Environmental Change and Food Security in China

ADVANCES IN GLOBAL CHANGE RESEARCH

VOLUME 35

For other titles published in this series, go to
www.springer.com/series/5588

Jenifer Huang McBeath · Jerry McBeath

Environmental Change and Food Security in China

Springer

Jenifer Huang McBeath
University of Alaska
Fairbanks, AK
USA
jhmcbeath@alaska.edu

Jerry McBeath
University of Alaska
Fairbanks, AK
USA
gamcbeath@alaska.edu

ISBN 978-94-007-3203-2 ISBN 978-1-4020-9180-3(eBook)
DOI 10.1007/978-1-4020-9180-3
Springer Dordrecht Heidelberg London New York

Softcover reprint of the hardcover 1st edition 2010

Printed on acid-free paper

Springer is part of Springer Science+Business Media (www.springer.com)

To our ancestors and to our children,
Bowen and Rowena

Acknowledgements

We began this study in spring 2004, when Jenifer Huang McBeath was a Science Fellow at the American Embassy in Beijing, and Jerry McBeath was Fulbright Professor at the China Foreign Affairs University, also in Beijing. We returned to China each year to collect research materials and conduct interviews, until the conclusion of research in mid-2009. For assistance in conducting research we are particularly grateful to Dean Dai Changzheng and Professor Wang Bo of the China University of International Business and Economics (CUIBE). In the 2006–07 academic year, graduate student Lu Na of CUIBE's School of Law helped us greatly by chasing down sources and facts. For assistance in other ways in Beijing, Shanghai, Hangzhou and Yunnan Province, we thank Stanley Huang; and for assistance in Yunnan Province, we thank Tian Rui-yun, RaiLongDa Trading Co.

A number of Chinese scholars were very helpful by sharing their collections of materials, their own research studies, and by reviewing preliminary drafts of our work. We thank: Dong Jiahong, Institute of Biotechnology and Genetic Resources, Yunnan Academy of Agricultural Sciences; Fu Chun-Hsu, Taiwan Forestry Research Institute, Council of Agriculture; Han Chenggui, Department of Plant Pathology, China Agricultural University; Huang Jikun, Center for Chinese Agricultural Policy; Li Zhou and Li Chenggui, Rural Development Institute, Chinese Academy of Social Sciences; Li Mingju, Institute of Plant Protection, Yunnan Academy of Agricultural Sciences; Li Zongzeng, Agriculture Department, Yunnan Province; Ma Jun, Institute of Public & Environmental Affairs; Nie Shanming, Chinese Academy of Fishery Sciences; Peng Yufa, Center for Biosafety Research, Institute of Plant Protection, Chinese Academy of Agricultural Sciences; Peng Youliang, Department of Plant Pathology, China Agricultural University; Ren Guoyu, National Climate Center; Sun Changjin, Forest Trends; Wang Xiaoming, Institute of Crop Science, Chinese Academy of Agricultural Sciences; Wang Jinxia, Center for Chinese Agricultural Policy; Alex Wang, China Environmental Law Project, NRDC; Xu Yinlong, Institute of Environment and Sustainable Development in Agriculture; Xue Dayuan, Nanjing Institute of Environmental Science and Central University for Nationalities; Yu Xuezhong, China Institute of Water Resources and Hydropower Research; Zhang Zhongkai, Institute of Biotechnology and Genetic Resources, Yunnan Academy of Agricultural Sciences; and Zhou Xiaoping, Beijing Normal University. In the United States, we thank reviewers of chapters of the manuscript: TJ Cheng of the College of William & Mary, Elizabeth Larus, University of Mary Washington, John Walsh, Center for Global Change and Arctic System Research, University of Alaska, and anonymous reviewers for the *Journal of Chinese Political Science*.

A number of other scholars, government officials, and NGO representatives were helpful to us in this research. We thank: Cai Lei, Cui Xiaoli, Guo Beihai, Han Nianyong, Huang Dafang, Angus Lam, Li Junqing, Li Yue, Li Zhihong, Liang Congjie, Liu Guihuan, Lo Szeping, Lu Houlin, Lu Xuedu, Luo Jinwen, Luo Youqing, Pan Jiahua, Pei Xiaofei, Qi Ye, Shen Gongming, Sun Yuehua, Wang Bin, Wang Gairong, Wang Sung, Wen Bo, Wu Zhongze, Xie Yan, Xu Yamin, Xuan Xiaowei, Yan Dongquan, Yuan Wenchuan, Zhang Dehui, Zhang Huimin, Zhang Li, Zhang Lei, Zhang Wenguo, Zhao Lei, Zheng Yisheng, Zhu Chunquan, Zhu Guangqing, and Eddie Zhu.

For assistance in manuscript preparation we thank University of Alaska Northern Studies graduate students Rudy Riedelsperger and Jesse Logan, the administrative assistant for the Department of Political Science, Courtney Pagh, and student assistants Kacie Take and Matthew Van Atta.

We are particularly appreciative of the assistance Margaret Deignan of Springer's Environmental Science unit gave us throughout the writing and publication process.

Finally, Jenifer McBeath expresses her deepest gratitude to her mother, Zhou Yuming, father, Huang Longjin, and maternal grandmother, Liu Yukung, from whom she learned the true meaning of dedication, sacrifice, bravery and honor.

Notwithstanding the assistance and support of so many, we cheerfully absolve all from any errors of fact and interpretation remaining in the manuscript.

Note on Names

We use pinyin throughout, except for commonly recognized names such as the Yangtze River. We introduce Chinese surnames first in the text, for example Deng Xiaoping; in the endnotes, Chinese surnames, like those in other languages, are listed last, as in Xiaoping Deng.

Contents

List of Figure

List of Tables

Chapter 1
Introduction

Abstract This chapter defines food security as the condition reached when a nation's population has access to sufficient, safe, and nutritious food to meet its dietary needs and food preferences. It stresses China's importance to global food security because of its population size. The chapter introduces the contents of the volume and then treats briefly food security in ancient and dynastic (211 BC–1912) China. It examines environmental stressors, such as population growth, natural disasters, and insect pests as well as imperial responses (for example, irrigation, flood control, storage and transportation systems). The chapter also briefly introduces the Republican era (1912–1949) and compares environmental stressors and government responses then to those of the imperial period.

Keywords Food system • Food security • Food production regions • Environmental stressors (Population growth • Natural disasters • Insect pests and Plant diseases • Deforestation • Climate change) • Irrigation systems • Flood control • Grand Canal

1.1 The Problem of Food Security and Environmental Change

Food is the material basis to human survival, and in each nation-state, providing a system for the development, production, and distribution of food and its security is a primary national objective. Many forces have influenced the food security of peoples since ancient times, with particular challenges from natural disasters (floods, famines, drought, and pestilence) and growing populations globally. From the late twentieth century to the early twenty-first, however, analysts have riveted their attention on environmental change and crises, for example pollution of arable land and water, climate change, insufficiency of water (and competitive pressures for water use), deforestation, desertification, and over-fishing among others. Our focus in this volume is on the food security of the world's most populous nation, China, and the impact on food security of vast environmental change in the last 60 years. First, however, we define important terms and explain why China must be considered in any global discussion of food security.

J.H. McBeath and J. McBeath, *Environmental Change and Food Security in China*,
Advances in Global Change Research 35, DOI 10.1007/978-1-4020-9180-3_1,

1.1.1 Definitions and Global Dimensions

There is a relationship between a nation's *food system* and that nation's *food security*. Typically, a food system comprehends three dimensions: (1) the *availability* of food (influenced in turn by its production, distribution, and trade); (2) *access* of the population to food (including such aspects as the affordability of food and preferences for different kinds of food); and (3) the *use* of food (whether, for example, it has value beyond fulfilling nutritional requirements, and food safety).[1]

Food security is co-extensive with food systems and a specification of their dimensions. The United Nations' Food and Agricultural Organization (FAO) defines food security as a condition attained within nations when food systems operate in such a way that:

> (A)ll people, at all times, have physical and economic access to sufficient, safe and nutritious food to meet their dietary needs and food preferences for an active and healthy life.[2]

These definitions may seem to imply that food is produced and consumed entirely within a national context, and that nations are autarkic with respect to their food systems. Of course, this is not what is meant: no large group of humans, whether a village, a city-state, nation-state, or empire has been entirely self-sufficient with respect to its food systems. Moreover, from the end of World War II to the early twenty-first century, food systems and food security have become increasingly globalized. On the one hand, the production, distribution, and marketing of food have become more intensive and complex. The distance between farmers and ultimate consumers has increased greatly, while different actors (such as food retailers) have become more prominent in the food chain. On the other hand, increased population, urbanization, and economic interdependence[3] globally have reduced barriers to the flow of food products. First, the General Agreement on Tariffs and Trade (GATT) increased trade flows exponentially; then, after 1995, the World Trade Organization (WTO) expanded membership of the global trade community to include most nation-states and focused interest on removing remaining restrictive trade practices (such as agricultural subsidies) within the system.

Availability of food, access to it, and its utilization always have been subject to environmental conditions. Throughout human history, floods, famines, drought, and pestilence have directly affected crop production and food security. Natural disasters are major examples of environmental change, but in the last generation they have been eclipsed by disasters humans have caused. There is a long and growing list of human-induced environmental changes: deforestation, desertification, over-fishing, biodiversity loss, pollution of air, lands, and water by industrial as well as agricultural activities, and most recently climate warming. These human

---

[1] P. J. Gregory, S. I. Ingram and M. Brklacich, "Climate Change and Food Security," *Philosophical Transactions of the Royal Society*, 360 (2005), 2139–48.

[2] FAO, *Report of the World Food Summit*. Rome: FAO, 1996.

[3] Economic interdependence remains the chief measure of globalization.

causes of environmental change interact with natural disasters, exacerbating them. For example, in August 2007, flooding in northern India and Bangladesh killed hundreds and made millions homeless. Flood conditions were worse than previously, experts opined, because extensive deforestation to increase agricultural land as well as to provide housing and industrial tracts had the effect of eroding soils and removing natural obstacles to water movement.

As food security has global dimensions, so too does environmental change. Although climate warming is most obvious in the polar regions, it will affect each of the globe's regions and have effects (not necessarily all malign) on agricultural production. Trans-boundary air pollution also influences growing conditions and food production. Both natural and human-induced disasters pose large problems to the food security of nations, with economically less-developed countries (LDCs) more vulnerable to adverse effects than economically-developed countries (EDCs). Also, vulnerabilities vary within individual countries, and the poor are especially vulnerable because they are most likely to lack the economic and social capabilities to cope with environmental change.[4] Too, the poor are least able to purchase high quality foods (those free of toxins, pesticide residue, and the like).

1.1.2 The Importance of China to Global Food Security

China has 22% of the global population but just 7% of the world's arable land. Food security has been a chief mission of the Chinese state since early in the dynastic era. It remains a primary state objective in the early twenty-first century. China in 2009 is largely self-reliant in food supplies, and its farmers produce about 95% of the staples consumed. Yet, any large disturbance in supply would have global ramifications, for example, by increasing world food prices.

This became an issue in 2008 as the FAO's food price index increased by 57%. On the international market, the price of cereals such as wheat and rice more than doubled. The director of the World Food Program in China noted: "At a moment when world food security is facing unprecedented challenges from rising prices, China's role is fundamental."[5]

China's environmental conditions directly impinge on its food security. Many observers believe China's environment is in crisis.[6] Population increases reduce arable land and water sufficiency; indirectly, population stress increases deforestation

---

[4] See, for example, B. Wisner, P. Blaikie, T. Cannon, and I. Davis, *At Risk: Natural Hazards, People's Vulnerability and Disasters*, 2nd ed. London: Routledge, 2004.

[5] Anthea Webb, "Why China is Crucial to World Food Security," *China Daily*, May 15, 2008, 9.

[6] For early studies, see Lester Ross, *Environmental Policy in China*. Bloomington, IN: Indiana University Press, 1988, and Vaclav Smil, *China's Environmental Crisis: An Inquiry into the Limits of National Development*. Armonk, NY: M. E. Sharpe, 1993.

and desertification as well as over-fishing. New environmental stress such as climate warming has an impact on plant diseases, pests, and invasive species too.[7]

China is a developing country, and its food security and environmental protection regimes are relatively new and untested. It was this combination of factors – a huge population with limited agricultural land, severe environmental challenges, and political, social, and economic systems in the process of modernizing – which prompted Lester Brown's 1995 book *Who Will Feed China?*[8]

Brown focused on what he believed was stagnating grain production in China of the early 1990s because of reduced arable land, lack of significant productivity gains, and environmental problems such as water insufficiency and large-scale soil erosion. He contended that China would need to import massive quantities of grain in future years to feed its population.

Brown's provocative thesis initiated debate about food security and environmental change in China. As we shall see below, much of the discussion focused on whether central government statistics were accurate. For the purposes of this study, the debate has been healthy by directing attention to the global consequences if China is unable to feed its large population.

1.2 Plan of the Book

After we outline the history of food security and environmental change in China, from early times, through the long dynastic era and the short Republican epoch, the volume unfolds in eight substantive parts. Chapter 2 treats the food security situation under rule of the Chinese Communist Party (CCP), from 1949 to the present. The new regime developed a command economy, influenced by the planning model of the Soviet Union; and this had important implications for the organization of agriculture. Early policy choices and radical development strategies, represented in the Great Leap Forward and the Cultural Revolution, left indelible marks on agricultural practices. Important changes in economic direction under Deng Xiaoping (starting in 1978) opened food production to market-based incentives. This chapter concludes with discussion of China's current food system, and it points out changes in food choices as the population grows in wealth.

In Chapter 3, we consider the immediate environmental stressors on food security. This chapter begins by describing the current status of China's arable land, forests, and grasslands, and how they have been measured. Then, it examines causes of loss to arable land through population growth, urbanization, and economic development.

---

[7] Nevertheless, China's ecological footprint is below the world average. In 2003, it was 1.6 ha per person (and ranked 69th globally), while the average was 2.2 ha per person. With 22% of the world's population, China uses just 15% of global total biological capacity. See Jing Fu, "Ecological Resources Use below Average," *China Daily*, June 11, 2008, 4.

[8] Lester Brown, *Who Will Feed China? Wake-up Call from a Small Planet*. Washington, DC: Worldwatch Institute, 1995.

We investigate the effects of socioeconomic change in terms of both land and water degradation. We also discuss the responses of the state to environmental stressors, and their overall impact on food security.

Chapter 4 examines one large near-term environmental stressor – climate change. To explain the climate warming situation in China, we consider energy use and policy and climate change impacts on traditional agriculture. We also consider atmospheric oscillations in East Asia and China specifically. Then, we look at the interconnections between these elements, leading to discussion of the current state of knowledge on climate change effects. In this chapter, we explore the correlation between climate warming and natural disasters. The chapter concludes with analyses of the impacts of climate change on food production.

In Chapter 5, discussion turns to the role that plant diseases and pests play in food security. We give a non-technical introduction to fungal, bacterial, and viral diseases on traditional food crops, and also introduce the major traditional pests and nematodes. The environmental stressors discussed at length in this volume have an impact on plant diseases and pests, as does climate warming. Throughout this chapter, we consider the traditional as well as modern means employed to mitigate plant pests and diseases.

Chapter 6 turns from native to alien and invasive species, which present challenges to China's food security. About 400 alien plant species threaten sections of China's food production, joined by alien insect and other species. Immediate environmental stressors are the cause of the increase in invasive species, particularly transportation improvements, explosive growth in China's international trade, and degradation of lands and waters through industrial, commercial, and agricultural development. The chapter also considers the impact of climate change on invasive species, and the difficulties in mitigating their adverse effects.

In Chapter 7, discussion turns to biotechnological responses to food security needs in China. We first track the evolution of biotechnology in China, and emphasize the success story of Bt cotton. We survey the other genetically-modified (GM) crops and plant species being tested for commercialization, and introduce the special case of Bt rice. The chapter then treats biosafety concerns regarding GMOs in China, and it concludes with a discussion of the biosafety regime formed to monitor agricultural biotechnology.

Chapter 8 evaluates the legal and institutional framework to address food security and environmental threats in China. The chapter begins with discussion of constitutional and legal provisions, and it introduces the several authorities supervising China's food system. Since 1978, the central government has devolved administrative functions to provinces, cities, and other local governments, and this has allowed an "implementation deficit" to form, which frustrates efficient administration of agricultural and environmental programs. We briefly review the nongovernmental organizations (NGOs), which monitor both food security and environmental change. The chapter concludes with the most recent case of food security: tainted food products China has exported to other countries, and the construction of a new food safety regime.

In Chapter 9, we consider six large issues in the implementation of China's food security regime. The first concerns the basis in knowledge of China's food system. The second issue is the large problem of governmental coordination, both horizontally and vertically. The third challenge concerns modernization of the infrastructure for food production, and treats transportation and food storage. The fourth issue concerns access to food and poverty alleviation. The fifth challenge is public participation with a focus on media reportage and citizen protests. The last issue treated is changes in agricultural trade after China joined the World Trade Organization in late 2001. The volume concludes, in Chapter 10, with a summary of the argument and our observations on what China's experience promises for the future.

1.3 Food Security in Traditional China

1.3.1 Food Production Regions

To understand the challenges of agricultural development and food production, scholars have adopted different analytical systems. In *A Biodiversity Review of China*, Mackinnon et al. divide the Chinese landmass into seven biogeographical regions.[9] A review of the regions informs us of the great variety in land forms (Fig. 1.1):

1. Northeast China: forested hills and the Changbai and Da Xing'an Mountains
2. North China: plains and low hills (with deciduous forests) of the heavily populated regions north of the Yangtze River
3. Inner Mongolia-Xinjiang: the northern and northwestern third of China, including steppes and deserts north of the Tibetan Plateau and also the Tianshan and Altai mountain systems
4. The Tibetan Plateau and Himalayan highlands
5. Southwest China: transitional mountains from the eastern Himalayas to the Sichuan and Yunnan plateaus, including deep river gorges
6. Central China: divided into the Guizhou plateau, the Sichuan basin, and southeastern provinces
7. Tropical South China: including the South China rain forest, Hainan Island, and Taiwan[10]

[9] John Mackinnon, Mang Sha, Catherine Cheung, Geoff Carey, Zhu Xiang and David Melville, *A Biodiversity Review of China*. Hong Kong: World Wide Fund for Nature (WWF) International, 1996.

[10] Ibid., 39.

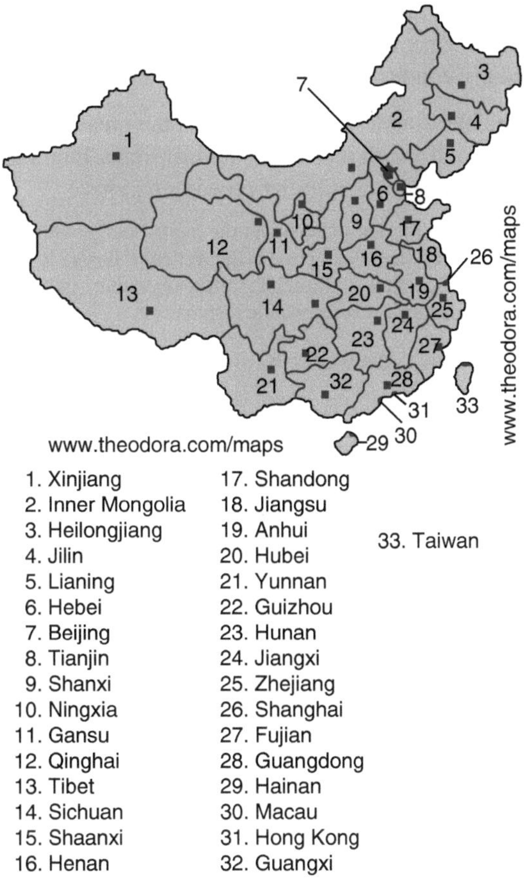

Fig. 1.1 Map of China

With a special focus on agricultural production, anthropologist E. N. Anderson divides China into five great regions:

1. North China: an area of dry farming including the Yellow River plains and adjoining areas, the heartland of Chinese civilization
2. Northeast China: an area of fertile river valleys but also of subarctic forests and cold winters
3. Central Asian China: including the deserts and dry lands of Inner Mongolia, Xinjiang and other border regions, with farming only in irrigated areas and most lands used to support nomadic herding
4. South China: an area south of the Yangtze River, with both hilly and mountainous lands as well as deltas and plains, where rice is the staple food

5. Tibet: an area of high and desolate plateaus and mountain ranges, including all of the province of Tibet as well as Qinghai and parts of Sichuan and Yunnan, where cold and dry conditions restrict farming[11]

For many observers, the simplest classification is the most serviceable: division of China into the North and South. For example, historians John Fairbank and Merle Goldman separate rice from wheat/millet growing regions:

> The dry wheat-millet area of North China and the moist rice-growing areas of the South divide along a line roughly halfway between the Yellow (Huang) River and the Yangzi River on the thirty-third parallel ... Rainfall, soil, temperature, and human usage create striking contrasts between these two economic regions.[12]

In most of this volume, we contrast the dry regions of northern China with the moist areas of the South.

1.3.2 Food Production in China's Pre-history

Archaeological finds in China date human settlements from 500,000 BC. At this early age, the people organized themselves into clans and tribes, and engaged in subsistence hunting, fishing, and gathering pursuits. Evidence of a millet-growing Neolithic culture in northern China is traced to the approximate period of 6500–5000 BC. Chinese farmers grew rice along the eastern coast as long as 7,700 years ago.[13] By 4000 BC, sheep and horses had been domesticated, and there were large farming villages in the North. In southern China, rice was grown from 3500 to 3000 BC. Because the South is close to the South China Sea, and is favored by much more precipitation than the North, rice plants were irrigated. The combination of ample water and human labor led to bountiful harvests and productivity per *mu* (about 1/15th of an hectare) of land that was nearly twice that of the dry-farming regions of the North. Wheat and barley were introduced to the North and South near the start of China's bronze age.[14]

The legendary three dynasties before the unification of China – the Xia, Shang, and Zhou – centered in different regions along the Yellow River in North China, and co-existed for the second and first millennia before the Christian era. These were feudal societies, and they practiced slavery. By the Eastern Zhou era (771–256 BC), the outlines of the food system were clear: reliance upon production of grains (millet, rice, wheat, barley), fiber crops, vegetables, fruits and nuts. Late in Zhou-era China, cast iron was developed and used in farming, which had a significant and positive impact on agricultural productivity.

---

[11] E. N. Anderson, *The Food of China*. New Haven, CT: Yale University Press, 1988, 2.

[12] John King Fairbank and Merle Goldman, *China: A New History*, 2nd ed. Cambridge, MA: Harvard University Press, 2006, 5.

[13] Xinhua, "Rice Cultivation Dated 7,700 Years," *China Daily*, October 3, 2007, 1.

[14] Anderson, 1988, 9–43.

Agricultural production was influenced by thought systems emphasizing protection of nature and especially natural resources. Philosophers from the pre-Qin to the Han Dynasties believed everything should be done "strictly according to the rules of nature." This early environmental ethic was combined with a social ethic, identifying the rule of nature with the social order. Thought systems after the Wei and Jin Dynasties paid more attention to the relationship between damages to forestry, land, and water resources and the environment.[15]

1.3.3 Primary Environmental Stressors in the Dynastic Era (211 BC–1912)

Five forces affected food security throughout the two millennia of dynastic rule in China: population growth, natural disasters, insect pests and plant diseases, deforestation, and climate change. We examine each in turn (Table 1.1).

1.3.3.1 Population Growth

The first census of China was taken during the Han Dynasty (about AD 2), and it revealed a population of 60 million. By the end of the later Han, approximately AD 280, the population had shrunk to 16 million (because of poor responses of the state

Table 1.1 Major periods in Chinese history (Adapted from John King Fairbank and Merle Goldman, *China: A New History*. Cambridge, MA: Harvard University Press, 2006, 24)

Eastern Zhou	771–256 BC
Warring States	403–221 BC
Qin Dynasty	221–206 BC
Earlier Han	206 BC–AD 8
Later Han Dynasty	25–220
Disunion in North and South	220–580
Northern Wei	386–535
Sui Dynasty	589–618
Tang Dynasty	619–907
Northern Song (Liao empire on northern border)	960–1125
Southern Song (Jin empire in North China)	1127–1279
Yuan (Mongol) Dynasty	1279–1368
Ming Dynasty	1368–1644
Qing (Manchu) Dynasty	1644–1912
Republican Era	1912–1949
People's Republic of China	1949–present

[15] Jinmin Zhang, "Nature Protection in Ancient China," (in Chinese), *China Agricultural History*, Vol. 18, no. 1 (1999), 71–77.

to natural disasters and wars), but then grew to 100 million at the time of the Northern Song. In the late Ming Dynasty, China's population had grown to 150 million, and then was larger than the population of Europe. By the early nineteenth century, in the Qing Dynasty, China's population stood at 360 million (which compared to a European population of about 200 million).[16]

China's population rose during periods of peace and stability, and it fell, often times precipitously, during periods of internal war and external threats. Since the late imperial period, however, the large population has brought great pressure on the environment. Food production in recent times barely has kept pace with population growth.

1.3.3.2 Natural Disasters

Natural disasters also had a limiting effect on growth in agricultural production. Virtually every year, drought, floods, famines, and pestilence wreaked havoc on human populations. China's pattern of rainfall was a "natural" factor influencing disasters. Most precipitation occurred consequent to the summer monsoon, which brought warm air, loaded with moisture, from the South China Sea to South China. This large region received heavy rainfall on a regular basis, and was subject to regular flooding.

North China received less rainfall, and the rain was much less predictable than in the South. Annual variation of rainfall in North China has been as large as 30% from 1 year to the next.[17] The variability of rainfall in the North often produced conditions of drought and famine. This particularly influenced the Yellow River, which flooded countless times in Chinese history (and repeatedly changed its course), resulting in the loss of agricultural land. Warmer and wetter conditions in the South also made it possible to double- and in some areas triple-crop rice production.

1.3.3.3 Insect Pests and Plant Diseases

The intensive nature of agricultural cultivation in China meant that crops were highly subject to infestation by bacterial, viral, and fungal diseases as well as by insect pests. For example, annual yield losses to the rice blast fungus in the late nineteenth century were estimated to exceed 20% (before development of effective pesticides). Rusts and scabs affected millet and wheat as well as soybeans.

Perhaps the most destructive pest for China's traditional agriculture was the locust. Records since AD 900 (near the end of the Tang Dynasty) show a consistent

---

[16] Anderson, 1988, 71. Fairbank and Goldman note a "doubling if not actually a tripling of population" in late imperial China from 1600 to 1911 (p. 163).

[17] Fairbank and Goldman, 2006, 5.

relationship between periods of drought and locust plagues, with the incidence particularly high at the end of the Ming Dynasty and beginning of the Qing Dynasty (AD 1601–1650) and the end of the Qing Dynasty (1851–1911).[18]

Government officials blamed diseases and insect plagues on people's lack of virtue, while the common people blamed them on supernatural forces. Two types of temples in traditional China acknowledged this. The first was dedicated to protecting against locust plagues; the second, usually aligned with a goddess, sought to protect crops and the people against all insect and plant disease threats.

1.3.3.4 Deforestation

Population growth brought increased pressure on China's forests, as farmers cut down trees to open up new areas for cultivation. Also, starting in the Song Dynasty, the demand for wood rose to support the iron, ceramics, and printing industries.[19] The loss of large once-forested areas increased greatly the amount of soil erosion, and this led to silting of rivers (such as the Yellow River), compounding flooding problems.

By the end of the Ming and start of the Qing Dynasty, in most parts of China, cultivation had expanded to the uplands and other marginal lands, because of continued population pressures. Unrelenting deforestation, especially in the Northeast, as well as reclamation of lakes and wetland areas intensified soil erosion and exacerbated flooding of rivers and delta areas.

1.3.3.5 Climate Change

A final factor influencing food security during the traditional period was change of weather and climate conditions. Historians note that the three dynasty period of the second and first millennium, BC, was wetter and warmer than in the early twenty-first century. By the Tang Dynasty, the weather had become colder and drier; large fluctuations in climate occurred from then and continued through the northern Song era.[20]

China's Little Ice Age corresponded roughly to the reign dates of the Ming Dynasty, and during this period winters were quite cold; summer rains were erratic and on occasion torrential.[21] This had adverse impacts on agricultural production. By the early Qing, the Little Ice Age waned, and China's weather became warmer and wetter,[22] which facilitated agricultural development.

---

[18] Shu-yan Yin, Chun-chang Huang, Li-hui Gun, Xin-yan Li and Yao-feng Jia, "The Statistics of Locust Disaster Frequencies and Analysis of its Influences in Historical Period in the Guanzhong Region (in Chinese)," *Journal of Arid Land Resources and Environment*, Vol. 20, no. 5 (September 2006), 159–162.

[19] Anderson, 1988, 79.

[20] Anderson, 1988, 71.

[21] A. F. Harding, ed., *Climate Change in Later Prehistory*. Edinburgh: Edinburgh University Press, 1982.

[22] Anderson, 1988, 109.

1.3.4 Imperial Responses

The unification of China under the short-lived Qin Dynasty in 221 BC facilitated large-scale change and many improvements in food production. The succeeding emperors of the early Han Dynasty created the rudiments of China's first comprehensive agricultural development policy, consisting of six elements:

1. Moderate taxation of agricultural land
2. Development of a class of independent small farmers (or stable tenancy arrangements)
3. Agricultural extension and compilation of textbooks and encyclopedias on agriculture
4. Support of irrigation systems and government granaries
5. Famine relief
6. Attention to herbs and medicine as well as to food[23]

These elements did not develop simultaneously. Moreover, the effectiveness of policy over time varied by wealth, political, and especially military stability, and the nature of the land tenure system established by the dynasties. For example, during the Northern Wei Dynasty, rulers devised the "equal field" system, a form of socialist redistribution of land. Although this equalized land holdings and was a good foundation for independent farmers, it also made the state the ultimate owner of land. Then, in the Song Dynasty, politically powerful elites established large estates, and this threatened the independence of small farmers.[24]

The second set of agricultural innovations occurred in the Song Dynasty, which historian Mark Elvin calls revolutionary in impact. He outlines four elements of this change:

> (1) Farmers learned to prepare their soil more effectively as the result of new knowledge, improved or new tools, and the more extensive use of manure, river mud and lime fertilizers. (2) Strains of seed were introduced which either gave heavier yields, or resisted drought better, or else by ripening more rapidly made it possible to grow two crops a year on the same land. (3) A new level of proficiency was reached in hydraulic techniques, and irrigation networks of unprecedented intricacy constructed. (4) Commerce made possible more specialization in crops other than the basic foodgrains, and so a more efficient exploitation of varying resource endowments.[25]

Few other major changes occurred in China's food production system from the Song Dynasty to the twentieth century. By that time, rice and wheat had become China's main staples. In fact, Harvard economist Dwight Perkins maintains that

---

[23] Anderson, 1988, 44.

[24] Anderson, 1988, 62, 64–65.

[25] Mark Elvin, *The Pattern of the Chinese Past*. Stanford, CA: Stanford University Press, 1973, 118.

little change occurred during the Ming and Qing dynasties, other than the extension in area of cultivated land.[26]

Throughout the dynastic period, the central role of agriculture in the state was signified by emperors, China's supreme authorities in the traditional era. In each dynasty, one of the functions of emperors was to perform auspicious ceremonies focusing on farming. For example, in the Ming Dynasty, the ceremony occurred as follows. The emperor and a group of high officials, wearing ceremonial robes, congregated at a site near the Temple of Agriculture. Assisting the emperor hold the plow were two elders, selected from farm communities around the capitol. Surrounding the emperor were designated farmers carrying farm tools and acting out agricultural tasks. Actors, selected from government-owned theatrical troupes, dressed in costumes representing wind, rain, thunder, and lightening gods. Villagers sang songs calling for a bumper harvest. The ceremonial plow was pulled by a cow led by two high officials. The emperor, holding a whip in his right hand and the ceremonial plow in his left, supported by the two elders, walked three paces, thereby inaugurating the new agricultural year.[27]

Governments responded to food production needs in six ways: through establishment of irrigation systems, flood control projects, granaries, transportation systems, trade, and mobilization of the population.

1.3.4.1 Irrigation Systems

By the first millennium BC, farmers had developed localized irrigation systems, which were based on simple ditching to move water along trenches. Fan Sheng-chih's agricultural manual of the first century BC describes several means of irrigating crops: rice land was leveled and rice paddies were irrigated; and water channels were rerouted by seasons to change the pattern of water circulation. In dry fields of the North, a water-trapping system was used. In areas where drainage was a problem, farmers cultivated crops along ridges.[28] By the time of the Yuan Dynasty, the development of a foot-operated water pump had improved irrigation.

In short, before the onset of centralized dynastic rule, Chinese farmers had local irrigation systems. By the third century, BC, large-scale irrigation works were established by the state, and the maintenance of these systems was a state responsibility (typically involving use of corvee labor).

---

[26] Dwight Perkins, *Agricultural Development in China, 1368–1968*. Chicago: Aldine, 1969, 17. This point is disputed by Evelyn Rawski, who argues that the introduction of high yield crops from the West, new fertilizers, and new agricultural practices were significant changes. See her *Agricultural Change and the Peasant Economy of South China*. Cambridge, MA: Harvard University Press, 1972.

[27] Renyu Huang, *1587: A Year of No Significance* (in Chinese). Taipei: Food Products Publishing Co., 1985.

[28] Quoted in Anderson, 1988, 49.

1.3.4.2 Flood Control

Beginning in the Yuan Dynasty, a central policy of the state was regulation of water flow on the Yellow River to avoid flooding. The Mongols claimed to have "tamed" this river, but throughout Chinese history rulers attempted to erect dikes, alter channels, and control the course of the river and particularly its lower course.

The most noteworthy (and also the oldest) example of flood control is found in Sichuan Province at Dujiangyan. The Dujiangyan irrigation system lies on the Min River to the northwest of Dujiangyan City in Sichuan. It was built more than 2,200 years ago under the direction of Li Bing, governor of the Shu Prefecture in the Qin State. The major purpose of construction was to manage water flow of the Min River, to facilitate irrigation of the Chengdu Plain, and to enhance water conservation.

After a careful survey of the areas, Li and his son directed that a channel be cut through Mt. Yulei to the west of Sichuan, creating a man-made river. A dike divided the river into two parts: the inner river and the outer river. In order to control floods and discharge silt, two spillways were built at the end of the dike. Following geographic characteristics of this region, the dike distributed 60% of the water to the inner river, and 40% into the outer river during the dry season. During the flood season, 40% of the water entered the inner river, and the rest the outer river. The curved shape of the dike matched the shape of the river. The bottom waters of the Min River, rich in sand, flowed into the outer river (helping to prevent flooding) while surface clear waters flowed to the inner river and were used for agricultural irrigation and water conservation.

The dike consisted of bamboo cages filed with egg-sized stones; spaces among the stones reduced the speed of river currents. The linkages and accumulation of bamboo stone barrels made the dike a solid and intact construction (and model) of civil engineering for over 2,000 years. Since the Tang Dynasty, villagers in the Dujiangyan region have maintained the system.

For more than 2,000 years, the Dujiangyan irrigation system has brought large benefits to agricultural production and people's lives in the Sichuan plain. The irrigated area has expanded gradually from 126,000 ha to nearly 660,000 ha of land, covering 36 counties. Dujiangyan was listed as a United Nations Educational, Scientific and Cultural Organization (UNESCO) World Cultural Heritage site in 2000.[29]

1.3.4.3 Imperial Granaries

It was during the Han Dynasty that leaders developed China's first granaries. The amount of grain stored varied by region and strength of dynasty. Wright notes, in a discussion of grain storage in the Sui Dynasty, that as many as ten million (Chinese) bushels were stored in public granaries.[30] Not only did the storage make it possible

[29] See Gerald A. McBeath and Tse-Kang Leng, *Governance of Biodiversity Conservation in China and Taiwan*. Cheltenham, UK: Edward Elgar, 2006, 23–4.

[30] Arthur Wright, *The Sui Dynasty*. London: Twitchett Publishers, 1979, 93–94.

for the regime to administer famine relief, but the granaries were a means to stabilize grain prices. By its large-scale purchases in years of good production, and by large-scale sales in years of poor production, the state reduced variation in price levels.

1.3.4.4 The Grand Canal

The Sui Dynasty rulers are also responsible for the development of China's first Grand Canal. Using local canals, rivers, and lakes as the base, the canal connected cities on the Yellow River with those on the Yangtze. Early rulers of the Tang Dynasty connected these canals to the Beijing and Tianjin areas.[31]

The Grand Canal was used for both military and economic purposes. It joined the rich agricultural regions of southern China to the northern regions, and provided food for soldiers defending the capital city. It had a critical role in food security by moving food products from south to north. Sections of the Grand Canal are still in use, and it has been incorporated in the South–North Water Diversion Project, discussed in Chapter 3.

1.3.4.5 Trade Flows

China's overland and marine transportation activities played a critical role in food security during the traditional era. Caravans of merchants moved goods on the Silk Road from Lanzhou and Yumen along the Tarim Basin and across the mountains into Farghana, Samarkand and other western points. This was the primary transportation linkage between China and Central Asia during most of the traditional era.

Maritime transportation was important for movement of food products to China's coastal cities, but less important in the trade of food products with South and Southeast Asia, and points west. Nevertheless, Chinese navigators, such as Cheng Ho in the early Ming Dynasty, brought Chinese food products in large vessels to ports in the South China Sea and Indian Ocean.

Trade flows introduced new products to China, for example, cotton, corn, sweet potatoes, peanuts, tobacco products and, as previously mentioned, wheat, barley, sorghum, and alfalfa.

1.3.4.6 Mobilization Capacity

The dynastic era in China's history demonstrates the sophistication and strength of the state, as compared to European states before modernization in the sixteenth and seventeenth centuries. Although most dynasties went through cycles of waxing and

[31] Edwin O. Reischauer and John K. Fairbank, *East Asia: The Great Tradition*. Boston, MA: Houghton Mifflin, 1960, 161–63. The first Grand Canal connected Hangzhou and Changan, while the second or modern Grand Canal lies between Hangzhou and Beijing.

waning power, from the time of the early Han Dynasty, regimes emphasized both Legalist and Confucian themes valuing agriculture as the root of state prosperity and power. For example, China's most eminent political philosopher, Mencius, emphasized the important role of agriculture:

> If you do not interfere with the busy seasons in the fields, then there will be more grain than the people can eat; … .When the people have more grain … than they can eat … then in the support of their parents when alive and in the mourning of them when dead, they will be able to have no regrets over anything left undone. For the people not to have any regrets … is the first step along the Kingly way.[32]

Strong rulers also developed capacity in the state to build large projects, such as irrigation and flood control projects, granaries, and the Grand Canal mentioned above. In fact, one of the important measures of the effectiveness of a regime was its ability to transport food to areas hit by natural disasters.

Karl Wittfogel suggested in the 1950s that the centralized strong state system of China largely was the inevitable result of the need to construct large irrigation projects.[33] The theory has been disputed by many scholars,[34] because the evidence points to the formation of a strong state before most large water control projects were established. Nevertheless, an insight in Wittfogel's analysis is the extent to which authoritarian rule and the ability to mobilize mass numbers benefited food production and food security.

1.4 Food Security in the Republican Period: 1912–1949

From the fall of the Qing Dynasty in the Shinhai Revolution of 1911, to the establishment of communist power in 1949, China was a republic, and for most of the period controlled by the Guomindang (Chinese Nationalist) Party. We discuss first the difference of environmental stressors in this period, as compared to the late Qing and earlier dynasties. Then we examine changes from imperial policy respecting food security.

1.4.1 Difference of Environmental Stressors

Most of the environmental stressors discussed above – population growth, natural disasters, plant diseases and insect pests, and deforestation – continued through the Republican period. However, two forces, war and industrial development, exacerbated the stressors.

---

[32] *Mencius*, translated by D. C. Lau. Hong Kong: The Chinese University Press, 1984.

[33] Karl Wittfogel, *Oriental Despotism*. New Haven, CT: Yale University Press, 1957.

[34] See, for example, Wolfram Eberhard, *A History of China*, 4th ed., Berkeley, CA: University of California Press, 1977.

Conflict and wars were not unusual events during the dynastic era; in fact, they punctuated Chinese history. Yet in the early twentieth century, three sources of conflict overlapped, putting immense pressure on the Republican government. First, the fragmentation of power occurring in the late Qing Dynasty intensified in the Republican era. From 1912 until the conclusion of the Northern Expedition in 1928, the central government lacked power to corral separatist tendencies of most provinces. Warlords could challenge the authority of the regime, and they made independent alliances with one another and foreign powers.

Second, from the formation of the Chinese Communist Party in 1921, the Nationalist government's authority was challenged by a rival linked to the Soviet Union, and during two periods, 1928–1937 and 1945–1949, warfare broke out pitting Chinese against Chinese. Third, China had lost sovereign control of its territory through the development of the Western treaty port system and was subject to predatory attacks by western imperial powers. The largest foreign threat, however, came from Japan. This power seized Manchuria in 1931 and then invaded the Chinese mainland in 1937, occupying all the eastern coast until its defeat by allied powers in 1945.

Altogether, conflict and war disrupted food production and the trade in food products. It intensified the effects of famine, drought, floods, and pestilence. Mass movement of armies as well as people fleeing from conflict put huge pressure on lands and resources.

Although industrial development in China began haltingly with the Self-strengthening Movement (from the 1860s), it accelerated in the Republican period. This too intensified pressure on food systems. Establishment of factories and new transportation routes took land out of agricultural production. Factory pollution soiled landscapes, making many infertile for food production.

1.4.2 Changes from Imperial Policy

Several changes occurred in agricultural and food production policy in the Republican era. First, the very limited capability of the regime meant that agricultural infrastructure and centralized famine relief were impaired. Yet, the flip side of weakened capacity was exposure of China to greater penetration of global forces, particularly international trade in food products. This penetration increased the commercialization of China's food system, and increasingly drew peasants into market relationships.

The second large-scale change was modernization of agricultural practices, and this occurred in several ways. Historically, Chinese agriculture has been highly labor-intensive, but during the Republican era, machines with diesel engines were introduced in several southern regions. Traditional Chinese agriculture employed natural means to fertilize crops, particularly use of human and animal manure. In the early Republican era chemical fertilizers were introduced to many Chinese farms, and by the end of the Republican era, fertilizer plants were springing up along the eastern coast. Both mechanization of agriculture and chemical fertilizers and pesticides improved agricultural productivity. Also, improved crop seeds were developed with disease- and drought-resistant properties.[35]

The third change is related to commercialization and modernization of agriculture, and this was the development of agricultural science in China. The government established a Ministry of Agriculture and several agricultural research institutes, such as the National Agricultural Research Bureau (founded in 1932). Several universities established schools and departments of agricultural science, and a few initiated extension work to apply research findings. A number of western universities established collaborative relationships with Chinese universities, for example Cornell University, which aided the modernization of Chinese agriculture.

1.5 Conclusion

In this chapter, we have introduced the concepts of food system and food security, their importance in China, and China's importance to global food production. No brief treatment can do justice to the extraordinary sweep of China's history before the mid-twentieth century. In very broad strokes, we have outlined the themes followed in this volume: the multiple forces of environmental change and the ways in which they have affected food sufficiency and security, as well as responses of the state to threats to agricultural production.

Although the dynastic system ultimately could neither counteract the force of western imperialism nor easily adapt to external pressures while undergoing internal turmoil, by the mid-twentieth century China still had a relatively well developed system of food production. In the next chapter we take the story from the founding of the People's Republic of China in 1949 to the early twenty-first century.

[35] John Lossing Buck, "Land and Agricultural Resources," in Yuan-li Wu, ed., *China: A Handbook*. New York: Praeger, 1973, 66–67.

Chapter 2
Communist Rule and the Food Security Situation

Abstract Since 1949, the Chinese Communist Party (CCP), a Leninist political organization leading a party-state, has ruled China. In this chapter we narrate the development of the planned economy in China, including agricultural growth targets, and then focus on the early choices of the regime, which influenced agricultural production: land reform and collectivization, the Great Leap Forward (1958–1960), and the Cultural Revolution (1966–1976).

After the death of Mao Zedong in 1976, and a brief succession struggle, Deng Xiaoping assumed leadership of the party and state. Deng radically changed China's economic development policy, opening the Chinese economy to competition and gradually engaging the Chinese marketplace with the world. The chapter also treats the current food system (focusing on production and use) in 2009, as well as changes in food choices in China.

Keywords Command economy • Collectivization • Communes • Great Leap Forward • Cultural Revolution • Household responsibility system • Grain reserves • "Opening to the world," Plant foods • Animal foods • Food sufficiency • Poverty • Changes in food preferences

2.1 Establishment of the Command Economy, 1949–1978

China's new rulers did not instantly control all economic decisions upon their ascension to power. Initially, they allowed "national" capitalists, as well as middle- and rich-peasants, who had supported the revolution, to retain their assets and lands. But the Soviet Union's support for the development of party rule in China, the Sino-Soviet Alliance, and the desire of China's rulers for rapid growth led to selection of the Stalinist model of economic development.

J.H. McBeath and J. McBeath, *Environmental Change and Food Security in China*, Advances in Global Change Research 35, DOI 10.1007/978-1-4020-9180-3_2,

2.1.1 Central Planning of Agricultural Production

About 80% of the population was involved in agricultural pursuits in 1949. As we note below, in the first few years after taking power, leaders disagreed on how land and production should be organized. Nevertheless, leaders shared the objective of rapidly developing China after the devastation of Japanese invasion and civil war, and the Soviet command economy model was intuitively attractive.[1]

A command economy is the antithesis of the market system. In the latter, prices – signals for allocation of resources – are not set by the state or by individual or collective producers or consumers. Instead, in ideal conditions prices reflect the underlying logic of the laws of supply and demand. And in market systems, most factors of production – land, labor, and capital – are privately held. In command economies, on the other hand, the state (or party-state, in the case of communist powers) sets prices, and centralized planning bureaus and ministries determine production targets, expressed in 5-year plans. Also, in command economies, the factors of production ideally are socially-owned (owned by the state on behalf of the people).

China was at the early stage of industrialization in 1949, and for this reason, development of centralized control of industrial enterprises met few obstacles. The state was able to control the production and distribution of industrial goods through establishing state-owned enterprises (SOEs). The SOEs fell under one of four levels of government: central, provincial, municipal, or county. Enterprise directors attained their posts through the elaborate nomenclatural system dominated by the CCP, not because of their qualifications as entrepreneurs; and they held control irrespective of the performance of the factories or firms. No SOEs went bankrupt, because they had access to the deep pockets of the state. Inasmuch as the state controlled all significant enterprises, it could influence their actions. This became the objective for agricultural production. The First Five Year Plan, with a baseline date of 1952, included agricultural as well as industrial production.

2.1.2 Ideological Control of Plan Targets and Methods

Plan targets and methods were not determined by objective analysis of stages in development. Instead, Chinese leaders and pre-eminently Mao Zedong applied Marxist and Leninist ideology, based on their experience of the Chinese revolution. Mao's style of governance emphasized three elements: mass line and mass movements, class struggle and continuous revolution, and superiority of the ideologically correct to expert opinion.

Mao developed the strategy of the mass line as the primary thought instrument in struggle with the Guomindang during the civil war. After the establishment of

[1] See Edward E. Rice, *Mao's Way*. Berkeley, CA: University of California Press, 1972, 120–35.

the People's Republic in 1949, the mass line continued, reflected in large-scale movements (a campaign psychology) such as the Anti-Rightist Campaign, the Great Leap Forward, and the Cultural Revolution. From Mao's perspective, the mass line was essential in order to mobilize support from the general population and to keep communist ideology pure. Mass movements and a collectivized economy also conferred some legitimacy on the communist regime.[2]

From Mao's perspective, class struggle and class conflict would continue for a long period after the establishment of New China. Communist leadership needed to pay attention to the transformation from "contradictions within the people" to "contradiction with the enemies."[3] Party leadership was responsible for identifying objects of struggle and launching investigation and struggle campaigns. Continuous class struggle, Mao believed, was the best way to maintain dynamism within the communist party (and avoid bureaucratic rigidity) and among the masses.

Politics commanded all economic decision-making processes during the Maoist period. Ideology set the directions and the methods for policy implementation. The centralized socialist command economy neglected local differences and reduced economic incentives, resulting in declining productivity. Those who deviated from the party line, such as intellectuals, were subject to thought reform and imprisonment.[4]

2.2 Policy Choices and Impacts on Agricultural Production

Unified in their objective to rapidly develop China, leaders of the People's Republic also learned from the experience of the Soviet Union that direct and abrupt changes to rural organization would be counter-productive. Thus, initially they moved slowly, but did so in fits-and-starts, largely determined by ideological divisions among the leadership. We treat three aspects of change in agricultural organization and in campaign strategies, which had an impact on food production as well as environmental change.

2.2.1 *Early Land Reform and Collectivization*

The communist party had engaged in land reform in areas under its control during the civil war, but at the establishment of the People's Republic, areas outside the party's orbit of power had not undergone land redistribution. An early decision was

[2] Stuart R. Schram, *The Political Thought of Mao Tsetung*. New York: Praeger, 1970, especially 294–330.

[3] Ibid., 301–12.

[4] See also Brantley Womack, ed., *Contemporary Chinese Politics in Historical Perspective*. New York: Cambridge University Press, 1991; Roderick MacFarquhar, *The Politics of China: Eras from Mao to Deng*. Cambridge: Cambridge University Press, 1990; and Tony Saich, *Governance and Politics in China*. New York: Palgrave, 2001.

to reduce rents in these areas, while developing a legal basis for change, in a new land reform law, promulgated in mid-1950s.[5] By 1952 land reform was concluded, and this campaign, which involved considerable class struggle (and liquidation of landlords), consolidated the power of the new government.

It was Mao's position that upon completion of land reform, agricultural collectivization should begin. However, Liu Xiaoqi urged a moderate policy; he sought to harness capitalistic initiatives of the peasants, thereby greatly improving agricultural productivity, before collectivization began in the near-term. Other leaders sided with Liu, and a compromise policy of forming cooperatives first, with the goal of collectivizing agriculture over a 15 year period, was adopted.[6]

Although agricultural productivity increased after land reform, the amount of grain sold to the state did not meet expectations of leaders (who were concerned, among other things, about feeding the growing urban and industrial population). Mao used this concern to bolster his ideological argument, and the regime proceeded to order collectivization of agriculture, which was accomplished by 1957–1958. By 1958, too, communes had formed, nearly autonomously, in several parts of China, and Mao then called for their spread nation-wide. Forming communes altered the distribution system and led to elimination of private plots. In the process peasants engaged in the widespread slaughter of livestock to prevent their being communized, which later led to a shortage of meat.[7]

In the period 1958–1978, the rural population was organized into three levels. At the base were the advanced producers' cooperatives (the stage reached by the end of collectivization), corresponding in size to small traditional villages and relabeled "production teams." These were formed into larger units called "brigades," which in turn composed the communes – of which there were about 70,000 throughout China.[8]

In order to appropriate the agricultural surplus to invest it in industrial production, the regime established a grain monopoly, and ordered the closing of all free grain markets. Through this means, it was able to obtain and distribute the basic food supply of the nation. Essentially, party cadres told peasants what to produce and in what quantity. Also, the government regulated the agricultural tax and determined what grain was "surplus" (with the level set below consumption needs), thereby determining the price of staples. Also, by establishing a comprehensive household registration system connected to grain rations, peasants were tied to the land and to their production team.[9]

---

[5] Kang Chao, *Agricultural Production in Communist China: 1949–1965*. Madison, WI: The University of Wisconsin Press, 1970, 14.

[6] Ibid., 17–18.

[7] Ibid., 24–26.

[8] John King Fairbank and Merle Goldman, *China: A New History*, 2nd ed. Cambridge, MA: Harvard University Press, 2006, 352–53.

[9] Ibid., 354, 356. Also see Jean Oi, *State and Peasant in Contemporary China: The Political Economy of Village Government*. Berkeley, CA: University of California Press, 1989.

2.2.2 *The Great Leap Forward (1958–1960)*

The stimulus for the Great Leap Forward (GLF) was the realization that grain production was barely keeping pace with population growth (of 9% in the countryside and some 30% in cities); the government's take in grain collection had not increased sufficiently to support rapid industrialization. Mao formulated another campaign, in this case to mobilize rural labor power on perhaps the largest scale in Chinese history.

Mao called his strategy "walking on two legs," by which he meant intensive use of rural labor in both agricultural and industrial development. To develop agriculture, peasants were organized to increase reclamation of land (which in many areas increased deforestation and erosion of lands unsuitable for agriculture), to improve irrigation systems, and increase flood control efforts.[10] Some of these efforts were salutary, but others – for example the campaign to eradicate the four pests – nearly eliminated sparrows, with deadly consequences for the ecological balance in rural areas.

To foster rural industrial development, with the slogan of overtaking Britain in steel production within 15 years among other goals, peasants were encouraged to construct backyard furnaces to produce steel. Nearly 100 million peasants were involved in this primitive form of steel smelting by late 1958, consuming all available iron products including pots and pans, and producing little of value.[11] To fuel the furnaces, peasants cut down trees where they could find them. The deforestation resulted in erosion, sedimentation, desertification, and changes of microclimates.[12]

The first year of the Great Leap campaign saw bumper crop production, but because party cadres shifted farm labor to steel production, many crops rotted in the fields. Then, over-reporting of agricultural gains increased demands on peasant communities to supply grain to the state in 1959, in a year when harvests turned poor. This led to man-made famine throughout much of rural China, and the loss of between 20 and 30 million lives to starvation and malnutrition.[13] The colossal failure of the GLF was a testament not only to the limits of ideological direction of agricultural policy but also to the failure of command economies.

2.2.3 *The Cultural Revolution (1966–1976)*

China's Cultural Revolution (CR) was partly a reaction of Mao to the aftermath of the Great Leap Forward. At the Lushan Plenum of the CCP in 1959, Marshall Peng Dehuai criticized Mao's direction of the state, leading to Peng's purge from leadership.

[10] Fairbank and Goldman, 2006, 370.

[11] Fairbank and Goldman, 2006, 371.

[12] See Judith Shapiro, *Mao's War Against Nature: Politics and the Environment in Revolutionary China*. Cambridge, UK: Cambridge University Press, 2001, ch. 2.

[13] Fairbank and Goldman, 2006, 368.

Mao withdrew from active party affairs, as different leaders, notably Liu Xiaoqi, Deng Xiaoping, and Chen Yun among others, sought to increase incentives for agricultural and industrial production, and return the state to normalcy.

As with many complex historical events, there were other stimuli to the CR. Mao objected to de-Stalinization in the Soviet Union and Khrushchev's leadership, as the Sino-Soviet split deepened, and as he saw similar tendencies in the CCP. He objected to what he thought was increased bureaucratic rigidity in the CCP, as well as to "capitalist roaders" (such as Liu and Deng), who seemed to have departed from communist principles. Also, he wondered how China's revolutionary spirit could be sustained, as years unfolded after the Long March without revolutionary campaigns to test the mettle of new party cadres. These ruminations occurred during the buildup of the American war in Vietnam from 1963 to 1965, when the United States seemed to press China extremely tightly.[14]

The CR was a time of extreme instability in China, both politically and economically. The party was internally divided, and external forces, such as the Red Guards, played major roles. (Chinese often describe the CR era today as "ten lost years" or the "lost generation," referring to opportunities lost by youth, especially in higher education.) Ultimately, the People's Liberation Army (PLA) was called on to preserve order in the state. Under these conditions, agricultural productivity did not keep pace with population growth. Threats to the Chinese environment were particularly severe during this period. We discuss three especially destructive campaigns of this era, as recounted in Judith Shapiro's *Mao's War Against Nature.*

The first example was the national campaign to "Learn from Dazhai," which was applied uncritically throughout China. The Dazhai model was artificially constructed to be an example of Mao's favorite parable of "The Foolish Old Man Who Removed the Mountain," in order to demonstrate that humans could conquer nature and bend it to their will. Among the most egregious applications of the Dazhai model were attempts to plant wheat on Mongolian grasslands, despoilation of wetlands, and encroaching on lakes and rivers to expand arable land. To purge the lakes, Shapiro notes, was to purge and rebuild the mind: "(T)he battlefield of the lake was an arena for urgent struggle: against nature, against political enemies, and against the limits of human will."[15] Mao's population policy further compounded the problem.

A second example was the large movement, occurring from 1964 to 1971 and prompted by the Vietnam War, to establish a "Third Front" in the western and

---

[14] For good studies of the Cultural Revolution, see Hong Yung Lee, *The Politics of the Chinese Cultural Revolution: A Case Study*. Berkeley, CA: University of California Press, 1978; Lowell Dittmer, *Liu Shao-ch'i and the Chinese Cultural Revolution: The Politics of Mass Criticism*. Berkeley, CA: University of California Press, 1974; Anita Chan, Richard Madsen, and Jonathan Unger, *Chen Village: The Recent History of a Peasant Community in Mao's China*. Berkeley, CA: University of California Press, 1984; and William A. Joseph, Christine P. W. Wong, and David Zweig, *New Perspectives on the Cultural Revolution*. Cambridge, MA: Council on East Asian Studies, Harvard University, 1991.

[15] Shapiro, 2001, 129.

southwest China hinterland. The irrational distribution to this area of strategic industrial plants and facilities caused severe air, water, and soil pollution as well as deforestation.

The third example was the movement to forcibly relocate "educated youth" (called the rustification movement) to rural areas. Their reclamation work damaged wetlands and forests, destroyed the ecology of steppes, and led to desertification. Transformation of the rainforest in Yunnan province into rubber plantations denuded the soil and encouraged overhunting and deforestation. Not only were these examples of the disastrous consequences of large, poorly planned, and hurried mobilization campaigns; they also represented the adverse impacts on the environment of those (such as the lost generation of rusticated youth) who had lost their sense of place.[16]

In summary, for the first generation of its governing of China, the regime made several disastrous policy choices, which imperiled China's food security and its ecological stability.

2.3 Policy Reforms, 1978–2009

Mao Zedong died in 1976, and after a short and nasty aftermath (the attempted seizure of power by the Gang of Four), Deng Xiaoping consolidated his grasp on the apparatus of state power. Until his death in 1997, Deng promoted reforms of China's economy and social system[17] (and some limited political liberalization), reforms continued by his successors. The plans for change crystallized ideas of China's earlier reformers, including Premier Zhou Enlai shortly before his death in January 1976, to engage in long-term industrial, agricultural, military, and scientific/technological development. Called the "Four Modernizations," the reforms unfolded gradually and often haphazardly.[18] Their overall objective, however, was ambitious: to create a prosperous and powerful China by the early twenty-first century.

The initial decade of reforms focused on increasing incentives for production, especially agricultural development, increasing trade with the world, attracting foreign investment to China, and improving China's infrastructure – all of which pertain to China's food security. The development of environmental policy fell behind this initial reform impulse, developing only in the late 1980s, and we consider it in later chapters.

---

[16] Shapiro, 2001, 147–53.

[17] Deng's reforms liberated the Chinese people from restrictions of classes and groups artificially created by Mao, and made them all productive members of society.

[18] For an early analysis of the Four Modernizations, see Richard Baum, ed., *China's Four Modernizations: The New Technological Revolution*. Boulder, CO: Westview Press, 1980.

2.3.1 *Introduction of Production Incentives*

Deng Xiaoping's practical orientation was reflected in statements such as "It does not matter whether the cat is white or black, so long as it catches mice," [the implication being that any person could be productive, whatever his or her class background] and "feeling for stones as one crosses the river" [meaning that Chinese should try new thinking and policies to solve old problems, in other words, be pragmatic]. His approach as China's preeminent leader was flexible, pragmatic, and non-ideological, yet he never recanted his belief in the leadership role of the CCP and that China would remain socialist (but with "Chinese characteristics").

Deng's initial reforms focused on China's rural and agrarian population, who still comprised nearly 80% of China's population in the late 1970s. The rural economy remained relatively stagnant. In the view of party cadres who worked in rural areas during the CR, conditions had not improved much since the establishment of the regime in 1949. Certainly, agricultural productivity lagged very far behind the "miracle economies" of China's capitalist neighboring states: Japan, South Korea, and Taiwan.

Thus there was general elite support for agricultural reform. First, communes were dismantled, with authority devolving to the production team (a small village or part of a village) as the primary agricultural work unit. Increasingly, however, decollectivization proceeded to the level of the household, the family farm. This soon became known as the "household responsibility system," by which was meant that the family had greater control over production decisions. Land still was owned by the state but leased on relatively long terms – as long as 30 years[19] – to farmers.

Farmers gradually have been freed from the requirement that they grow cereal crops at the direction of the state. They could switch to higher-value crops, and they also were free to engage in sideline production and sell products to local markets. Farmers could not alienate the land, as they lacked complete property rights, which is a continuing issue in China's development. The land each household used also was variable; it could increase or decrease based on changes in family size as well as different estimates of land productivity. Recent surveys by Kung and Liu indicate that in the counties they studied, farmers preferred a situation of periodic reassignment of land to a situation of de jure land ownership.[20]

The important point is that the rural household became the center of agricultural production. It had entered the market, as it had production contracts with the state

[19] Jean Oi notes that in the case of marginal lands – barren hills, slopes, ditches and beaches – lease rights could extend to 50 years or more. See Oi, "Two Decades of Rural Reform in China: An Overview and Assessment," *China Quarterly,* 159 (September 1999), 619.

[20] James Kai-sing Kung and Shouying Liu, "Farmers' Preferences Regarding Ownership and Land Tenure in Post-Mao China: Unexpected Evidence from Eight Counties," *The China Journal*, No. 38 (July 1997).

including all means of production except land-use rights and access to irrigation facilities (remaining under the control of the collective). The contract also included all raw materials used in production.[21]

2.3.2 Grain Sales and Changing State Intervention

During the Maoist era, as noted above, the state exercised a monopoly on grain procurement, and it used agriculture to benefit heavy industry by paying farmers low prices for their products. After filling their quotas, farmers often had little left to sustain their families. The economic reforms in agriculture initially tolerated the opening of private markets in grain, which effectively established a dual-market in grain prices: the private, market rate, and the government rate. Yet during the first of what can be identified as six changes in purchase and distribution of staples during the 30-year reform era from 1978 to 2008,[22] the system remained a state monopoly.

Then, beginning in 1984, it became difficult for the government to store all the staples it purchased (as discussed in the next section). Owing to the development of the household responsibility/contract system, farmers were motivated to produce, and there was a large scale increase in production. The inability to store all grain produced led the state to implement a contract purchase system, under which the government signed purchase agreements with farmers. The state purchased only the amount of grain specified in the contract, and remaining grain could be sold on the free market.

The third stage, beginning in 1995, was a reaction to Lester Brown's *Who Will Feed China?* that drew attention of China's leaders to food sufficiency issues and led to research by Chinese scientists on food production and consumption. In 1996, to insure food security, the State Council established a new policy to motivate farmers to grow crops. It established a price support system for staples. The price ceiling was based on the price of each central commodity during the previous 2 years. This policy was based on the assumption that, given the government's control of approximately 60% of the harvest of cereals, it would be able to control the price overall. The government sold cereals (from storage) at a slightly higher price than it purchased cereals, assuming this would avert a fiscal burden to the state. As it eventuated, this assumption proved to be incorrect.

From 1996, the production of staples increased, and the government was not able to avoid losing money when selling grains. Because of production increases, and because the remainder of the harvest was sold on the market, the officially-set price was higher than the market price, causing a drain on the central treasury.

---

[21] Tony Saich, *Governance and Politics of China*, 2nd edn. New York: Palgrave, 2004, 245–46.

[22] This section is based on communications from the State Grain Administration, in 2007–2008, as well as a detailed interview with grain policy experts in the Development Research Center of the State Council, May 19, 2008.

Every year, state fiscal policy allocated $50 billion RMB (US$7.2 billion) in subsidies, simply to pay the interest and storage fees, and to dispose of old (the previous year's) stored cereals.

The fourth stage, from 1998 to 2000, was the peak of grain production in China to that point. In order to reduce its fiscal burden, the central government decentralized the pricing of food to provincial grain bureaus. Also, it increased the quality standards of cereals it purchased, limiting state purchases to high quality goods (without defining clearly what "high" quality meant).[23] This lack of clarity led to disputes and contradictions between central and local authorities.

At the fifth stage, starting in 2000, it was clear that China did not have a food deficit and instead had surplus cereals, available for export. The net export of food in that year exceeded 20 million kilograms. Due not only to excess food production but also to environmental deterioration, the central government loosened its policy on food production. One of the tactics initiated at this time was the *Tuigeng Huanlin* or Sloping Land Conversion Program (SLCP, discussed in Chapter 3), which converted marginal agricultural land to forests or grasslands.

Simultaneously, the state increased flexibility in the structure of the agriculture system. No longer did it rigidly regulate what crops farmers could grow. They could decide themselves whether to plant cereals or other economic crops. Under these circumstances, in the 3–4 years before 2003, the production of cereals declined yearly. By 2003, cereal production had dropped to the level of 1986–1987.

It was in response to this decline in grain production that policies of the current, sixth stage, were developed. In 2004, Premier Wen Jiabao announced that the agricultural tax would be reduced annually and eliminated within 5 years. This was a highly popular proposal for farmers, and in most parts of China, the tax was completely eliminated within 3 years. Other minor, nuisance taxes on agriculture, with the exception of the tobacco tax, were eliminated as well. This reduced the burden on farmers by $50 billion RMB (about US$7.2 billion), and nearly 800 million farmers benefited from the changes.

Second, the state began to provide subsidies for farmers to purchase high quality seeds (in addition to other subsidies for staples production), and it provided subsidies to assist farmers purchasing capital equipment (such as farm machinery) needed to modernize their agricultural practices. The central government took funds from the crop insurance fund (with more than $10 billion RMB [US$1.45 billion]) to subsidize farmers in main crop production areas, and then added an additional $7 billion RMB (US$1.1 billion) more to insure that the system would motivate farmers.

The prices of grains are not completely set by the market in China today. In the early twenty-first century, the central government carefully monitors both the quantity and the price of food staples. To insure grain sufficiency and to provide enough grain for urban consumption, the state enters the market. The state employs three

[23] Ibid.

methods to adjust prices: (1) It uses grain reserves under its direct or indirect control to affect prices; (2) it adjusts the state purchase price (the price support) for major cereals; and (3) it provides other incentives (such as tax relief and subsidies) to encourage greater cultivation of cereals. Food price inflation (about one third of China's general price index or CPI) is watched carefully, and grain prices are paid special attention, for they comprise one third of food prices (or over 10% of the weight of CPI).[24] For example, when cereal prices rose in late 2006, the State Grain Administration and National Development and Reform Commission (NDRC) auctioned off 3.8 million tons of grain reserves.[25]

As prices continued to rise due to the spike in energy as well as food costs (the CPI rose to an 11-year high of 8.7% in February 2008), the state raised the minimum purchase price of rice[26] and wheat. It acted first in early February, and then in late March 2008 raised state purchase prices again. NRDC officials stated their objectives were to encourage grain production and ease inflation. They added that stable food prices would curb costs for livestock farmers and related industries.[27] In addition, the central government again released supplies from state reserves of wheat, rice and pork, and it also increased subsidies to spur further growth of grain.[28] Some economists questioned whether these measures would have a long-term impact, and thought they might instead distort market-based pricing, which would discourage farmers from growing more cereals. Agricultural economist Huang Jikun argued: "China should let grain prices rise now so farmers have a greater incentive to plant more and earn more as well."[29] He wondered how long grain reserves would last were global inflation in food prices to continue.

---

[24] Yali Guo, "Food Prices Push Up CPI, Say Analysts," *China Daily*, May 3, 2007, 2.

[25] Huanxin Zhao, "Ample Food Reserves Can Feed Market Needs," *China Daily*, December 13, 2006, 1.

[26] Rice prices are watched particularly closely. As international rice prices doubled in early 2008, Premier Wen Jiapao said: "China has an abundant supply of rice" and had stockpiles of 40–50 million tons. See: Jize Qin and Zhiming Xin, "World Rice Price Hikes 'Will not hurt Supply,'" *China Daily*, April 1, 2008, 1. Also see: Diao Ying, "Cost of Rice 'Stable' in China," *China Daily,* April 5–6, 2008, 10; and Qiwen Liang, "Claims of Grain Shortage Dismissed," *China Daily*, April 15, 2008, 4.

[27] Zhiming Xin, "Nation to Raise Purchase Prices for Grains," *China Daily*, March 29–30, 2008, 10.

[28] For discussion of grain reserves and their use in stabilizing prices in 2007 and 2008, see these Internet materials: "Chinese State Administration of Grain Refutes Empty Granary Rumor" at http://rss.xinhuanet.com/newsc/english/2008-05/08/content_8131350.htm (retrieved May 18, 2008); "China Says It Is Able to Maintain Stable Grain Supply and Price" at http://news.xinhuanet.com/english/2008-05-06/content_8116728.htm (retrieved May 18, 2008); "Chinese Grain Reserves Sufficient" at http://news/xinhuanet.com/english/2008-05-05-content_8106686.htm (retrieved May 18, 2008); "China's State-Owned Grain Companies Shrug Off Losses" at http://news.xinhuanet.com/english/2008-02/10/content_7586454.htm (retrieved May 18, 2008); and "Time to Reform China's Grain Trade Management Regime" at http://www.caijing.com.cn/20050404/13819.shtml (retrieved May 18, 2008).

[29] Xinhua, "Importing Inflation," *China Daily*, May 19, 2008, 3.

In 2007 the China Grain Reserves Corporation (Sino Grain), controlling the bulk of the wheat crop, easily stabilized market prices for wheat.[30] When global prices rose for wheat, rice, and corn, and several LDCs experienced food riots,[31] Chinese officials attempted to calm consumers and allocated additional funds to subsidize farmers' purchase of seed, diesel, fertilizer and other production costs.[32] Experts in the State Grain Administration (SGA) said grain reserves were 150–200 million tons (if private reserves are included, the total rises to around 250 million tons), about 30% of annual production. Nearly 70% of reserves are rice. Overall, China's reserves are nearly double the 17% safety level set by the FAO. Yet, reflecting concerns mentioned by many agricultural policy analysts, the SGA head Nie Zhenbang referred to shrinking arable land and water shortages, saying in 2008 "We now have less room to increase grain planting acreage, and it's becoming more and more difficult to raise yields."[33]

2.3.3 *Improvements in Infrastructure*

Since the onset of economic reforms, the state also has initiated comprehensive improvements in infrastructure. Given the importance of grain reserves to the control of prices, the state has invested heavily in granaries, but on at least two occasions this was insufficient to meet demand. In 1984 there was the first "granary crisis" as central state storage facilities could not hold all the cereals produced in that year.[34] Then, in 1990 China confronted another granary crisis. Owing to the shortage of strategic warehouses, a large amount of staples had to be stored outside, covered by reed mats. Using this method, it was possible to store grain only for one year, and there were problems of rotting of some stored foods.[35] From available accounts, it appears that the maximum storage capacity was reached in 1998, a peak production year, when some 512.3 million tons of grain were stored in state facilities.[36] As we discuss in Chapter 9, this may not be sufficient for predicted much larger harvests of the future, nor may the facilities be adequate to avoid spoilage.

---

[30] Zhiming Xin, "Bumper Summer Crop on the Cards," *China Daily*, June 18, 2007, 3. Also see, Xinhua, "Bumper Summer Harvest Expected," *China Daily*, June 17, 2008, 4.

[31] See Keith Bradsher, "High Rice Cost Creating Fears of Unrest," *New York Times*, March 29, 2008, A5.

[32] Jing Fu and Lan Wang, "Rising Prices Eating into Farmers' Income," *China Daily*, July 25, 2008, 2.

[33] Jiao Wu, "Ample Grain to Keep Food Prices Stable," *China Daily*, May 7, 2008, 1.

[34] Personal interview with State Grain Administration official, Beijing, May 16, 2008.

[35] Personal interview with grain policy expert, State Council Information office, Beijing, May 19, 2008.

[36] See "Chinese State Administration of Grain Refutes Empty Granary Rumor," http://rss.xinhuanet.com/newsc/english/2008-05/08/content-8131350.htm (retrieved May 18, 2008).

Government investments in transportation have been large and continuing. Along the eastern coast, the central government, provincial and municipal governments have invested in high-grade highways and expressways, often in association with outside investors. In the late 1980s, the state announced planned construction of the national trunk highway system (NTHS). This network of highways will, at its completion in 2010, include 12 high-capacity toll roads, five north–south highways and seven east–west highways. The network will extend 36,000 km and connect all cities with populations greater than one million.[37] Construction of roads in rural provinces has vastly improved transportation of products to markets. Construction and improvement of ports and harbors, particularly in the Shanghai suburban area of Pudong, has made this China's premier international economic and commercial center.

The development of special economic zones in South China, such as Shenzhen and the integration of the economies of Guangdong, Hong Kong, and Macao (in the Pearl River Delta) have made South China one of the most important regions of China. A new Beijing-Kowloon Railway also has opened up the large hinterland for South China and redirected the flow of goods and services.[38] (China has more than doubled the length of its railways in the reform period.)

In the late 1990s, the state launched a "Great Opening of the West" campaign to reduce disparities between this region and the much more highly developed east and south coastal provinces. Development efforts focused on Shaanxi, Gansu, Qinghai, Sichuan, Yunnan, and Guizhou provinces, the autonomous regions of Ningxia, Tibet, Inner Mongolia, Guangxi, and Xinjiang, and the newly formed municipality of Chongqing. Infrastructure improvements included roads, railways, airports, and electric power transmission facilities.[39]

Notwithstanding these significant improvements in infrastructure, much remains to be done, particularly if China applies the principle of comparative advantage to food production and imports grains from countries which have more of an advantage in their production. For example, Li and Wang note that "In 1995, China imported 20 million tons of grain, which took about 6 months to clear from the ports, and some grain vessels had to wait in port for 2 months, causing serious economic losses."[40]

A more important need for improvement in infrastructure is to reduce the imbalance in supply of grains and demand in regional markets of China. In previous years, "transportation bottlenecks impeded the ferrying of quality rice from the

[37] Samuel P. S. Ho and George C. S. Lin, "Non-Agricultural Land Use in Post-Reform China," *China Quarterly*, 179 (September 2004), 764.

[38] Max Lu, "Changes in China's Space Economy Since the Reform," in Chiao-min Hsieh and Max Lu, eds., *Changing China: A Geographic Appraisal.* Boulder, CO: Westview Press, 2004, 116.

[39] See Elizabeth Economy, *The River Runs Black: The Environmental Challenge to China's Future*. Ithaca, NY: Cornell University Press, 2004, 210.

[40] Chenggui Li and Hongchun Wang, "China's Food Security and International Trade," *China & World Economy* (November 4, 2002), 33.

Northeast to the South." Rice shortages in Guangdong in 2008 prompted the Ministry of Railways to order railway authorities in northeastern provinces to improve efficiency and send 10 million tons of rice to the South within 90 days.[41]

2.3.4 Opening to the World

An important and all-encompassing reform of Deng Xiaoping was the opening of China to foreign trade, and the invitation of foreign corporations and business firms to participate in the economic development of China. The success of this policy can be seen in the impressive growth in number of foreign firms operating in China, China's foreign trade, and foreign direct investment (FDI) in China.

Statistics tell part of this story. From 1980 to 1997, more than 160 foreign financial institutions gained government approval to develop businesses in China.[42] Over 80% of the world's largest 500 multinational corporations have now invested in China.[43] The growth in China's foreign trade has been immense. From 1978 to 2002, China's exports grew from about $10 billion to $325 billion; its imports grew from about $11 billion to $295 billion. Actual FDI in China during this period increased from $916 million to more than $52 billion, and China for most of this period drew more FDI than any other country in the world.[44] By the turn of the twenty-first century, China had become the seventh largest trading country in the world. Its foreign exchange reserves in September 2008 at their peak were $1.9 trillion, the largest in the world.[45]

When the reform era began, China was not a member of the General Agreement on Tariffs and Trade (GATT), and it needed to negotiate trade agreements individually with its major trading partners. In the case of some countries, notably the United States, this was an humiliating process for China. After the normalization of diplomatic relations with China in 1979, the USA did extend special trading preference for Chinese imports. However, trade ties were subject to annual review by the executive and the Congress. Indeed, after the Tiananmen crisis in June 1989, the Congress regularly sought linkage of the extension of most-favored nation status to improvement in China's human rights record.

---

[41] Xinhua, "Stabilize Grain Supply," *China Daily*, April 30, 2008, 8.

[42] Bian Kou, "Foreign-funded Banks Land in Chinese Market," *Beijing Review*, December 29, 1997–January 4, 1998, 15–16.

[43] See Guangsheng Shi, Chinese Minister of Foreign Trade and Economic Cooperation, "Remarks at the Reception Hosted by EU-China Business Association and Belgium Chinese Economic and Commercial Council," December 9, 2002, cited in Rex Li, "Security Challenge of an Ascendant China," in Suisheng Zhao, ed., *Chinese Foreign Policy: Pragmatism and Strategic Behavior*. Armonk, NY: M.E. Sharpe, 2004, 38.

[44] Nicholas R. Lardy, *China in the World Economy*. Washington, DC: Institute for International Economics, 1994, 303, and Li, 2004, 38.

[45] Xinhua, "Forex Reserves Drop for First Time since 2003," *China Daily*, December 12, 2008, 1.

From the late 1980s through the 1990s, China underwent the long negotiations process to join the global trade organization (after 1995, the World Trade Organization or WTO). China's acceptance into the WTO required changes in the nature of its economic structure, centering on trimming back the state-owned enterprises established during the command economy phase, and on increasing competitiveness throughout the economy. China's major trading partners (the USA, EU, and Japan) allowed it several years to make a smooth transition, but entrance to the WTO, finally approved by all parties in late December 2001, required the leadership to step up reform efforts, including privatization. In the case of state intervention in agriculture, WTO entry forced changes in operations of the state grain procurement bureaus. In 2003, the government announced it would end hidden subsidies for agriculture. Instead, farmers would plant based on market circumstances and receive state subsidies if the market performed poorly. The state grain procurement offices and trading companies would need to compete with private grain buyers.[46]

Altogether, the economic reforms discussed above have been comprehensive, gradual, and far-reaching in their effects. Although a number of socialistic characteristics remain in the food production system, it is considerably more capitalistic in its approach, efficient and productive than in the Maoist era.

2.4 China's Current Food System

As noted in Chapter 1, a nation's food system includes the dimensions of availability, access, and utilization of food. In this section, we first consider the food products on China's tables, and then review estimates on food sufficiency. We also look at differences in food access of population groups, and then treat changes in food preferences.

2.4.1 The Food on China's Tables

The basis of the Chinese diet is boiled grain, and this provides most of the calories in the average diet. Reliance on plant foods is heavy, but increasingly animal foods are supplying more protein in the diet.[47]

---

[46] Saich, 2004, 250–51.

[47] This section relies extensively on E. N. Anderson, *The Food of China*. New Haven, CT: Yale University Press, 1988, 137–81.

2.4.1.1 Plant Foods

The primary grain crop for about half to two-thirds of China's population is rice, and its major area of cultivation is southern China, where typically paddy fields are extensively irrigated. China is the world's largest producer of rice, followed by India and then Indonesia.[48] Of the several varieties of rice, the most prevalent in China are *indica, japonica*, and glutinous rice. In the North, dry rice is cultivated. Rice is double-cropped throughout southern China, and in some areas even triple-cropped. Yields of rice in China are among the highest in the world, but diseases, especially rice blast, influence productivity.

Among all the types of rice, paddy rice is the most important and has the largest acreage. Paddy rice was domesticated from wild rice by the BaiYue tribe in southern China. It is the main staple for people in the South. Because of its greater yield, cultivation of paddy rice gradually expanded to the North, as rice varieties developed which were more tolerant to low temperatures and improvements in irrigation systems. Recent successful cultivation of paddy rice in Heilongjiang Province (46° latitude) was heralded by the government as a major accomplishment. In China, to make better use of limited agricultural land, rice is cultivated by first raising the seedlings in a nursery. When seedlings (in a small bundle of two or three) reach 25 cm in height, they are transplanted to a prepared paddy field in a row of 20–30 m long with spacing of 10–20 cm. The field is flooded which is also an effective means to reduce weeds.

China's second most used grain crop is wheat. China also is the world's largest producer of wheat, followed by India, the USA, and Russia.[49] This crop was introduced to China from the Middle East, possibly through Xinjiang.[50] Evidence found during the Neolithic period (3,000 BC) indicates that wheat was under cultivation west of the Yellow River in what today is Gansu Province. Cultivation of wheat became common in the Qin Dynasty. Today, wheat is grown mostly in North, Central, and western China. Typically, wheat is used to produce flour, and the flour in turn is the basis for dumplings and noodles. Wheat is grown in both summer and winter.

The earliest description of maize (corn) in China was in the *Diannan Benciao* (Southern Yunnan Botanical Record), several decades prior to Columbus' voyage to the Americas. Initially, maize was cultivated in the southwestern areas of Yunnan and Sichuan, and then spread to the entire country. Large scale cultivation of corn in China began in the Qing Dynasty, as the population increased. Corn is very adaptable to different environmental conditions, and it grows well on marginal lands. Compared to other crops such as sorghum, corn is much more economical. By the nineteenth century, corn cultivation expanded to the plains of Huabei (Hebei, Henan, Shanxi, Shaanxi) and Dongbei (Liaoning, Jilin, Heilongjiang).

---

[48] See http://www.fao.org/es/ess/top/commodity.html?lang+en&item27&year=2005, retrieved March 11, 2008.

[49] See: http://www.fao.org/es/ess/top/commodity.html?lang=en&item=15&year=2005, retrieved March 11, 2008.

[50] Genpan Li, *China's Ancient Agriculture* (in Chinese). Taipei: Taiwan Commercial Press, 1994, 58.

Maize is now the third most important grain in China, and China is the world's second largest producer, following the USA.[51] In the early twenty-first century it also is the most controversial. About 7% of corn is consumed as food – eaten directly in the central North, and in mountainous areas of the West and South, or used to make steamed bread. Most corn (about 68%) is used as animal feed. Increasingly, however, corn has become an industrial product to produce starch and ethanol,[52] and by 2007–2008 China had become the world's third largest producer of ethanol.

The dramatic increase in use of corn to make ethanol brought about a 40% increase in corn production from 2000 to 2006. This growth responded to global demand for biofuels to meet energy shortages created by rising oil prices in 2004–2008, as well as the search for cleaner energy sources to combat global warming. Corn production earned higher profits than other staples.[53] Corn price increases from 2006 to 2008 fueled concerns about food price inflation. Officials blamed China's price surges on international corn market inflation, and particularly the diversion of US corn production from food to ethanol markets.

Corn constituted three fourths of the raw material producing ethanol in China, and the state had planned to increase biofuel production several fold in the next decades.[54] However, in mid-2007, the State Council responded to rising corn prices by imposing a moratorium on projects making ethanol from corn and other basic food crops. Authorities decided to phase out use of corn for ethanol, and replace it with substitutes of non-cereals such as batata, cassava, grasses, and sweet potatoes.[55] To make production of these non-cereal crops more attractive, the Ministry of Finance proposed subsidies for each hectare of non-food products that could be used to make biofuels, dependent on crude oil prices.[56]

Returning to plant foods, the fourth most important are species of the millet family. Foxtail millet was used traditionally throughout northern China to make porridge. Today, the most prevalent millet species is sorghum (*gaoliang* in Chinese), and it grows well in arid areas of China. China is the world's third largest producer of millet, following India and Nigeria. It is the seventh largest producer of sorghum.[57]

---

[51] See http://www.fao.org/es/ess/top/commodity.html?lang=en&item=56&year=2005, retrieved March 11, 2008.

[52] Jiao Wu, "Industrial Demand for Corn Increasing," *China Daily*, April 10, 2007, 1.

[53] Dingding Xin, "Corn Output Reaches Record High," *China Daily*, March 13, 2007, 2.

[54] Chuanjiao Xie, "Ethanol Output has Corn Prices Popping," *China Daily*, December 6, 2006, 1.

[55] See Xiaohua Sun, "Non-Staple Crops New Source for Ethanol," *China Daily*, June 13, 2007, 14; Tian Le, "Ban on Use of Corn for Ethanol Lauded," *China Daily*, June 22, 2007, 2; Xiaohua Sun, "Corn Won't Be Used for Biofuel in 5 Yrs." *China Daily*, July 17, 2007, 1; and Zhuoqiong Wang, "Biofuel Expert Allays Food-Shortage Worries," *China Daily*, October 12, 2007, 3. Agriculture minister Sun Zhengcai concluded this discussion by stating: "China will never develop biofuels at the cost of grain supplies or arable land." Jiao Wu, "Food Before Biofuels," *China Daily*, May 19, 2008, 3.

[56] Xinhua, "Bio-fuel Makers to Get Subsidy," *China Daily*, December 8–9, 2007, 1, 2.

[57] See http://www.fao.org/es/ess/top/commodity.html?lang=en&item=79&year=2005 and http://www.fao.org/es/ess/top/commodity.html?lang=en&item=83&year=2005, retrieved March 18, 2008.

Both white and sweet potatoes are considered grains for statistical purposes in China. Although rarely are potatoes the sole staple, they are an important part of the diet locally in many parts of China. Potatoes were introduced to China at the end of the Ming Dynasty, by the Portuguese from the South (through Taiwan) and by Russians from the North.[58] By the early twenty-first century, China had become the world's largest producer of potatoes, followed by India.[59] In the judgment of experts, China cultivates fewer hectares of potatoes than optimal. The potato is vitamin-, protein-, and mineral-rich and provides most of daily nutritional requirements. Also, drought-resistant varieties grow well in the half of China's arable land lacking sufficient water.[60] However, China's potato yields are about one-third those of developed countries, largely because of the prevalence of potato diseases (as well as poor growing conditions). Other root crops, such as manioc and taro, play a less important part in the diet.

The soybean is China's best known pulse crop. High in protein, it is typically used to make fresh or dried bean curd (*doufu*), cooking oil, and soya cakes for animal food. China is the world's largest importer of soybeans. The largest producers in 2009 were Brazil, the United States, and Argentina, and increasingly South American producers have competed heavily with the USA in the China market.[61] Other beans include the mung bean (called *ludou* in Chinese), peanuts, red beans, and the yam bean.

Both leafy and non-leafy vegetables (*cai*) play important parts in the Chinese diet. Leafy vegetables include cabbages, mustard greens, kale, collards, spinach, lettuce, and watercress. The most important non-leafy vegetables are root crops, such as the white radish, carrots, onions, scallions, and garlic.

Besides potatoes, other plants of the nightshade family also are important in the Chinese cuisine – for example, eggplant, tomatoes, chili and other peppers. Melons, zucchini, bitter melon, squash, pumpkins, and cucumbers are standard items of the diet too.

Fruits prevalently found on China's tables include apples, pears, peaches, plums, grapes, apricots, sweet melons, watermelons, cherries, loquat, and jujube (dates). Citrus fruits such as oranges and lemons are prevalent, as are lichi, papaya, pineapple, pomegranates, and pomelos. Nuts are a much smaller part of the Chinese diet than in other countries, but herbs and spices are essential parts of diets based on grains.

Animal Foods

Aquatic species play a larger role in Chinese diets than land animals, and a diverse set is consumed. Perhaps the most popular water species are sea cucumber and

[58] Li, 1994, 102.

[59] See Qingbin Wang and Wei Zhang, "China's Potato Industry and Potential Impacts on the Global Market," *American Journal of Potato Research*, Vol. 18 (Mar/Apr 2004), 101–09.

[60] Weifeng Liu, "Potato Could Rival Rice, Expert Says," *China Daily*, October 11, 2007, 4; and Xinhua, "Potato Seen as Viable Alternative," *China Daily*, August 15, 2007, 5.

[61] See "Soybeans and Oil Crops: Market Outlook," at http://www.ers.usda.gov/briefing/soybeansoilcrops/2008baseline.htm; retrieved June 24, 2008.

shark fins (for the affluent), shrimp, crab, carp, groupers, pomfret, and oysters. Increasingly, aquatic species and especially fish are farmed and not taken wild from fresh or marine waters.

The primary meat animals in China historically and today are pigs and chickens. China produces half the world's pork, and it is the second largest poultry producer. The third animal species used in diets is sheep. Cattle, particularly water buffalo, were too precious as draft animals to be used as food in traditional China; moreover, Buddhists avoided beef in their diet. Today beef is gaining in popularity, but still is less important in the standard diet than pork, mutton, and chicken. Nevertheless, China ranks third in global beef production. Ducks and geese also are eaten. Dairy products have gained in popularity, as incomes have risen, and as awareness of the health benefits of milk has grown.

The Chinese diet is exceptionally varied; when food security is discussed, however, what most observers first consider is cereal crops – rice, wheat, corn, sorghum, and potatoes.

2.4.2 Degree of Food Sufficiency

Grain crops comprise about three-fourths of the total growing area and one-third of the overall value of China's agricultural output. As mentioned, rice is the most important cereal crop; it makes up roughly one-third of the area planted to cereals and between 40% and 45% of total grain production. Wheat and maize together comprise nearly one-fourth of total grain production,[62] and as mentioned, maize production areas are increasing rapidly.

In the Maoist era as well as in the period of economic reform, self-sufficiency in grain production has been a major objective of the state. Before the early 1990s, imports of grain exceeded exports. In 2000, China was the world's ninth largest exporter of food products ($16.38 billion or 2.9% of world exports) and the eighth largest importer ($19.54 billion or 3.2% of world imports).[63] After entering the WTO, China's food exports increased by 11% and its imports rose by 22%. Yet, with imports and exports comprising less than 15% of agricultural GDP, China has a relatively low dependence on the international market. Throughout the history of the People's Republic (except in the GLF), China has been mostly self-sufficient in grain production. However, for two cereals, corn and soybeans, trade dependence is likely to increase. China's self-sufficiency in corn is expected to decline from 97.4% in 2000 to 74% in 2020. For soybeans, the ratio is expected to change from

---

[62] Shaolian Liao, "Food Production in China," in Aileen S.P. Baviera, Shaolian Liao, and Clarissa V. Militante, *Food Security in China & Southeast Asia*. Manila, Philippines: Phlippine-China Development Resource Center, 1999.

[63] Peter Oosterveer, *Global Governance of Food Production and Consumption: Issues and Challenges*. Cheltenham, UK: Edward Elgar, 2007, 3.

Table 2.1 Grain production, selected years (Adapted by authors from Shaolian Liao, "Food Production in China," 1999, 10; and USDA, Foreign Agricultural Service, "Facts about Agriculture in China – 2005" retrieved October 11, 2006 from http://www.fas.usda.gov/remote/china_countrypage/chindex.htm; National Bureau of Statistics, *China Statistical Yearbook 2007*. Beijing: China Statistics Press, 2007, 462, and Xinhua, "Grain Trade in Deficit for Jan, Feb," *China Daily*, April 16, 2008, 3)

Year	Grain production (million tons)
1957	200
1978	305
1984	407
1985	379
1990	446
1995	465
1997	495
2001	453
2002	457
2003	431
2004	469
2005	484
2006	497
2007	502

46.1% in 2000 to 20.6% in 2020. By that time, value of imported grains is expected to total 15% of agricultural GDP.[64] Table 2.1 presents some statistics on food production.

As we note in Chapter 3, the amount of arable land in China has declined since 1949. Food production, however, has kept pace (but in some years, just barely) with population growth. As Liao notes, China is the world's largest grain producer and "has basically solved the issue of feeding its huge population."[65] Indeed, China has fed its population while increasing domestic food availability, from 1,500 cal per capita per day at the start of the 1960s to 3,000 cal per capita per day at the onset of the twenty-first century.[66]

Other specialists in food production issues emphasize the challenges to self-sufficiency, including insufficient agricultural investment, declining arable land, inadequate application and use of modern agricultural technology, among other factors. Of course, sufficiency depends not only on population size but the average consumption of grain. Production, in turn, varies by natural factors (natural disasters), disease and pest problems, as well as state incentives. For example, responding

[64] Jiao Wu, "Overseas Food not China's Staple," *China Daily*, April 26, 2007, 3.

[65] Ibid., 11.

[66] Edward Anderson, "Annex 4: Food Security in China," in Gerard J. Gill, John Farrington, Edward Anderson, Cecilia Luttrell, Tim Conway, N.C. Saxena and Rachel Slater, *Food Security and the Millennium Development Goal on Hunger in Asia*. London: Overseas Development Institute, Working Paper 231, December 2003, 1.

to the low production figures for 2003, the state made provisions for tax relief, direct payments to grain farmers, and guaranteed prices – all of which had an impact, leading to much larger productivity in 2004.

One cannot assume that grain consumption will remain constant in the future. Were per person consumption 400 kg of grain a year, with a population of 1.3 billion, total demand in 2008 would be 520 million tons. However, were Chinese to consume on average 500 kg of grain yearly, total demand would be 650 million tons. Demand for grains has increased faster than population growth. However, counteracting this is changing food preferences, discussed below. To the present, food security in China has been interpreted almost exclusively as grain security, and this may no longer be the most appropriate measurement.

2.4.3 Regional and Income Variations

Access to food depends on geographic variables (in developing nations, this particularly means distance from food producing regions) as well as income. Both factors influence access to grains in China. Part of the population, especially inhabitants of peripheral areas – Inner Mongolia, Xinjiang, Tibet, sections of Yunnan – are distant from grain processing regions. Their access depends on infrastructure, which is less well developed in these regions. Overall food costs are higher than those in China's central and eastern areas, and incomes are lower. Although state intervention in grain markets provides some uniformity of price for staples nationwide, other food costs are likely to be higher.

Where one lives in China also influences one's income-making opportunities, and food access correlates with income as well. While there are urban residents with limited access to food – and these are largely migrant workers from rural areas – the largest number of people with concerns about food costs are in China's vast rural areas. Inequality of rural and urban regions in China has greatly increased since the onset of economic reforms. Saich reports that the income gap between rural and urban areas in 1986 was 2.33:1,[67] worse than it had been at the start of economic reforms and much greater than in any period of the Maoist era. Poor people in rural areas earn low wages, and spending on food products (beyond payments available as non-cash income, if they work on farms) comprises a larger percentage of income. The rural poor frequently supplement their income by collecting wild mushrooms and wild vegetables, selling them in markets and restaurants, which often causes disturbance to the ecosystem and loss of biodiversity.

Official figures show that more than 200 million Chinese have been lifted out of absolute poverty since economic reforms began. In 1978, government statistics showed 250 million rural people living in poverty, compared to just 28 million by 2002. However, Saich points out that the government measurement of poverty was

[67] Saich, 2004, 248.

$0.66 per day, whereas the usual World Bank standard is $1.00 per day. By the World Bank benchmark, about 23% of rural residents, or 18.5% of the total population of China in 2002 – some 200 million – were still quite poor. Adopting the World Bank's standard of $2.00 per day for a subsistence lifestyle, more than half, or 54% of the rural population, would be included.[68] As we note in Chapter 9, there is no consensus on the number of impoverished Chinese because of differences in definitions. The most authoritative (and recent) figure is the World Bank's: 135 million Chinese fall below the poverty line, or about 10% of the population.[69]

Government programs to alleviate poverty (and thereby improve food access) include subsidized loan programs, the food-for-work program, a budgetary grant program, and in 2004 elimination of the main agricultural tax. Saich comments about the subsidized loan and food-for-work programs that "(T)he impact of these programmes on helping the poorest has not been significant and there have been serious distortions in implementation."[70]

A recent study of family income uses household survey data collected by the Economics Institute of the Chinese Academy of Social Sciences (CASS). These survey data are available for 1988, 1995, and 2002, allowing a comparative analysis of poverty alleviation and income inequality in China. Authors Khan and Riskin find that from 1995 to 2002, inequality in the distribution of rural income fell almost 10% as measured by the Gini ratio (the internationally accepted measure of income inequality in national populations).[71] They attribute this reduction to several factors. First, inequality among the different provinces declined and within each province, due to a reduction in regressive taxes as well as rises in farm and wage incomes. A second important factor was the decline in rural population (which peaked, at about 750 million, in 1995). Effectively, surplus rural labor migrated to urban areas in search of work. While they were less likely than urban residents to have jobs, they were no longer dragging down rural income charts.

Despite declines in rural/urban inequality rates, China's overall Gini ratio remained virtually unchanged, at 0.45,[72] which is less than that of the most unequal countries in Latin America – Argentina (0.52), Chile (0.57), and Brazil (0.58), but above the ratio in the United States (0.41).[73] Khan and Riskin suggest that government policies have had some beneficial effects, particularly the "Great Western Development Strategy," programs targeted at both rural and urban poor, and restrictions

---

[68] Saich, 2004, 298.

[69] World Bank, *From Poor Areas to Poor People: China's Evolving Poverty Reduction Agenda.* March, 2009, iii.

[70] Saich, 2004, 300.

[71] This discussion follows Azizur Rahman Khan and Carl Riskin, "China's Household Income and Its Distribution, 1995 and 2002," *China Quarterly*, 182 (June 2005), 356–84.

[72] A 2007 report by CASS listed China's Gini coefficient at 0.496. The institute's data showed that the richest 10 percent of Chinese households owned more than 40 percent of private assets; the poorest 10 percent, on the other hand, controlled less than 1 percent. Cited in Robert Ash, "Quarterly Chronicle," *China Quarterly*, 190 (June 2007), 537.

[73] Khan and Riskin, 2005, 358.

on discriminatory treatment of migrant workers, but they believe that little clear progress has been made in reducing the large gap of income between rural and urban areas.[74]

We return to the topic of poverty alleviation in Chapter 9. Suffice it to say here that access to food remains uneven in China today, with greater difficulties of access in rural than in urban areas.

2.4.4 Changes in Food Preferences

An extremely important change to food consumption has been the increased wealth of China's population, and the significant reduction in poverty. Chinese in 2009 have more disposable income to use on food purchases than at any time in the previous 60 years. A large middle class has grown in tandem with the rapid increase of economic development. Indeed, many newspaper stories in 2006–2009 featured the obesity problem, without emphasizing the large increase in household wealth and changes in standard of living (including consumption of fast foods) that has produced it.

Changes in food consumption patterns can be observed in the shopping behavior of Chinese, in meals cooked at home, in street stalls, even in restaurants. A sign of change is that McDonalds has about 500 restaurants in China, which usually are crowded. Foreign food shelves in supermarkets bulge with goods.[75] A recent study goes one step further by examining through survey research the food consumption pattern in Qingdao, a coastal city in Shandong Province. The study by rural sociologist Jussaume notes two overall changes in preferences.

First, consumers are less reliant on grains and more likely to increase their use of "high quality" foods, such as meat and fresh produce.[76] (In the last 30 years, per capita direct consumption of grains [by humans] has steadily declined, matched by increased grain used as feed for animals.[77]) Survey respondents reported they were more likely to eat pork and chicken (but not beef) than previously. Also, they spent more of their food budget on fresh produce (and we could add, on fruit products as well). Meat products (but to a lesser extent vegetables and fruits) are income elastic, meaning that the higher one's income, the more able one is to afford them. Thus, as the population has become richer, it has shifted from the traditional diet of rice, noodles, or steamed bread, to a diet that is more varied, richer in protein, and more expensive to consume.

[74] Khan and Riskin, 2005, 383–84.

[75] Erik Nilsson, "Hunger for the West," *China Daily*, June 20, 2007, 20.

[76] Raymond A. Jussaume Jr., "Factors Associated with Modern Urban Chinese Food Consumption Patterns," *Journal of Contemporary China*, Vol. 10, no. 27 (May 2001), 229.

[77] Nuo You and Jiao Wu, "Looking Behind the Global Food Crisis," *China Daily*, July 29, 2008, 8.

Reports from other scholars of food point to increased use of dairy products, particularly milk, among middle and upper-class Chinese.[78] Too, there has been increased attention to food safety, which is represented in the appearance of "green food," organically-grown products in marketplaces.[79]

A second change is the increased preference for shopping at supermarkets and other large retail establishments. This change in food retailing practices follows the rise of market penetration by large food industry firms, such as Carrefoure. Because supermarket prices tend to be higher than those of street markets (but not necessarily higher than those in smaller retail outlets), this behavior is related to income level as well.

Jussaume did not examine the preference of his respondents for western foods. When considering the motivation behind the change in food preference patterns, the most interesting association he found was between preference for meat, fresh produce, and fruits and inclination to shop at supermarkets.[80] Of course, both are income-related, which suggests that it is change in income that drives change in food preference. As Chinese economic development proceeds, and more Chinese are able to purchase "high quality" foods, we can expect that grain sufficiency alone will be even less good a measure of food security.

2.5 Conclusions

We have recounted the large changes in China's food system since the establishment of the People's Republic in 1949. Initially, China's new rulers developed a command economy. They centrally determined grain production targets, and were obsessive in their desire to insure that China was self-sufficient in grain. Although leaders began agricultural change in modest steps, soon Maoist ideological goals took over planning efforts. Within a short space of years, peasant plots were collectivized and by 1958 organized into communes.

Then there followed two disastrous socioeconomic and political experiments: the Great Leap Forward (1958–1960) and the Cultural Revolution (1966–1976). The GLF exacerbated the effects of natural disasters and abrupt change in agricultural

[78] For additional information on China's nutritional transition, see Barry M. Popkin et al., "The Nutrition Transition in China: A Cross-Sectional Analysis," *European Journal of Clinical Nutrition,* Vol. 47 (1993), 333–46; Du et al., "A New Stage of the Nutrition Transition in China," *Public Health Nutrition* (Vol. 5(1A) (2002), 169–74; and Vaclav Smil, "Feeding China," *Current History*, Vol. 94, no. 593 (1995), 280–84.

[79] Observations of the authors. Also, see F. J. Simmons, *Food in China: A Cultural and Historical Inquiry.* New York: CRC Press, 1991; and Hsin-Hui Hsu, Wen S. Chen, and Fred Gale, "How Will Rising Income Affect the Structure of Demand," in Fred Gale, ed., *China's Food and Agriculture: Issues for the 21st Century*. Washington, DC: USDA Economic Research Service, 2002, 10–13.

[80] Jussaume, 2001, 232.

practices and organization, leading to one of the greatest famines in Chinese history and the starvation of 20–30 million people. Both the GLF and CR were serious assaults on the agricultural ecosystem, whose adverse effects continue to the present.

The death of Mao Zedong ushered in another pattern of change: Deng Xiaoping's economic reforms. Communes were dismantled and the household responsibility system created incentives that led to increased agricultural efficiency and productivity. The regime intervened somewhat less in setting agricultural prices, which had both positive and negative effects. Overall, however, production of staples kept pace with population growth. The state also invested heavily in infrastructure, and, by opening China to the world market, increased external competitive pressures on food production in China.

China's food system in 2009 is still reliant on rice, wheat, maize, and sorghum (and to a lesser extent, potatoes). In the 1990s and early twenty-first century, China has imported more agricultural products than it has exported, but is nearly self-sufficient in cereals (about 95%). Yet economic reforms beginning in the late 1970s introduced large disparities in income, both within rural and urban areas and across them. For poor Chinese, access to food in 2009 is not much better than during the 1950s and 1960s. There are fewer poor Chinese than in the 1950s, though, and a new middle class has risen. Its food preferences are different from those in traditional Chinese society, making grain security alone an unreliable indicator of food security.

In the next chapter we turn to the immediate environmental stressors on food security.

Chapter 3
Immediate Environmental Stressors on Food Security*

Abstract This chapter considers the immediate forces influencing China's food system and food security. By immediate is meant events of the reform period, from the late 1970s to 2009. It begins by asking the question that has preoccupied specialists since the publication of Lester Brown's *Who Will Feed China?* in 1995: How much arable land does China have? Is that land area sufficient to insure grain sufficiency? To insure food security?

The chapter focuses on the human pressures on the food production environment, and then treats the effects of socioeconomic change: land, air, and water degradation. The core of the chapter examines seven responses of the state to both perceived and actual environmental stressors: policy restricting arable land conversion, China's one-child policy, investment in irrigation systems, large-scale dam construction, the South-North Water Diversion Project, large-scale afforestation and reforestation campaigns, and the program to convert marginal agricultural lands to forests and grasslands.

Keywords Arable land • Urbanization • Economic development • Erosion • Deforestation • Desertification • Land pollution • Air pollution • Water sufficiency • Water pollution • Ocean pollution • One-child policy • Three Gorges Dam • South-North Water Diversion Project • Afforestation • Reforestation • Slope Land Conversion Program ("Grain to Green")

3.1 How Much Arable Land Does China Have?

China has a total land area of 960 million hectares, making it the world's third largest country. Three terrestrial systems cover 82% of this surface area: grasslands cover 42%, forests 27%, and croplands only 13%.[1] China's cropland is small for the size of the population, and the productivity of the land is low.

---

*An earlier version of this chapter was published in the *Journal of Chinese Political Studies*, a Springer publication, entitled "Environmental Stressors and Food Security in China

[1] State Statistical Bureau. *China Statistics Yearbook*. Beijing: China Statistics Press, 2005.

J.H. McBeath and J. McBeath, *Environmental Change and Food Security in China*, Advances in Global Change Research 35, DOI 10.1007/978-1-4020-9180-3_3,

Lester Brown's prediction in 1995 that China would have to import 200 million tons of grain by 2020[2] initiated a debate among scholars as well as government officials on grain sufficiency. This debate focused on the amount of arable land in China, and whether it was sufficient to sustain agricultural production of staples. In the late 1990s, the official government estimate (now revised upward) was approximately 95 million hectares; on a per capita basis, this would equal 0.08 ha per person, making China's land availability about one-fourth of the global average.[3]

The major critic of the Brown hypothesis has been Vaclav Smil, a geographer at the University of Manitoba. Smil makes a convincing argument that grain sufficiency pessimists underestimate the amount of China's arable land by at least 50%.[4] He points out the several reasons why official statistics are wrong: (1) a non-standard accounting unit is used for the areal measurement of land – the *mu* (there are about 15 mu to the hectare[5]); (2) there were large incentives to underreport land in the Maoist era, for underreporting reduced land taxes and also allowed peasants (and collective leaders) to claim higher harvests per mu; and (3) under the somewhat more privatized system of land use in China today, underreporting land allowed fairer apportionment of marginal, less productive land; it reduced the quotas required for delivery to the state at fixed prices; and it reduced taxes as well.[6]

In the last 2 decades, analysts have made two improvements in land measurement: remote sensing and detailed land surveys. These have produced a consensus among researchers that the range of arable land is between 131 and 137 million hectares.[7] (As we note below, government officials in 2008 most often used the figure of 121.5 million hectares.) Smil finds confirmation for the recent estimates of Chinese researchers and officials in results of the MEDEA study, a multi-disciplinary scientific program using US intelligence satellites and a methodology employing stratified, multi-stage area estimation.[8]

Even this approach is too conservative, in the view of Smil, because it omits measurement of non-traditional land uses, which nevertheless produce goods serving

---

[2] Lester R. Brown, *Who Will Feed China?* New York: W.W. Norton, 1995.

[3] World Resources Institute, *World Resources 1998–99*. New York: Oxford University Press, 1999.

[4] Vaclav Smil, "Research Note: China's Arable Land," *China Quarterly*, 158 (June 1999), 414–29.

[5] The hectare is approximately 2.47 English acres. The *mu* (also spelled *mou*) is approximately one-fifteenth of a hectare. However, historically the mu has not been standardized. See Clifton W. Pannell and Runsheng Yin, "Diminishing Cropland and Agricultural Outlook," in Chiao-min Hsieh and Max Lu, eds., *Changing China: A Geographic Appraisal*. Boulder, CO: Westview Press, 2004, 36.

[6] Smil, 1999, 417. See also China Development Brief, "Statistics: Seeking Truth from (Tonnes of) Facts." Retrieved February 15, 2006 from: http://www.chinadevelopmentbrief.com

[7] Smil, 1999, 419; also see Chuanchun Wu, "Land Utilization," in Geping Qu and Woyen Lee, eds., *Managing the Environment in China*. Dublin: Tycooly International, 1984; and Gerhard K. Heilig, "Anthropogenic Factors in Land-Use Change in China," *Population and Development Review*, Vol. 23, no. 1 (1997), 142.

[8] Smil, 1999, 419–20.

nutritional needs of modern Chinese. Specifically, traditional land measurement does not include fish ponds and orchards, and both farmed fish and fruits play an increasingly important role in Chinese nutrition (discussed in Chapter 2). By adding these surfaces, Smil estimates that land devoted to intensive food production was in the range 146–160 million hectares in 1997 – an average of 153 million hectares or 63% higher than official estimates (and on a per capita basis higher than figures for Japan, South Korea, or Taiwan).[9]

In 2007, China listed the area of cultivated land as 130,039,200 ha. This is based on the situation surveyed as of late 1996. Estimates of the National Bureau of Statistics for 2001 show 127,082,000 ha. Of this amount, "regularly cultivated land" comprised 105,826,020 ha and "temporarily cultivated land" was 21,256,000.[10] As mentioned throughout this chapter, attention focuses both on loss of arable land to other purposes and attempts to increase arable land. For example, in 2006 China lost 307,000 ha, mostly for new construction.[11] In the National Agricultural and Rural Economic Development Program for the Eleventh Five-Year Plan (2006–2010), the Ministry of Land and Resources predicted that grain-producing land would decline by 0.18% annually (based on loss of 8 million hectares of arable land from 1999 to 2005). It estimated the need for at least 103.33 million hectares in 2010 to reach a target production of 500 million tons of grain.[12] Simultaneously, the Ministry of Land and Resources announced that between 1999 and 2006, China gained 2.4 million hectares of arable land (and during this period, grain production increased by 10–20% in pilot areas).[13]

As we note below, this information does not close the debate, which has refocused on the ways to increase production of both plant and animal foods. Moreover, it is abundantly clear that whatever the areal measurement of China's land in 1949, since the late 1970s, China has lost lands formerly used for production of food crops. We now seek an explanation for this loss.

3.2 Causes of Arable Land Loss

Three interrelated factors are the source of pressure on arable land in China: population, urbanization, and economic development. As Ho and Lin note, they explain about three-fourths of the variation in the share of land employed for non-agricultural uses.[14] We treat each in turn.

---

[9] Smil, 1999, 423–24.

[10] National Bureau of Statistics, *China Statistical Yearbook 2007*. Beijing: China Statistics Press, 2007, 464.

[11] Fangchao Li, "Arable Land Bank Continues Decline," *China Daily*, April 13, 2007, 3.

[12] Huanxin Zhao, "Bottom Line Set for Grain Production," *China Daily*, August 4, 2006, 1.

[13] Chuanjiao Xie, "1.7m Hectares of Arable Land 'by 2020'" *China Daily*, June 22, 2007, 2.

[14] Samuel P. S. Ho and George C. S. Lin, "Non-Agricultural Land Use in Post-Reform China," *China Quarterly*, 179 (September 2004), 776.

3.2.1 Population Growth and Pressure

China is the world's most populated nation-state and has been so since the dynastic era. Table 3.1 reports the growth of population, in selected years:

China's population more than doubled in the 50-year period from 1949 to 1999. Only since 1979 has a clear population limitation strategy been in effect. It was obvious from the results of the first national census in 1953, when population increases clearly exceeded rises in agricultural productivity, that some form of national birth control planning effort would be needed. However, Mao Zedong was at best ambiguous on the subject of birth control.[15] He made several statements to the effect that he favored birth control, but he also said (in 1958) that a population of more than one billion would be "no cause for alarm."[16]

Mao's actions on population questions, however, sent a clear message. In 1957, demographer and Peking University president Ma Yinchu warned, based on the 1953 census, that China's rapid population growth would jeopardize development if not checked. For his forthright views, which contradicted state policy and Mao's many statements that China's strength lay in her huge and growing population, Ma was silenced, forced to resign from the university, and stripped of his academic and government posts.[17]

The size of China's population, which is now expected to peak at 1.6 billion in 2030,[18] puts immense pressure on the land, but this pressure is uneven. In the

Table 3.1 China's population, selected years (China Statistical Yearbook, 2007, 105)

Year	Population (in million)
1949	541.67
1954	602.66
1959	672.07
1964	704.99
1969	806.71
1974	908.59
1979	975.42
1984	1,043.57
1989	1,127.04
1994	1,198.50
1999	1,257.86
2004	1,299.88
2006	1,314.48

[15] Susan Greenhalgh & Edwin A. Winckler, *Governing China's Population: From Leninist to Neoliberal Biopolitics*. Stanford, CA: Stanford University Press, 2005, 74.

[16] Ibid.

[17] See Judith Shapiro, *Mao's War Against Nature:* Politics *and the Environment in Revolutionary China*. Cambridge, UK: Cambridge University Press, 2001, 37.

[18] This is an estimate only. Several sources predict that population will not peak at the 1.6 billion level until 2050.

deserts of western China, population pressures are slight; Tibet, too, is lightly populated. The eastern coastal provinces, however, while occupying only 15% of China's expanse have 41% of China's total population. These differences are summarized in population density statistics. China's population density in 2000 was 351.3 persons per square mile, by no means the highest in the world (in Bangladesh, density was then 1,520 persons per square mile). Yet, in Jiangsu province, the most densely populated region, the statistic was 1,567 persons per square mile, as compared to Tibet and Xinjiang, with fewer than four persons per square mile.[19]

Growth of China's population brings a corresponding increase in use of land for housing and human settlements. Although population growth has slowed, it is still increasing. Moreover, the improvement in economic conditions has released a pent-up demand for more, better, and larger housing. The housing construction boom, noticeable in cities as well as in the countryside, has used a large amount of land, including cultivated land.[20]

3.2.2 *Urbanization*

The fast pace of urbanization in China has swallowed up huge areas of arable land. In the first 20 years of economic reforms, the number of cities in China increased from 193 to 666.[21] In 1995, the rural population of China peaked at about 750 million, while the urban population continued to grow. By the early twenty-first century, China's urban population was greater than 500 million. As cities became more populous, they expanded into the countryside, consuming land once used for agricultural purposes. One estimate is that urban sprawl and transportation networks took up 1.4 million hectares annually, just in the period of the Eighth Five-Year Plan (1991–1995).[22]

Li Diping reports results of landsat mapping for Chengdu, the capital of Sichuan Province. From 1988 to 1996, the city doubled in size.[23] Ho and Lin discuss changes in Guangzhou, which illustrates the combined impact of rural-urban migration, rapid industrialization, and urbanization. From 1988 to 2000, the amount of land used for non-agricultural purposes doubled, from 35,000 ha to nearly 70,000 ha. They remark: "With a negligible amount of unused land in the region, the increase in

---

[19] Chiao-min Hsieh, "Changes in the Chinese Population: Demography, Distribution, and Policy," in Chiao-min Hsieh and Max Lu, eds., *Changing China: A Geographic Appraisal*. Boulder, CO: Westview Press, 2004, 204.

[20] Ho and Lin, 2004, 762.

[21] L. Wang, "The Degree, Characteristics, and Trends of China's Urbanization," *People's Daily*, October 31, 1995, 4.

[22] Liping Di, "Land-Use Patterns and Land-Use Change," in Chiao-min Hsieh and Max Lu, eds., *Changing China: A Geographic Appraisal*. Boulder, CO: Westview Press, 2004, 19.

[23] Ibid.

non-agricultural land was largely at the expense of agricultural land."[24] The same pattern can be observed not only in other cities, but also in rural towns.

Urban residents have more disposable income than most rural residents in China (a global pattern). Members of China's growing middle class expect to be able to use their leisure time in recreational activities. Much farmland near cities and towns has been converted into golf courses, parks, and other recreational uses. Yet, critics of conventional wisdom regarding urbanization's adverse impact on cultivated areas suggest that under certain conditions urbanization may save arable land, and produce more efficient land uses than if rural residents were left in rural areas (or if the central government promoted development of small cities and towns, with less concentrated populations).[25]

3.2.3 Economic Development

In Chapter 2 we noted the large pressures of people on the land during the Great Leap Forward and Cultural Revolution. These events had disastrous consequences for China's environment, but they were of limited duration. The economic reforms unfolding since 1978 have spurred economic development in all parts of China, at the cost of China's arable land.

Factories, office buildings, hotels and resorts, and shopping centers consume space in China's cities and suburbs. They are as important as human settlements in accounting for loss of arable land. One estimate is that loss of agricultural land to industrial development has been underreported by as much as 61%.[26] Perhaps the clearest example of land loss is to what is called "development zone fever (*kaifa qu re*)."

At the outset of reform, central planning authorities established experimental special economic zones (SEZ) on the south coast of China. Planners established four SEZs in Guangdong and Fujian provinces and in 1988 declared Hainan Island a SEZ. In 1989 the state created two additional special zones in the Xiamen SEZ (Haicang and Xinglin) for investors from Taiwan. In 1990 SEZ status was extended to the Pudong New District in Shanghai. Then in 1994, the China–Singapore

[24] Ho and Lin, 2004, 766. For a case study using different methods, see Guifen Liu and Chengtai Diao, "Current Situation of Cultivated Land Resources and Food Security in Jiangjin City," in Yancui Liu, ed., *Study of the Strategy of Land Resources and Regional Coordinated Development in China* (in Chinese). Beijing: Meteorology Publishing Co., 2006.

[25] See Jikun Huang, Lifen Zhu, Xiangzheng Deng and Scott Rozelle, "Cultivated Land Changes in China: The Impacts of Urbanization and Industrialization," *Society of Photo-Optical Instrumentation Engineers*, Vol. 584 (2005), 58840I 1–15.

[26] Anthony Gar-On Yeh and Xia Li, "An Integrated Remote Sensing and GIS Approach in the Monitoring and Evaluation of Rapid Urban Growth for Sustainable Development in the Pearl River Delta, China," *International Planning Studies*, Vol. 2, no. 2 (1997), 193–210.

Suzhou special zone for industrial development was established. By 1995, some 422 zones had been approved by the central government.[27]

The state gave SEZs several privileges and advantages in order to spur rapid development, and high officials visited them, such as in Deng Xiaoping's highly publicized "Southern Tour" of 1992. One measure of the impact of industrial development zones is seen in the 1997 provincial survey of Fujian Province. It revealed that in the previous 7 years, more than half of the 700,000 ha being used for industrial development had been arable land.[28] Cartier reports the egregious example of a large seaport industrial zone associated with Liem Sioe Liong, the head of the Salim Group and Indonesia's leading industrialist. Local authorities allege that the provincial government sold him 800 ha of farm land for "virtually nothing."[29]

An unintended consequence of the SEZ model was widespread copying of the concept in rural counties and towns.[30] The chief force has been the Township and Village Enterprise (TVE, *siangjen qiye*). In the initial stages of economic reform, TVEs were owned collectively by all rural residents of the township or village in which the enterprise was located. Many are now owned by local governments (or their subsidiaries), and a large number have been privatized and are under the control of owner-managers. By the early 1990s, there were 1.3 million TVEs, and they then produced about 30% of national industrial output (rising to 40% by the late 1990s).[31] Most of the TVEs are small factories, and they have taken up land once used for farming in rural areas. A large number of TVE factories sit in industrial parks covering more than a hectare of land. We discuss the effects of these and other types of economic development in the next section.

Illegal land acquisition, which implicates local governments throughout China, is perhaps the gravest threat to China's diminishing arable land. Gan Zangchun, the deputy state land inspector-general of the Ministry of Land and Resources stated "Violations of land laws and regulations have cropped up recently in some areas." He directly accused local governments, remarking that "Some local governments have arbitrarily expanded development zones in violation of the master plan for land use, and encroached on land using various pretexts."[32] The root of the problem is the lack of a property right to land of farmers. Local governments illegally lease land, the prices of which have become inflated due to a booming land and property market, which makes land sales and leases a lucrative business for local government.

---

[27] Carolyn Cartier, "'Zone Fever,' the Arable Land Debate, and Real Estate Speculation: China's Evolving Land Use Regime and its Geographical Contradictions," *Journal of Contemporary China*, Vol. 10, no. 28 (August 2001), 454–56.

[28] Cartier, 2001, 452.

[29] Ibid.

[30] Dali L. Yang, *Beyond Beijing: Liberalization and the Regions in China*. London: Routledge, 1997, 54–55.

[31] Jean Oi, *Rural China Takes Off: Institutional Foundations of Economic Reform*. Berkeley: University of California Press, 1999, 37.

[32] Fangchao Li, "Illegal Land Use Poses Major Threat," *China Daily*, September 18, 2007, 1.

Corruption has become rampant through officials' siphoning off land sale proceeds and abusing land use powers to improperly allot land. Xu Shaoshi, Minister of Land and Resources, said "The illegal acquisition of arable land (for purposes other than agriculture) has endangered food safety and social stability both." He emphasized in a pessimistic voice: "Given the growing population and fast industrialization and urbanization, illegal land acquisition will probably continue."[33]

Senior researcher Li Guoxiang of the Chinese Academy of the Social Sciences (CASS) Rural Development Institute, said: "Local governments don't get any incentive for protecting arable land, even though the central government wants them to do so." Establishing industrial units produces higher revenues than what could be derived from agriculture. Local officials see urbanization and industrial production as a solution to poverty, low rates of literacy, economic backwardness, and other ills of the countryside. "Agriculture isn't considered important in their policymaking because it can't bring quick returns that can improve their careers."[34]

3.3 Effects of Socioeconomic Change

Increased population, urbanization, and economic development have had some benign effects on food production in China. Certainly, rapid economic development has made China the world's third largest economic power, and earned it the foreign exchange to purchase whatever food it cannot produce to sustain the population. Yet as noted throughout, the environmental cost of rapid economic growth may have outweighed many of the gains. A Chinese Academy of Sciences report in late 2008 indicated that the cost of exploitation of natural resources, ecological degradation and environmental pollution in 2005 was 13.9% of GDP, while growth in that year was 11.3%.[35]

Our focus is on domestic food security in China, and economic development and industrialization in particular have had mostly adverse impacts on food production. Our two large topics here are degradation of land and of water.

3.3.1 Land Degradation

By degradation of the land, we mean reducing or eliminating its ability to generate plant life and sustain humans and animals. The immediate causes of such despoilation are erosion, changes to the nutrient balance of soils, and pollution of the land with toxic substances. Erosion occurs naturally in most ecosystems, but our concern is

[33] Jiao Wu, "Land Loss Threatens Food Safety," *China Daily*, December 26, 2007, 1.

[34] Zhiming Xin, Jing Fu, and Ping Zhu, "Security of Food Calls for Serious Thought," *China Daily*, January 16, 2008, 7.

[35] Xu Wang, "Report Shows Real Price of Growth," *China Daily*, September 12, 2008, 3.

with erosion caused by human actions, such as through deforestation. Changes in nutrient balance of soils occur through weather and climate changes but also through excessive use of chemical fertilizers and other poor farming practices. Pollution of soils occurs primarily through human action.

Several reports in 2006 pointed out the serious extent of land degradation. The Ministry of Water Resources stated that 37% of China's total territory suffered from land degradation. This despoliation included soil erosion, deforestation, salinity, reduced fertility and sand storms, affecting 3.56 million square kilometers.[36] The study *China Ecological Protection* issued by the State Environmental Protection Administration (SEPA, in 2008 upgraded to the Ministry of Environmental Protection) in this year reported continued deterioration of China's ecology. Major problems included excessive logging, degradation of natural pasture land, shrinking wetlands, overuse of pesticides and fertilizers in farmland and contaminated coastal areas. The study reported these specific findings:

- The ecology of 60% of China's territory was considered fragile.
- About 90% of natural pasture land (accounting for more than 40% of China's territory) faced degradation and desertification; desertified pastures had become major sources of sand and dust storms.
- Only about 40% of China's wetlands were under effective protection.[37]

The most recent report indicates that 10% of China's farmland was contaminated with pollution from chemical plants, steel factories, crude oil storage tanks, and other industrial facilities. This resulted in contamination of 12 million tons of grain and losses of $2.8 billion.[38]

We treat erosion, deforestation, desertification, and land pollution, giving examples of each form. (Air pollution and particularly acid rain are somewhat less important factors in agricultural production, and for this reason we make only passing reference to them.[39])

3.3.1.1 Erosion

Erosion of soils is a general problem of ecological degradation in China. We provide examples from three regions: the Northeast, the Northwest, and the South. Northeast China – including the provinces of Heilongjiang, Liaoning, Jilin and part

---

[36] Xiaofeng Guan, "Experts Discuss Answers to Land Degradation," *China Daily*, August 28, 2006, 2.

[37] Fangchao Li, "Ecological Degradation Continuing," *China Daily*, June 5, 2006, 1.

[38] Jiao Wu, "Govt Targets Land Pollution," *China Daily*, June 20, 2008, 3.

[39] One example of a recent study is: Yizong Huang, Zhixian Li, Xiangdon Li, Wenmiao Yang, Zhaoyong Liang, Huafeng Li, Dinglan Liu, and Bisheng Lu, "Effects of Acid Deposition and Atmospheric Pollution on Forest Ecosystem Biomass in Southern China" (in Chinese), *Ecology and Environment*, Vol. 16, no. 1 (2007), 60–65.

of the Inner Mongolia Autonomous Region – is China's breadbasket. It is an area of black soil, covering more than 35 million hectares, and one of the world's three largest black soil regions (the others being the Ukraine and the USA). The black soil belt accounts for 30% of China's total grain output, and its yields feed 10% of the population. Research institutes of the Chinese Academy of Sciences in Harbin and Shenyang recently demonstrated that the thickness of the soil had dropped radically from more than 80 cm to less than 30 cm in the past 6 decades. The density of organic substances in the soil declined from 12% in the 1940s to less than 2%; about 85% of the soils lacked sufficient nutrients. Causes of soil erosion and degradation included excessive farming, overuse of fertilizers and unsustainable logging. Soil erosion, in turn, has brought about more frequent drought, floods and sandstorms. Zhang Xudong, an expert with the Shenyang-based Institute of Applied Ecology, commented that "Soil erosion and degeneration will jeopardize the nation's grain security."[40]

Soil erosion has become a large problem in northwestern China's Xinjiang Uygur Autonomous Region as excessive herding and farming have outpaced state conservation efforts. Remote sensing surveys show 1.03 million square kilometers of land degraded by soil erosion. Xinjiang itself accounts for about 30% of China's total acreage of soil erosion. A local soil conservation official remarked: "The region has a vulnerable ecology. Besides natural factors, human activities (excessive herding on pastureland and farming along the Tarim River) are largely to blame for the deteriorating soil erosion."[41] As noted below, this is a primary cause of desertification, which affects most of Xinjiang's counties and cities and nearly two-thirds of its territory.

Soil erosion is even a problem in prosperous Guangdong province, ranking second on the mainland for this form of land degradation. The provincial water conservancy bureau reported that more than 2,200 km^2 of soil had eroded during the Tenth Five-Year Plan period (2001–2005) alone, with expected worsening during the next plan period. Experts feared that soil erosion would spread to 5,748 km^2 of land (or about 3.2% of the province's farmland) by 2010. In the case of this province, industrial developments have been the primary factor damaging soils.[42]

A 3-year survey released by the Ministry of Water Resources in late 2008 reported that China how has more than 3.5 million square kilometers of eroded land. It warned that land erosion increased the risk of floods and was a growing threat to food supply. Sun Honglie, a survey team participant, said: "Agricultural and forestry exploitation, and highway, railway and urban construction projects are the major causes of land erosion, accounting for 78% of the total."[43]

---

[40] Xinhua, "Erosion Threatening Grain Security," *China Daily*, August 28, 2007, 2.

[41] Xinhua, "Xinjiang Losing Land to Soil Erosion," *China Daily*, July 3, 2007, 4.

[42] Caixiong Zheng, "Soil Erosion Targeted in Guangdong," *China Daily*, November 20, 2007, 4.

[43] Yu Xie, "Land Erosion 'Threat to Food Supply'," *China Daily*, November 23, 2008, 2.

3.3.1.2 Deforestation

Population growth and the timber industry are the major factors causing a substantial reduction in forests. About half of China's forests have been destroyed since 1949. Today, forests cover 134 million hectares, 14% of the land area, but few virgin forests remain.[44] In recent years, they have decreased at an annual rate of 5,000 km^2. Mining and logging have deforested mountains, which causes erosion, reduced water storage capabilities, and severe sandstorms in northern China.

Agricultural development and housing settlements also have reduced forest and vegetative cover. As will be noted below, government policies of afforestation, reforestation, and converting cropland to grassland and forests have ameliorated some of the deleterious effects of deforestation, but because they replace natural forests, they have "altered the variety, quality, and the pattern of delivery of plant and wildlife habitats that had been provided previously."[45] The massive reforestation and afforestation programs have not yet curbed soil erosion, which as discussed threatens more than one-third of China's territory.[46]

Starting in the late 1980s, the central government developed a natural forest protection program (also known as the National Greening Campaign). After massive flooding of the Yangtze River in 1998, this program was strengthened. It included a complete logging ban in the upper reaches of the Yangtze River and the upper and middle reaches of the Yellow River. Also, the program called for a reduction or adjustment of timber output in state-owned forest farms of the Northeast and Inner Mongolia, as well as rehabilitation and development of natural forests in other regions.

However, some illegal logging continues, notwithstanding the ban. One of our respondents, a forest ecologist, estimated that one-third of industrial wood in China is harvested illegally. Most of this timber was harvested above the official quota level. He remarked:

> On the recommendation of the State Forestry Administration (SFA), the State Council fixes and approves a 5-year quota, which is adjusted annually. There is an extensive network of checking stations. At these stations, permits to log are checked. But the system is corrupted. Loggers give money to local officials to register a lower amount than that actually logged. Thus, they avoid the quota limit and also avoid paying heavy taxes. The end result is that 5 years later, at the end of that quota period, a national inventory will be taken. SFA will do surveys; they'll use remote sensing. Then they discover that the inventory doesn't match the quota.[47]

---

[44] This figure differs from the 27 percent mentioned above, because the State Forestry Administration uses several different definitions for forest coverage.

[45] Scott Rozelle, Jikun Huang, and Vince Benziger, "Forest Exploitation and Protection in Reform China," in Hyde, Belcher, and Xu, 2003, 20.

[46] Chao Liang, "Probe Launched into Erosion Threat," *China Daily*, July 5, 2005, 2.

[47] Personal interview with forest ecologist, Beijing, May 18, 2004.

In a highly publicized case of 2004–2005, the environmental NGO Greenpeace (see Chapter 7) attacked Asia Pulp & Paper (APP), a multinational pulp and paper production giant, for illegal logging in Yunnan Province. Greenpeace charged the firm with logging a large section of natural forests, violating the state's Forest Law as well as the national natural forest protection program.[48] The firm then replanted 183,000 ha with eucalyptus plantations. Local farmers claimed their land had been requisitioned at yearly rents of only $1.45 per hectare. In this case, Greenpeace enlisted the collaboration of the Zhejiang Hotels Association in a boycott of APP products, which led the corporation to modify its actions.[49]

3.3.1.3 Desertification

Sand and desert cover about 27% of China's land area. The expanse of deserts has increased dramatically in the contemporary period. Desertification annually claims an additional 3,400 km^2.[50] Desertification has dried up rivers and lakes (leading to salinization of the soil, which then cannot be used for growing crops), shriveled plants and vegetative cover, and led to dropping levels of ground water, posing a direct threat to more than 100 million people. Specifically, it degrades farmland and pastures, and leads to the reduction of crop production.[51] Desertification also has threatened national treasures such as the Great Wall[52] and the Mogao grottoes.[53]

Desertification in parts of China is attributable to deforestation as well as to poor protection and overutilization of water resources in arid and semi-arid regions of the North and West. Increased desertification in some parts of China also is attributable to agriculture, commercial, industrial, and residential development.

Grasslands degradation is a principal form of desertification. It refers not just to degradation of the grass but also the soils, with a despoilation of the entire ecosystem. China's many steppes have undergone extensive degradation as a result of land reclamation, grazing, and wood cutting. In the estimate of experts,

---

48 Greenpeace China, *Investigative Report on APP's Forest Destruction in Yunnan*. Hong Kong: November 2004.

49 China Development Brief, "Greenpeace, Zhejiang Hotels, Stand Firm against Paper Giant," March 31, 2005. Retrieved February 14, 2006 from: http://www.chinadevelopmentbrief.com/node/77/print Also, see Liang Chao and Cui Ning, "Authorities Crack Down on Illegal Logging," *China Daily*, March 31, 2005, 2.

50 Desheng Cao, "Nation Fighting Ever-Engulfing Deserts," *China Daily*, June 18, 2004, 2.

51 See Xiang-yun Li, Jun Yang, and Li-xin Wang, "Quantitative Analysis on the Driving Role of Human Activity on Land Desertification in an Arid Area: The Case of the Tarim River Basin (in Chinese)," *Resources Science*, Vol. 26, no. 5 (September 2004), 30–7.

52 Lie Ma, "Water Rule Will Protect Great Wall, Gansu Oasis," *China Daily*, August 7, 2007, 5.

53 Shanshan Wang, "Creeping Desert Threatens Mogao Grottoes," *China Daily*, November 6, 2007, 3. The Mogao caves are located in Dunhuang, Gansu province and are famous for their 1,000 year old Buddhist statutes and wall paintings. They are threatened by the encroaching Kumutage desert, China's sixth largest.

the productivity of plant communities on steppes has declined 60% since the 1950s, with particularly obvious degradation in the last decade.[54]

A special case of desertification is erosion around the world's highest and largest wetlands, the Sanjiangyuan Nature Reserve in Qinghai province. Sanjiangyuan means "source of the three rivers" (the Yangtze, Yellow River, and Lancang River), but in recent years the region has become increasingly sandy because of overgrazing, increased human activity, and climate change. Although ecological restoration work began in 2005, a conservative estimate is that efforts will need to continue for 10 years to bear fruit.[55]

Desertification is correlated with the increase of sandstorm activity in North China.[56] Most of the sandstorms blowing into Tianjin and Beijing originate in Inner Mongolia.[57] Inner Mongolia has China's four largest deserts – Otindag, Horqin, Moo-Os and Hulun Boir – with a combined area of 15 million hectares. The Otindag desert in Xilingol League once was mostly grassland and produced few sand storms (only one recorded from 1930 to 1960). However, population increases and over-grazing (an increase in livestock from one million in 1947 to 24 million in 2000) severely increased pressure on the land; by 2000, the region produced 14 sand storms alone.[58]

While it is clear that sandstorm activity has increased consequent to China's rapid industrial development, scientists suggest that spring sandstorms are a "fact of nature" in North China. Qin Dahe, director of the China Meteorological Administration, introduced a note of caution in the discussion of sandstorm activity: "The public should have a full understanding of sand and dust, which have complex effects on nature and society.... Without sandstorms, Chinese society would not have arisen." Qin maintained that sandstorms contributed to creation of nearly 1 million square kilometers of loess plateau. It was the Yellow River, he remarked, whose course over the plateau washed away huge amounts of dust, which formed the North China Plain.[59]

There is consensus that some of the causes of erosion should be addressed, and several mitigation strategies have been employed to reduce desertification. The regime's re- and afforestation policy was the original strategy to reduce soil erosion and desertification, but lack of water in desert regions frustrated this policy.

---

[54] See Nianyong Han, Gaoming Jiang, and Wenjun Li, *Management of the Degraded Ecosystems in Xilingol Biosphere Reserve* (Chinese and English). Beijing: Tsinghua University Press, 2002, 120–34. See also Biodiversity Working Group of China Council for International Cooperation on Environment and Development, *Restoring China's Degraded Environment: The Role of Natural Vegetation*. Beijing: 2001, 2–7.

[55] Xinhua, "10-year Plan Needed to Save Wetlands from Desert," *China Daily*, September 18, 2007, 4.

[56] R. Gluckman, "The Desert Storm," *Asiaweek*, October 13, 2000, 3–6. See also Lie Ma, "Desertification Threatens Northwest Areas," *China Daily*, September 6, 200, 3; Weiyu Jiang and Fanglin Chan, "Controlling Desertification in North China" (in Chinese), *Journal of Arid Land Resources and Environment*, Vol. 26, no. 5 (September 2004), 30–37; and Qianzhao Gao, Jianjun Qu, Run Wang, Yuan Li, Ruiping Zu, and Kecun Zhang, "Impact of Ecological Water Transport to Green Corridor on Desertification Reversion at Lower Reaches of Tarim River" (in Chinese) *Journal of Desert Research*, Vol. 27, no. 1 (January 2007), 52–58.

[57] Huanxin Zhao, "Inner Mongolia's Green Efforts aid Beijing," *China Daily,* July 26, 2007, 3.

[58] Yong Wu, "Playing with Nature or Helping It Is the Question," *China Daily*, August 22, 2007, 12.

[59] Xiaohua Sun, "Sandstorms a Fact of Nature," *China Daily*, March 15, 2007, 6.

A major recent method has been fencing in grasslands to protect vegetative cover against advancing deserts. While this has been superior to planting trees, which require more water resources than available in most desert areas, it has a large impact on the ecosystem. Migratory species are then restricted in their movement.[60] Another mitigation strategy used in Inner Mongolia has been erecting sand barriers and planting soil-stabilizing shrubs.

3.3.1.4 Land Pollution

Three types of pollution afflict agricultural lands in China: industrial plant waste, mining operations, and use of chemical fertilizers and pesticides. Chemical and other industrial facilities pollute land with toxic contaminants, diminishing or exterminating plant growth. Pollution caused by rural industries is far more severe than that caused by urban industries. Second, China has a large number of small-scale mining operations, particularly coal mines, for China is reliant on coal for about 70% of its energy needs. Mine waste dumps including sulfides as well as other toxic chemicals have had adverse impacts on the soil microbial communities in adjacent areas.[61] A third cause of land pollution is from excessive use of chemical fertilizers and pesticides by farmers. About 7% of China's agricultural land has been polluted through improper uses of pesticides and fertilizers, not to mention the number (50,000–120,000) of people who are poisoned each year by pesticides.[62]

Reports from the Ministry of Land and Resources in 2007 indicated that about 12.3 million hectares – more than 10% of China's arable land by current government estimates – is contaminated by pollution, and the situation is worsening.[63] The ministry acknowledged that heavy metals alone had contaminated 12 million tons of grain and caused losses of $2.6 billion each year, and that the contaminated grain would ultimately be a health hazard.

Land pollution concerns prompted SEPA to conduct the first soil survey of China's farmlands to insure food safety, beginning in mid-2006. The survey has focused on main grain-producing and industrial areas: Jiangsu and Zhejiang provinces in the Yangtze River Delta and Guangdong Province in the Pearl River Delta and also Liaoning Province in the Northeast and Hunan in central China. In addition to pollution of grain production regions by wastewater, solid waste and other pollutants, vegetables and fruits also have been polluted by excessive nitrates in the soils.[64]

---

[60] Personal interview with government-organized NGO representative involved in desert control activities, Beijing, May 18, 2007.

[61] Guolan Liao and Chao Wu, *Heavy Metal Pollution and Control in Mining Environment* (in Chinese). Changsha, Hunan: Central South University Press, 2005.

[62] Yang Yang, "Pesticides and Environmental Health Trends in China," Woodrow Wilson Center, China Environmental Forum, 2007.

[63] Fangchan Li, "Official: More than 10% of Arable Land Polluted," *China Daily,* April 23, 2007, 1. Also see Xinhua, "Bitter Harvest," *China Daily*, May 19, 2008, 3.

[64] Xiaohua Sun, "SEPA Sets Sights on Polluted Soil," *China Daily*, January 9, 2008, 3.

3.3.1.5 Air Pollution

Some 2,000 t of mercury, from more than 2 billion tons of coal burned every year, enter the soil and pose threats to agricultural production and human health,[65] and this is one indication of the serious impact air pollution has on agricultural land. In 2005, acid deposition affected one third of China's land mass; in some regions of the nation, all rainfall was acidic. With 26 million tons of sulfur dioxide discharged in 2005 – 27% greater than in 2000 – China became the world's largest sulfur dioxide polluter.[66] Coking plants and coal-burning power stations were primarily responsible for these emissions.

Air-borne pollution particles have cut rainfall in many regions of China, particularly in the Northeast and Northwest. Scientists studying mountain regions have noted a particular kind of precipitation called orographic. This occurs when moist air is deflected upwards by a topographic feature such as a mountain, which cools the air and causes cloud droplets and then raindrops to fall. Polluted air carries more particles that divide cloud droplets into smaller ones. The smaller cloud droplets are slower to combine into rain drops, reducing precipitation.[67]

Air pollution is a major cause of lung cancer, as harmful particulates enter the lungs and cannot be discharged. As noted below, water pollution also is a cause of cancer, which in recent years has been the most lethal disease for China's residents. A 2007 survey administered by the Ministry of Health (of 30 cities and 78 counties) indicated that death rates from cancer had risen to 19% in cities and 23% in rural areas.[68] In recent years, reports on "cancer villages," where residents describe high rates of deaths from cancer, have increased. The World Bank reports that deaths resulting from water-related pollutants and bad air reach 750,000 a year.[69]

Optimistic reports from government agencies in 2007–2008 said the high rate of air (as well as water) pollution might be dropping somewhat. SEPA noted that there were fewer sandstorms in 2007 than in the previous year, that key cities in North China were less seriously polluted, and that "good air" days were more frequent than in the previous year.[70] Further, SEPA noted that although the total amount of

---

[65] Xiaohua Sun, "Soil Survey to Monitor Pollution," *China Daily*, May 9, 2007, 1.

[66] Li Liu, "One Third of Nation Hit by Acid Rain," *China Daily*, August 28, 2006, 2. In Guangzhou, eight out of every 10 rainfalls were acidic in 2007; see Hong Chen, "Guangzhou Swamped by Acid Rain," *China Daily*, March 28, 2008, 4. China is a major source of transboundary air pollution reaching its neighbors. Even in Los Angeles, city officials estimate that on some days, one-quarter of the city's smog comes from China. See Doug Struck, *Washington Post National Weekly Edition*, February 25–March 2, 2008, 10.

[67] Chong Wu, "Researchers: Polluted Air Causes less Rainfall," *China Daily*, March 9, 2007, 3.

[68] Chuanjiao Xie, "Pollution Makes Cancer the Top Killer," *China Daily*, May 21, 2007, 1.

[69] Joseph Kahn and Jim Yardley, "As China Roars, Pollution Reaches Deadly Extremes," *New York Times*, August 26, 2007, A13.

[70] Xiaohua Sun, "SEPA Publishes Q1 Report," *China Daily,* May 22, 2007, 3.

pollutant discharges rose in 2006, the rate of increase was slower than in the previous year. Sulfur dioxide emissions fell 0.3% in the first quarter of 2007.[71]

Yet by mid-2008, MEP officials admitted that the target of cutting emissions of major pollutants 10% by 2010 likely was beyond reach.[72]

Annual surveys of public attitudes on the environment have been conducted since 2005 by the China Environment Culture Promotion Association. The most recent survey of attitudes in 2008 showed that 76% of 10,000 people surveyed in 31 provinces and municipalities thought environmental problems were "very serious" or "relatively serious." Pollution was the most serious problem after rising prices and food security, and air quality was a somewhat greater concern than water quality.[73]

3.3.2 *Degradation of China's Waters*

While degradation to China's land is serious and worsening, water degradation, in the view of many observers, has reached crisis proportions. We consider first the issue of water sufficiency in China, and then treat respectively pollution to fresh waters and to the oceans off China's coasts.

3.3.2.1 Water Sufficiency

Lester Brown directly connected the nature of China's water system to global food security when, in 1998, he commented: "As rivers run dry and aquifers are depleted, the emerging water shortages could sharply raise the country's demand for grain

[71] Jiao Wu, "Pollution Picture to Brighten," *China Daily*, June 6, 2007, 1. NGO activity targeting air pollution became more public in 2007–08. Ma Jun, author of *China's Water Crisis* and head of the Institute of Public and Environmental Affairs, set up a China Air Pollution Map to mirror the China Water Pollution Map. Its blacklist in late 2007 included 40 multinational corporations among 4,000 firms on the roll. See Zhuoqiong Wang, "Global Giants on Pollution Blacklist," *China Daily*, December 14, 2007, 1. Too, regulation of vehicle traffic and moving highly polluting industrial plants out of town did clean up Beijing's air for the 2008 Olympics; some of these measures may become permanent. See Zhe Zhu, "Nothing but 'Blue Skies' from Now On," *China Daily*, August 20, 2008, 3; Zhe Zhu and Ying Wang, "Clean Air Measures to Remain after Games," *China Daily*, August 24, 2008, 2; and Xiaohuo Cui, "Clear Days' Target Met Before Time," *China Daily*, December 1, 2008, 3.

[72] Jing Li, "Slow Progress," *China Daily*, July 20, 2008, 4.

[73] Xiaohua Sun, "Pollution Problems Serious: Survey," *China Daily*, January 19, 2009, 2. Respondents rated the effectiveness of environmental protection as 52 points (with 1 being lowest) as compared to 58 in 2007 and 68 in 2006. In other words, perceptions of protective efforts had declined.

imports, pushing the world's total import needs beyond exportable supplies."[74] As in his previous critique of China's loss of arable land, he maintained that if China did not address this problem, world grain prices would rise, creating instability in Third World cities.

China ranks fifth in the total water resources of nations in the world, but on a per capita basis, China's water supply is 25% below the global average. Future projections are more troubling. By 2030, per capita supply is expected to drop from 2,200 m^3 to below 1,700 m^3, and at this level would meet the World Bank's definition of a water-scarce country.[75] Agriculture consumes from 70% to 80% of China's water resources, but as supplies tighten, agricultural use of water is threatened by rising industrial and household consumption. However, most observers of China's water sufficiency believe that shortage of water has not yet led to a substantial loss of irrigated area or industrial production.[76]

Thus, on a national basis, China's water resources currently seem to be sufficient; however, water is not evenly distributed throughout China. It is relatively scarce in the North and West, and is abundant in the humid South. Although the dry North produces more than 40% of China's grain supplies, it has less than a quarter of China's water resources and parts have been subject to drought conditions, for example Shaanxi in 2007.[77] The North China Plain (*Huabei pingyuan*) is the heartland of Chinese civilization, and is traversed by three major rivers: the Yellow, the Hai, and the Huai. The Yellow River (affectionately dubbed the "mother river") is the most obvious example of water scarcity in the North. In 1972, for the first time in Chinese history, the Yellow River dried up before water could reach the sea. In 1997, for 330 days of the year, water from the Yellow River did not reach the ocean.[78] Causes of water loss included extensive upriver exploitation of water as China rapidly industrialized; future threats include melting of glaciers and depletion of underground water systems feeding the river.[79] Both the Hai and Huai rivers suffer

---

[74] Lester R. Brown, "China's Water Shortage Could Shake World Food Security," *World Watch*, July 1, 1998. See also, China Development Brief, "Thirsty Cities and Factories Push Farmers Off the Parched Earth," June 1, 2001. For a generic response to Brown's argument, see Jianxin Xu, Zezhong Zhang, and Fa Liu, "The Analysis of the Water Resource's Assurance of Grain Production Security" (in Chinese), *Water Conservancy Science and Technology and Economy*, Vol. 11, no. 10 (October 2005), 611–14; and Bryan Lohmar and Jinxia Wang, "Will Water Scarcity Affect Agricultural Production in China?" in *China's Food and Agriculture: Issues for the 21st Century*. Washington, DC: Economic Research Service, USDA, 2003, 41–43.

[75] Jennifer Turner & Kenji Otsuka, *Reaching Across the Water*. Washington, DC: China Environment Forum, Woodrow Wilson International Center for Scholars, 2006, 1.

[76] Bryan Lohmar, Jinxia Wang, Scott Rozelle, Jikun Huang, and David Dawe, "China's Agricultural Water Policy Reforms," *Economic Research Service, Agriculture Information Bulletin*, no. 782 (2003), 1. Also see Renee Vassilos, Tanya Franke and Jorge Sanchez, *Water Situation in the North China Plain*. Beijing: USDA Foreign Agricultural Service, Global Agriculture Information Network, 2008.

[77] Lie Ma, "Shaanxi Reels from Drought," *China Daily*, May 11, 2007, 50.

[78] Jun Ma, *China's Water Crisis*. Norwalk, CT: Eastbridge, 2004, vii.

[79] Jim Yardley, "China's Path to Modernity, Mirrored in a Troubled River," *New York Times*, November 19, 2006, A1, 14–15.

from depleted flow, leaving entire valleys short of water, notwithstanding construction of thousands of large- and small-scale reservoirs.[80]

With less available water (and because most water from rivers is polluted), the people of the North and West have turned to use of ground water. In recent years, however, the ground water tables through most of North China and parts of the South have dropped, making it necessary to drill deeper wells. A recent survey indicated that the water tables beneath much of the North China Plain have fallen an average of 1.5 m/year in the last 5 years.[81] The dropping water tables have caused large areas of subsidence.

Not only is water in limited supply in China, but it also is used inefficiently. One estimate suggests that only 43% of the water used in agriculture is used efficiently, as compared to 70–80% of irrigated water in developed countries. Moreover, about 25% of the water transmitted through pipes is lost through leakage, much higher than the 9% lost in this way in Japan, and 10% in the United States.[82] In China's irrigation systems overall, much water is lost through evaporation.

A final factor affecting water supply is the pricing system for water use. Until recently, prices for commercial, industrial, and household use were not well-differentiated. Moreover, prices for water in most of China's regions and cities did not vary in proportion to the amount of water used.[83] Lohmar et al. comment: "Despite increasing water prices, current pricing policies do not effectively encourage water saving and in fact contribute to China's water problems in other ways."[84] Also, researchers have identified challenges as well as opportunities in the transition from collective management of water to water user associations and contracting systems.[85] In a country that remains communist, with clear policy goals of egalitarianism, the transition to a market-based system for water use is especially difficult. Only in 2007, did the regime gingerly begin planning to deregulate prices of water, to reflect its scarcity.[86] For these reasons, the water use system encourages over-use of water instead of careful conservation.

---

[80] As a number of studies have indicated, the crop structure of regions in the North China Plain, affecting the amount of water consumed through irrigation systems, strongly influences water sufficiency prospects. See, for example, Yongsong Liao and Jikun Huang, "Impact of Crop Structure Change on Irrigation Water Demand in the Basins of Yellow River, Huahe River and Haihe River," *Journal of China Institute of Water Resources and Hydropower Research* (in Chinese), Vol. 2, no. 3 (September 2004), 184–88.

[81] Brown, 1998, 2.

[82] Turner & Otsuka, 2006, 3.

[83] Personal interview with water engineer, Beijing, May 15, 2007.

[84] Lohmar, et al., 2003, 17.

[85] Most of the analyses suggest that incentive systems and not the development of new water management institutions are most likely to produce water savings. See Jinxia Wang, Zhigang Xu, Jikun Huang, and Scott Rozelle, "Incentives in Water Management Reform: Assessing the Effect on Water Use, Production, and Poverty in the Yellow River Basin," *Environment and Development Economics*, Vol. 10 (2005), 769–99.

[86] Jing Fu, "High Prices to 'Persist' Next Year," *China Daily*, September 18, 2007, 2.

3.3.2.2 Water Pollution

The consensus of water specialists is that water sufficiency currently is less of a problem in China than water pollution.[87] One obvious indicator is that 16 of the world's 20 most polluted rivers are in China. Moreover, air and water pollution combined with widespread use of food additives and pesticides have made cancer the leading cause of death in China. Chen Zhizhou, a health expert with the cancer research institute of the Ministry of Health noted: "The main reason behind the rising number of cancer cases is that pollution of the environment, water and air, is getting worse day by day." He continued: "Many chemical and industrial enterprises are built along rivers so that they can dump the waste into water easily. Excessive use of fertilizers and pesticides also pollute underground water. The contaminated water has directly affected soil, crops and food."[88]

There are three major sources of water pollution: industrial contaminants spewed into rivers and lakes, chemical pesticide and insecticide run-off from crop fields, and human waste and garbage disposed into waterways. A 2006 study examined 30 of China's major rivers carrying processed water to the sea, accounting for 82% of the total run-off volume. Results showed a large increase over the previous year in levels of pollutants discharged via the Yangtze, Pearl, Yellow, Minjiang and several other rivers. Of the total volume of major pollutants, chemical oxygen demand accounted for 86.3%, nutritive salts some 12.5%, with the remainder distributed among oil, heavy metals, and arsenic.[89]

Industrial pollution events hit the news repeatedly in 2006–2009. In late 2006, a chemical spill caused by an explosion at the Jilin Petrochemical Corporation (in the Northeast, China's rustbelt), created a toxic slick on the Songhua River, forcing downstream cities in Heilongjiang to suspend their normal water supplies.[90]

[87] Based on personal interviews with officials of the Ministry of Water Resources, professors of hydrology, and NGO representatives, Beijing, March 15, 22, May 18, 22, 30, 2007.

[88] Chuanjiao Xie, "Pollution Makes Cancer the Top Killer," *China Daily*, May 21, 2007, 1. Environmental degradation also has been linked to birth defects, which increased by nearly 40 percent between 2001 and 2009. See Yinan Hu, "Rural Environmental Protection Plan Released," *China* Daily, November 22, 2007, 5. Also see Jia Chen, "Birth Defects Soar Due to Pollution," *China Daily*, February 1, 2009, 2. (The article estimated that 10 percent of the physical defects in Chinese infants were caused by environmental pollution.) The health effects of contaminated water are both direct and indirect. Farmed fish raised in contaminated waters may lead to higher rates of cancer as well as liver disease and other afflictions. See David Barboza, "In China, Farming Fish in Toxic Waters," *New York Times*, December 15, 2007. Eel farmers called the *Times* accusation that fish farmers had mixed illegal veterinary drugs and pesticides into fish feed "totally groundless." Local environmental protection bureau officials said strict regulations since 2003 had made drug use illegal, and that "The major pollutants in eel breeding are nitrogen, phosphorous and excrement which are found naturally" and that 97 percent of aquatic products met standards during random tests." See Yinan Hu and Jiao Wu, "Fishing in 'Troubled Waters' Leads to Big Losses," *China Daily*, January 3, 2008, 1.

[89] Xinhua, "Volume of Pollutants Exceeds 13m Tons," *China Daily*, May 11, 2007, 4.

[90] Fangchao Li, "Provincial Officials Promise to Close Loopholes," *China Daily*, May 11, 2007, 4.

In fact, the Ministry of Water Resources (MWR) labeled water quality at level 5, the poorest, equivalent to raw sewage. The basin of the 1,900 km Songhua River spreads to Jilin and Heilongjiang provinces, Inner Mongolia, and it flows into Russia (and its level of pollution nearly created an international incident). This event prompted the resignation of the minister. Government officials planned to let the river "rest in peace and rehabilitate itself" for 10 years,[91] but experts were not sanguine about the prospects of full recovery in this period, given lack of clean-up success in other rivers, such as the Huai. Pollution on this river, which authorities have been attempting to clean up for 10 years, has become significantly worse.[92] The pollution situation in the Liaohe River of the Northeast is equally bleak. In Shenyang, capital city of Liaoning Province, some 400,000 t of sewage are released into a major tributary of the Liaho River daily, and emissions of chemical oxygen demand have steadily increased since 2001.[93]

Pollution levels in the Yellow River have increased rapidly during the reform era. Between the 1980s and 2005, the volume of wastewater flowing into the river increased from about 2 billion tons to 4.3 billion tons.[94] Much of this pollution has been caused by industrial enterprises, which produce large amounts of sewage, but untreated household waste also has been released in the river. For example, the city of Lanzhou, capital of Gansu Province, had released millions of tons of household sewage into the Yellow River because the city's sewage treatment facilities were out-of-date and inefficient.[95] Tributaries of the Yellow River are similarly affected. For example, the Weihe River, the largest tributary, is seriously polluted. Cities along the Weihe such as Baoji, Xianyang and Weinan dump sewage into it daily; Xian, capital of Shaanxi Province, dumps nearly 1 million tons of sewage into the Zaohe River, a tributary of the Weihe. The Weihe River basin covers one-third of the province's total area. Home to 63% of the Shaanxi population, it contains 56% of the farmlands and generates about 89% of the province's GDP.[96]

---

[91] Fangchan Li, "Rehabilitation Effort for Country's Dirtiest Waterway to Last a Decade," *China Daily*, May 11, 2007, 4. For a review of the incident and analysis of the flawed response capability of central and regional authorities, see Nat Green, "Positive Spillover? Impact of the Songhua River Benzene Incident on China's Environmental Policy," Woodrow Wilson Center, China Environmental Forum, 2009.

[92] Xiaohua Sun, "SEPA Publishes Q1 Report," *China Daily*, May 22, 2007, 3. The Huaihe River is China's third longest and stretches through Henan, Anhui, Jiangsu, and Shandong provinces. In early 2008, wastewater treatment plants treated just 60 percent of daily wastewater from industrial plants and households. Its chemical oxygen demand level was still 80 percent higher than the accepted standard, and a Ministry of Environmental Protection report faulted provincial governments for inadequate pollution monitoring and enforcement. See Xiaohua Sun, "3 Provinces Fail on River Targets," *China Daily*, April 22, 2008, 4.

[93] Xiaohua Sun, "Polluters Face Stiff Penalties," *China Daily*, August 27, 2007, 1.

[94] Zhuoqiong Wang, "River is 10% Sewage: Official," *China Daily*, May 11, 2007, 4.

[95] Ibid.

[96] Lie Ma, "Shaanxi Tackles River Pollution," *China Daily*, May 15, 2007, 5.

Pollution levels also have increased enormously in the Yangtze River, which (including its tributaries) accounts for about one-third of China's total fresh water resources. In a 2007 report, the Yangtze River Water Resource Commission stated that one-tenth of the 6,211 km main course of the river was in critical condition. In addition, about 30% of the major tributaries of the Yangtze – including the Minjiang, Tuojiang, Xiangjiang, and Huangpu rivers – were heavily polluted by excessive ammonia, nitrogen, phosphorous and other pollutants.[97] High levels of pollution led to a shrinking of fish catch, and threatened endangered species.[98] Other industrial pollution reports included a cadmium spill along the Beijiang River in Guangdong Province, spills from factories in Gansu Province, and factories in Hunan Province flushing waste water with high concentrations of arsenide into the Xinqiang River.[99]

Both the Yangtze and Pearl River estuaries were listed as "dead zones" in a study released in late 2006 by the United Nations Environment Programme (UNEP).[100] Dead zones are water areas where nutrients from fertilizer runoff, sewage, human and animal waste and the burning of fossil fuels trigger algae blooms. The blooms require oxygen and remove it from the water, a condition called eutrophication, which endangers all water life. The number of deoxygenated areas has increased each decade since the late 1970s; they are a major threat to fish stocks and people depending on the fisheries.

In the summer of 2007, many lakes in China experienced major algae outbreaks. High concentrations of nitrogen and phosphorus in the waters (produced by adjacent chemical factories, waste water and agricultural emissions) caused spurts of blue-green algae (cyanobacteria) that threatened the safety of the water supply of Wuxi in Jiangsu Province, a city with a population of nearly six million.[101] Resembling a sheen of green oil paint, the canopy of algae covered 70% of the lake's surface. To dilute the polluted

---

[97] Xiaohua Sun, "Pollution Takes Heavy Toll on Yangtze," *China Daily*, April 16, 2007, 1.

[98] The white-fin dolphin, with a history of 20 million years along the Yangtze, has practically died out because of pollution and over-fishing. See Liu Xiao, "Swift Action Needed to Save Yangtze, Forum Says," *China Daily*, September 18, 2007, 3.

[99] Xinhua News Service, "Sepa to Get Tough on Gov't Violations," *China Daily*, March 2, 2007, 3. Industrial pollution even reached the Three Gorges Dam area in mid-2008, which was the first time a blue-green algae outbreak had occurred there. See Xiaohua Sun, "Algae Infests River near Dam," *China Daily*, July 22, 2008, 3 and Jing Li, "Concerns Remain over Water Safety," *China Daily*, December 25, 2008, 4.

[100] Xiaohua Sun, "Yangtze and Pearl River Estuaries Now 'Dead Zones'," *China Daily*, October 20, 2006, 2.

[101] Kun Zhang, "Experts Identify Algae Afflicting Taihu Lake," *China Daily*, June 13, 2007, 4; also see Fangchao Li, "Official Warns of Major Algae Outbreak," *China Daily*, July 14–15, 2007, 3; and Na He, "Blue Algae Hits City's Water Supply," *China Daily*, July 18, 2007, 4, for a report of an outbreak of blue algae in the water supply of Changchun in northeast China. For analysis of the problems in environmental governance reflected in the Taihu case, see Pei-yu "Catherine" Tai and Linden Ellis, "Taihu: Green Wash or Green Clean?" Woodrow Wilson Center, China Environmental Forum, 2008.

lake water, local officials diverted water from the Yangtze River and also seeded clouds to bring rain.[102] China's natural lakes are estimated to be disappearing at the rate of 20 a year because of eutrophication.

Pollution has increasingly affected groundwater supplies throughout China. A recent report found that 90% of the groundwater of China's cities is polluted to some extent, which poses huge problems because nearly three-quarters of the population of China relies on ground water for drinking. [103] SEPA Vice-Minister Pan Yue reported in 2007 that the quality of potable water in key cities had dropped by 5% age points as compared to the previous year; only 66 cities had source water meeting national environmental standards. [104] Groundwater contamination is a more serious problem in North than in South China, because abundant rainfall flushes out contaminants in the South. Yet even in the South, pollution remains a serious problem. A study of drinking water in Guangdong province by the Guangzhou Institute of Geography indicated that 17,000 m^3 of sewage were being discharged into rivers every year. At least 16 million residents of the province faced water shortages because of pollution.[105]

Most of these reports come from urban areas in China, but the situation in rural areas doubtless is worse. Primary pollutants in rural areas are poisonous fertilizers and discharge of untreated sewage water. China uses more than 360 kg of fertilizer per hectare, much higher than developed nations' usage rates, and fertilizer is used inefficiently. Fertilizer runoff after rains causes contamination of water and water life. Most of the 280 million tons of sewage generated each year is untreated and directly discharged into rivers. Some 9 billion tons of sewage water is discharged every year. Overall, about one-third billion rural Chinese use unsafe drinking water.[106]

Government officials, particularly in the national SEPA and MWR, as well as provincial and local environmental protection bureaus, addressed these problems

---

[102] Lifei Zheng and Huanxin Zhao, "Water Better, But Not Drinkable," *China Daily*, June 2–3, 2007, 1. An environmental activist who protested the impact of the chemical industry in the pollution of Lake Tai sought retribution from the government. See Joseph Kahn, "In China, a Lake's Champion Imperils Himself," *New York Times*, October 14, 2007. Five of the nine "Green Chinese," winners of environmental awards sponsored by seven ministries in 2007 won the prizes because of their service in reducing water pollution, including Zhang Xiaojian, who treated polluted water of Lake Tai in Wuxi. See Zhuoqiong Wang, "Turning Back the Tide," *China Daily*, December 31, 2007, 8. Despite these actions, water in Lake Tai remained polluted the following year. See Yanfang Qian, "Taihu Lake Water Safety Fears Remain," *China Daily*, May 30, 2008, 6.

[103] Turner & Otsuka, 2006, 4; also see Zhuqing Jiang, "Groundwater Quality 'Deteriorating'," *China Daily*, October 10, 2006, 2.

[104] Xiaohua Sun, "SEPA Publishes Q1 Report," *China Daily*, May 22, 2007, 3.

[105] Hong Chen, "Guangdong Water Woes 'to Worsen'," *China Daily*, November 28, 2007, 5. This region too was afflicted by algae blooms. See Qiwen Liang, "Algae Blooms Watched," *China Daily*, December 28, 2007, 4.

[106] Jiao Wu and Fangchao Li, "Rural Water Woes to be Addressed," *China Daily*, August 23, 2007, 1. Also see Jing Li, "China Cleans Up Countryside," *China Daily*, April 13, 2009, 6 and "Rivers Need to be Cleaned Up," *China Daily*, September 24, 2009, 3.

by tightening regulations and increasing inspections.[107] Yet the problems persisted and increased in frequency and severity.[108] A senior engineer working in an institute affiliated with the Ministry of Water Resources commented "The water environment is not good, and this influences the quality and quantity of cereals production in China."[109] The security of China's drinking water and purification of major rivers and lakes – together with major pollution and emissions control and urban waste treatment measures – were highlighted in the Eleventh Five-Year (2006–2010) Plan. Total investment in environmental protection will increase to 1.35% of GDP, and the new investment will focus on treating pollution sources.[110]

The 4 trillion yuan ($586 billion) economic stimulus plan announced in November 2008 included $41 billion to be invested in sewage treatment in 90% of China's counties over 3 years. An environmental official declared: "This is the first time that there is a nationwide sewage treatment program."[111] Xiaoqing Wu, a vice minister of MEP said the infrastructure was badly needed because annually some 280 million tons of household garbage, 9 billion tons of household sewage and 260 million tons of human excrement are not treated.[112]

It is difficult to know whether promised infrastructure will be constructed. Yet the policy response of government to water pollution crises has been noteworthy in the case of the Songhua River benzene spill, which focused domestic and international

---

[107] For example, in 2007 SEPA announced that it was launching an automated system to closely monitor key polluters, who account for 65 percent of the country's industrial waste (in response to environmental activists who complained that many industrial plants shut off expensive sewage disposal facilities after SEPA inspections and resumed dumping wastes into rivers). The agency claimed it had reached a "turning point" in this year because the rate of increase in pollutant discharges increased at a slower rate than in the previous year. See: Xinhua, "Big Polluters Face Strict Monitoring," *China Daily*, May 11, 2007, 4, and Jiao Wu, "Pollution Picture to Brighten," *China Daily*, June 6, 2007, 1.

[108] In 2007, SEPA reported that despite an increase in funding on pollution control, amounting to 1.23 percent of China's GDP, "China is under increasing pressure to cope with environmental pollution." Of 842 pollution accidents reported for 2006, more than half were water related. Moreover, half the country's population lived in an environment where sewage was not treated. Orders from SEPA to reduce pollution routinely were disregarded by some cities. See Xinhua, "Spending Failing to Solve Pollution Problems," *China Daily*, September 25, 2007, 4, and Dingding Xin, "278 Cities Suffer Untreated Sewage," *China Daily*, August 31, 2007, 3.

[109] Personal interview with Senior Engineer, Institute of Water Resources, Beijing, May 30, 2007.

[110] Xiaohua Sun, "Drinking Water Gets Top Priority in New Plan," *China Daily*, November 27, 2007, 3. SEPA official Shengxian Zhou gave an optimistic projection, claiming that "the quality of China's key drinking water resources should meet national standards by next year." He also believed that pollution of waterways would decline: "The ecological system should maintain a virtuous circle and all rivers should flow calmly along their natural course." See Wutai Zhu, "SEPA Chief Outlines Vision for Clean Drinking Water." *China Daily*, November 22, 2007, 5. It seems unlikely that either goal will be realized without greater government coordination and enforcement actions. The immediate plan was to inaugurate the first national pollution census to examine sources of industrial, agricultural and residential pollution without, however, linking results to either punishment or evaluation of derelict officials. See Xiaohua Sun, "1st National Pollution Census Starts," *China Daily*, January 5–6, 2008, 1.

[111] Jing Li, "41b Flows into County sewage Facilities," *China Daily*, November 18, 2008, 6.

[112] Jing Li, "China Cleans Up Countryside," *China Daily*, May 13, 2009, 6.

pressure on China's environmental regulatory regime. It prompted amendment to the Water Pollution Prevention and Control Law in early 2008. Li and Liu summarize four ways in which policy was toughened:

1. Strengthening of local governments' environmental protection responsibilities
2. Enhancing opportunities for public participation, including permitting class action suits
3. Increasing fines (including fining individuals of polluting industries) and expanding enforcement measures
4. Improving existing practices such as the requirement for discharge permits[113]

In Chapter 8 we consider implementation and enforcement issues raised by this case.

3.3.2.3 Ocean Pollution

China's coastline extends 18,400 km and abuts four seas: the Bo Hai (considered an "inland" sea), the Yellow Sea, the East China Sea, and the South China Sea. In 2006, China's seas generated $270 billion or just over 10% of GDP,[114] yet development of a booming regional economy along this coastline is jeopardized by increased degradation of the ocean. Threats to China's oceans include overfishing, destructive fishing methods, pollution, and the reclamation of coastal lands. Marine fisheries are nearly 75% of China's total fisheries, and overfishing has resulted in a serious decline of take in recent years. The mariculture industry has caused degradation of water quality as well as put pressure on fish fry, small crustacea, and shellfish.[115] Moreover, the use of dynamite and poison fishing has damaged coral reefs and mangrove forests. At least 50% of the coral reefs off China's coasts have disappeared in the past 20 years. Loss of coral reefs in turn increases the risk of typhoon damage to China's coasts.

Pollution from industries, agriculture, domestic sewage, oil and gas exploration, and fish farming has degraded China's ocean environment, as has extensive runoff of silt from rivers and seabed dredging. As one NGO representative remarked: "All the coastal cities of China dump their wastes in the sea."[116] A State Oceanic Administration official stated: "The coastal marine ecosystem is worsening, the quality of ocean water is deteriorating, and large amounts of pollutants are infiltrating from land to sea."[117] The loss of coastal wetlands to agriculture, aquaculture, and reclamation projects has

[113] Jingyuan Li and Jingjing Liu, "Quest for Clean Water in China's Newly Amended Water Pollution Control Law," Woodrow Wilson Center, China Environmental Forum, 2009.

[114] Xiaohua Sun, "10% of GDP Now Comes from Sea, Says Report," *China Daily*, April 10, 2007, 4.

[115] John Mackinnon, Mang Sha, Catherine Cheung, Geoff Carey, Zhu Xiang and David Melville, *A Biodiversity Review of China*. Hong Kong: World Wide Fund for Nature (WWF) International, 495.

[116] Personal interview with NGO representative, Beijing, January 11, 2005.

[117] Personal interview with official, State Oceanic Administration (SOA), Beijing, January 1, 2006. A SOA report issued in August 2007 and based on more than 500 pollution outlets monitored by the agency found that 77 percent of the outlets were discharging more pollutants than permitted, some 18 percent more than in the previous year. See Xiaohua Sun, "Offshore Water Quality Deteriorates," *China Daily*, August 4–5, 2007, 2.

devastated both wildlife and marine resources. Several species already have become extinct: sea cows, species of kelp, and the habitat of sea turtles has been threatened. Enforcement of existing regulations and laws on pollution remains problematical.

Red tides have increased in frequency and range, with 80 or more occurrences in recent years. The tides are caused by buildup of marine plankton that consume oxygen while releasing toxic substances into the water, killing fish and plant life in the coastal regions. The third appearance of a red tide in Shenzhen bay caused serious pollution in mid-2007, which marine experts said was the largest ever appearing off the city's coast.[118]

Problems of ocean pollution have attracted less attention even than those of land and freshwater systems in China. Director of the Guangdong Oceanic and Fishery Administration, Li Zhujiang, said problems of the Pearl River estuary were the product of years of ineffective protection measures: "The government has spent so much on cleaning the river, but it never set up a special financial foundation to deal with pollution near the sea."[119]

In 2008, Director of the State Oceanic Administration, Sun Zhihui, pledged to address these concerns. He said the agency would tighten monitoring of the ecology of waters offshore China, conduct experiments to restore the marine ecosystem, and establish seven special protection zones.[120] However, the agency is small and not well regarded for its enforcement capability.

3.4 State Responses to Environmental Stressors

The environmental challenges to China's food producing lands and waters have been huge, and the state has responded in kind with large-scale projects as well as standard bureaucratic routines. We treat the food safety system in Chapter 8, and consider implementation issues at length there and also in Chapter 9. Here, we examine seven different examples of state responses: policy restricting arable land conversion, China's one child policy, investment in irrigation systems, large-scale dam construction, the South-North Water Diversion Project, large-scale afforestation and reforestation campaigns, and the program to convert marginal agricultural lands to forests and grasslands.

3.4.1 Restriction on Arable Land Conversion

In the 1980s and early 1990s, the central government employed an hierarchical system to regulate conversion of agricultural lands to other purposes – primary industrial, commercial, and residential. The regulatory system had several loopholes, however.

[118] Jonathan Yeung, "Red Tide Returns to Shenzhen Coastal Area," *China Daily*, June 7, 2007, 4.
[119] Quanlin Qiu, "Pearl River Waste Harming the Sea," *China Daily*, July 25, 2007, 5.
[120] Xinhua, "Official Vows to Cut Offshore Pollution," *China Daily*, February 26, 2008, 4.

Moreover, local level officials had large incentives to bend the rules because of the benefit to provinces, counties, and municipalities of the conversion of collective land to commercial and industrial purposes. It increased tax revenues of the municipalities. In response, the State Council promulgated a revised Land Management Law in 1998.[121]

Under the revised law, the central government resumed decision-making control on land conversions from agricultural to other uses. The land utilization plan for the period 1996–2010 called for a reduction in land allocated to human settlements and industrial sites, and specified that very little agricultural land would be converted for any purpose in the coastal provinces. Also, the central government imported sophisticated remote sensing technology from France, which made it less dependent on provincial and local governments for information on land use.

Ho and Lin believed these goals were overly ambitious, given the expected pressure of population increases and increased industrial and commercial activity in the coastal areas.[122] Cartier had a more sanguine view of the revised law, believing that it might resolve some of the internal contradictions (between different interests in land conservation and development of different levels of government in China),[123] and that it might promote fiscal stability by cooling speculative fever in land transactions.

These measures did not reduce pressures on arable land, which reached a high point by the end of the Tenth Five-Year Plan in 2005. To curtail conversion and safeguard future food security the regime responded in three ways: (1) it set a limit on the minimum amount of cultivable land, (2) it tightened regulations on land conversion, and (3) it sought ways to increase the amount of arable land. We discuss each in turn.

At the 17th Congress of the CCP, Premier Wen Jiabao announced that China could not have less than 120 million hectares (about 1.8 billion mu) of arable land. Reiterating this point, Minister of Land and Resources Xu Shaoshi stated: "The red line of 120 million hectares of arable land cannot be crossed."[124] (At the end of 2006, official accounts reported that the arable land total was 121.8 million hectares, compared to 122 million hectares in 2005.)

In 2006, the Ministry of Land and Resources established a new classification system for lands. It strictly barred any construction of villas, golf courses, or race tracks taking up large amounts of arable land.[125] Then, it initiated a process to define lands into four different categories: those where urbanization was "prioritized," "encouraged," "limited," and "forbidden." At the completion of this national blueprint, provincial governments would be given greater freedom to plan their own development projects in accord with the national plan.[126]

---

[121] This discussion follows Ho and Lin, 2004, 776–78.

[122] Ibid., 778.

[123] Cartier, 2001, 466–69.

[124] Huanxin Zhao, "No Yielding on Arable Land: Minister," *China Daily*, July 13, 2007, 1.

[125] Fangchao Li, "Arable Land Bank Continues Decline," *China Daily*, April 13, 2007, 3.

[126] Jing Fu, "Land Plan to Preserve Countryside," *China Daily*, September 25, 2007, 1, 2.

Too, the land approval process was tightened in 2007 to force local governments to make better use of their available land and spur disposal of land already approved for use. In mid-2007, the Ministry of Land and Resources began a 3-year national land-use survey, to ascertain changes in land utilization and management. A previous survey had been done from 1984 to 1996, but a number of local officials camouflaged land status or fabricated data during the inspection, leading to many cases of illegal land acquisition, as mentioned above. The ministry planned random checks and strict penalties for manipulation of land data.[127]

The ministry also increased fees and penalties for illegal conversions. It doubled the land-use fee for arable land for new construction projects, which reduced the incentive for local governments to sell land (as they would receive less income for doing so). The ministry also set a minimum pricing standard for land sales for industrial use, as a means to stop local governments from attempting to attract investors with heavily discounted land prices.[128] Finally, the ministry announced a campaign to check land law enforcement, and to hold provincial governments responsible for diverting farmland to other uses in excess of quotas.[129]

China's land authorities also have made efforts to increase the amount of arable land. From 1999 to 2006, China added 2.4 million hectares of arable land, which was a greater amount than land made available for construction projects. In the expanded, pilot areas, grain production increased by 10–20%.[130] A second plan, announced in mid-2007, was to convert at least 5.5 million hectares into cultivable land through two forms of consolidation: (1) re-planning of random, scattered and

---

[127] Fangchao Li, "Minimum Land Bank to Stay in Effect," *China Daily*, June 26, 2007, 3. A State Council regulation of 2008 required greater cooperation among land survey participants, and it further tightened penalties for falsifying or distorting information. See Dingding Xin, "Authentic Land Data Promised," *China Daily*, February 28, 2008, 3.

[128] Fangchao Li, "Protection of Arable Land Banks a Joint Responsibility: Minister," *China Daily*, July 13, 2007,4.

[129] Fangchao Li, "Illegal Land Use Poses Major Threat," *China Daily*, September 18, 2007, 1. Nearly half of the rural protests in China were triggered by illegal land seizures or expropriations. The State Council ordered local governments to raise compensation for farmers losing land to development projects, as one means to address protests; increasing enforcement of land law violators including local government officials is another. See: Huanxin Zhao, "Farmers' Protests Decline Sharply," *China Daily*, January 31, 2007, 1. Some 22,000 cases of illegal land use covering more than 328,720 hectares were reported between January 2005 and September 2006. Late in the next year, land inspectors in the Ministry of Land and Resources ran a 100-day campaign to detect major rule violators, catching a few local government officials. See Jiao Wu, "Crackdown on Illegal Land Use," *China Daily*, December 11, 2007, 3. The ministry promised to station inspectors in every province as part of a pilot project to curb illegal land acquisitions involving local authorities. See Jiao Wu, "Illegal Land Use in Cross Hairs of New Nationwide Scheme," *China Daily*, February 2–3, 2008, 3.

[130] For example, authorities in Jilin Province expected to increase rice output through converting large areas of salina lands to paddy fields. The plan was to make the salt-encrusted land arable by flooding it with nearby river water. See Xinhua, "Salt Lands May Offer Solution to Land Shortage Problem," *China Daily*, December 27, 2007, 3.

small plots, and (2) merging villages and returning land used to build houses and other structures to farming.[131]

Altogether, these measures were designed to insure sufficient arable lands for production of staples in the near-term. They seemed to be having some effect, as the rate of arable land loss in 2007, a reduction of 40,700 ha to a total of 121.73 million hectares, was the smallest annual decrease since 2001.[132]

In August 2008, the central government consolidated these reforms in a land use policy for the next two plan periods (until 2020). A State Council executive meeting chaired by Premier Wen Jiabao announced the formal goal of maintaining at least 121.2 million hectares of arable land until 2010 and 120.3 million hectares until 2020.[133] A requirement of the policy was that real estate developers who obtained arable land would need to reclaim an area of comparable size in order to begin construction on the new property.[134]

Then, on November 12, 2008, the central government announced its $586 billion stimulus plan to offset the effects of the global recession. Virtually every part of the stimulus package required acquisition of land, and especially for improvement of rural infrastructure such as roads, railways and airports.[135] Officials of the Ministry of Land and Resources said no more than 40,000 ha of arable land would be acquired for the projects (and another 40,000 ha of non-arable land), and the amount was "not a big deal."[136] While some of the planned infrastructure projects would disturb the environment, stimulus funds also would be available for environmental projects such as remediation of pollution to waterways.[137] Indeed, the MEP did reject a number of heavy polluting projects such as coal-fired power plants and resource-intensive coal chemical plants. A spokesman for MEP said the agency would "set up a 'fire wall' to block projects that fall short on environmental standards."[138]

We mentioned above that uncertainty concerning land use rights has accelerated the loss of arable land. Clarification of property rights is a major issue in contemporary China, posing market-based arguments such as permanent land contracts and freer trading of land against those who fear that unregulated sales will concentrate land in the hands of the few. At the plenary session of the CCP Central Committee in October 2008, party leaders agreed to allow farmers to transfer their

[131] Chuanjiao Xie, "1.7m Hectares of Additional Arable Land 'by 2020'," *China Daily*, June 22, 2007, 2.

[132] Jiao Wu, "Arable Land Reserves Continue to Decline," *China Daily*, April 17, 2008, 3.

[133] Jiao Wu, "Policy Thrust on Saving Arable Land," *China Daily*, August 14, 2008, 2.

[134] Jiao Wu, "New Measures to Protect Farmland," *China Daily*, September 10, 2008, 2.

[135] Jiao Wu, "Enough Land Pledged for Booster Plan," *China Daily*, November 14, 2008, 2.

[136] Qian Wang, "Stimulus Sops Up Farmland," *China Daily*, December 31, 2008, 3.

[137] Yanfeng Qian, "Stimulus Spending 'Could Aid' Environmental Protection," *China Daily*, December 11, 2008, 4.

[138] Jing Li, "Green Light Given to 153 New Projects," *China Daily*, January 10, 2009, 10. Also see Xinhua, "Arable Land 'Must Not be Misused'" *China Daily*, December 4, 2008, 4.

land use rights, in order to boost rural incomes and agricultural productivity. The new policy endorsed the formation of markets to allow farmers to sub-contract, lease and exchange their land use rights, and they could pool their land in larger shareholding entities. To allay concerns that the policy would encourage conversion of arable land to non-agricultural uses, the policy reiterated language on enforcing "the most stringent farmland protection system."[139] This is the most significant expansion of privatization in rural China since the development of the household responsibility system. As of late 2009, however, the state has not developed an effective system for the arbitration of disputes over land use rights. This is the largest category of complaints received by the Ministry of Agriculture, and is a significant catalyst for protests, which have increased in the economic downturn as millions of unemployed migrant workers return home.[140]

3.4.2 *China's One-Child Policy*

Population pressures figure in each of the environmental stressors discussed above, and this is a problem to which the regime responded radically. In 1979, China introduced the one-child family policy, which is the single most important reduction of environmental stress to have occurred globally in the past generation. The policy was designed primarily for urban areas, where there were incentives for residents to have small families. In rural areas, the policy effectively was an "one child with exceptions" policy. The army of enforcement officials (at least one million) usually tolerated families with two, and sometimes three children. The policy also was not applied to minority households at all. Recently, an additional exception to the policy has allowed married couples, both of whom are single children, to have two children. The onus of policy implementation fell on women and led to horrible abuses such as forced abortions and sterilizations.[141] Preference for male offspring resulted in cases of female infanticide and under-reporting of births, as well as skewed sex ratios and large future problems as millions of men will lack marriage partners.

Notwithstanding these serious defects, the policy has sharply reduced the rate of growth in China's population as compared to relatively unconstrained population

---

[139] Xinhua, "Farmers Get Lee way on Use of Land," *China Daily,* October 20, 2008, 1; Jiao Wu, "Major Farmland Reforms Mulled," *China Daily*, October 8, 2008, 3; and Edward Wong, "China May Let Peasants Sell Rights to Farmland," *New York Times*, October 11, 2008, A1.

[140] Xinhua, "Move to Minimize Disputes over Land Use," *China Daily*, October 11, 2008, A1.

[141] Tony Saich, *Governance and Politics in China*. New York: Palgrave, 2004, 246–68. For a critical perspective see Susan Greenhalgh, "Missile Science, Population Science: The Origins of China's One-Child Policy," *China Quarterly*, 182 (June 2005), 253–76, and Greenhalgh and Winkler, 2005. Also, party officials, celebrities, and the rich have ignored the policy, and enforcement has had little effect in deterring this in recent years. See Xinhua, "Hubei Luminaries Fined for Flouting Family Rules," *China Daily*, January 2, 2008, 5 and Juan Shan, "Rich Flout Family Planning," *China Daily*, June 22, 2009, 1.

growth in other large developing countries such as India and Indonesia. Demographic experts estimate that the population would have reached 1.7 billion instead of the current 1.3 billion had the policy not been implemented.[142]

Repeated calls for a loosening of the one-child policy have had no effect. In March 2008, Zhang Weiqing, Minister of the State Population and Family Planning Commission, ruled out any policy change for the next decade. He noted that China faced a new birth peak in the upcoming 10 years, when nearly 200 million would enter child bearing ages:

> Given such a large population base, there would be major fluctuations in population growth if we abandoned the one-child rule now. It would cause serious problems and add extra pressures on social and economic development.[143]

3.4.3 State Investments in Irrigation Systems

One of the factors typically used to explain China's ability to achieve food self-sufficiency is the huge investments the state has made in irrigation infrastructure. In the 1960s and 1970s particularly, spending on water control played a very important role in rural development. Fan et al. note that government spending on irrigation was 30% of total expenditures in rural China in 2000.[144] Whether in poor or rich areas, spending on irrigation systems has been the most important form of agricultural investment. They mention that the state invests more than ten times as much in irrigation as it does in agricultural research.

Approximately 51% of the cultivated area in China is irrigated; nearly two-thirds of the irrigated areas use surface water, while the rest is irrigated with groundwater.[145] Several studies of the impact of irrigation on crop yields as well as household incomes report positive findings. For example, Huang et al. point out the "strong and robust" effect of irrigation on agricultural performance.[146]

3.4.4 Large-Scale Dam Construction

The regime consistently looks at the construction of dams as one means of stabilizing China's water supply (as well as producing hydroelectricity), and in the 50 years

---

[142] Chuanjiao Xie, "Baby Boom Set to Start Next Year," *China Daily*, December 12, 2007, 3.

[143] Zhigang Xing, "Population Policy Will Stay for Now," *China Daily*, March 10, 2008, 1.

[144] S. Fan, L. Zhang, and X. Zhang, "Reforms, Investment, and Poverty in Rural China," *Economic Development and Cultural Change*, Vol. 52, no. 2 (2004), 417.

[145] National Statistical Bureau of China, *Statistics Yearbook of China*. Beijing: China Statistics Press, 2007.

[146] Qiuqiong Huang, Scott Rozelle, Bryan Lohman, Jikun Huang, and Jinxia Wang, "Irrigation, Agricultural Performance and Poverty Reduction in China," *Food Policy*, Vol. 31 (2006), 30–52.

from 1949 to 1999 built 85,000 dams of different sizes.[147] The largest and most controversial recent dam construction has been at the Three Gorges of the Yangtze River. This is the largest major project in contemporary China, with an estimated cost of $37 billion. Standing 200 m tall, the dam will create a reservoir longer than 360 miles and 175 m deep. Planned energy production is 18,000 MW, about one-tenth of China's needs, when completed.[148]

Construction of the Three Gorges Dam aroused objections because of its destructive impact on people who would be displaced, on the environment of the Yangtze region, and on archaeological treasures in the area. About 1.3 million people have been displaced in the process of dam construction, and critics maintain that they have had difficulties relocating and have not been compensated adequately for loss of their homes and livelihoods. The dam will flood high quality arable land; moreover, industrial, mining, and human detritus will collect in the huge reservoir.[149] Finally, historic towns such as Fengdu (Ghost Town) have been inundated, along with relics from the past. For these reasons, when the party leadership presented the construction plan to the National People's Congress in 1992, nearly one third of the delegates of this once docile body either abstained or opposed it.

A test of the environmental effects of the Three Gorges Project occurred in 2006. Some observers blamed dam construction for the severe drought affecting Sichuan province and Chongqing Municipality. Government meteorologists refuted these charges, claiming the drought was attributable to both heat island effects around big cities and climate change.[150] Critics continued their attacks by pointing to severe erosion and landslides, as well as ecological destruction, saying that further relocations might be required when those already relocated had not been adequately compensated and were having difficulty finding employment.[151] The response to these criticisms (including from the foreign press) was predictable: Landslides had been common in the area before the project began, the dam presented no threat to the environment, most residents had been resettled and provided for, and both money and expertise had been allocated for any needed repairs to the project.[152]

---

[147] Jun Ma, 2004, ix.

[148] Economy, 2004, 205–08.

[149] Ibid., 208.

[150] Xinhua, "Three Gorges Project 'Not to Blame' for Severe Drought," *China Daily*, August 19–20, 2006, 2; and Xiaofeng Guan, "'Dam Not Responsible for Drought'," *China Daily*, October 24, 2006, 1.

[151] Jim Yardley, "Chinese Dam Projects Criticized for Their Human Costs," *New York Times*, November 19, 2007; see also Xinhua, "Gov't Tackling Three Gorges' Hidden Environmental Threats," *China Daily*, September 27, 2007, 3.

[152] See *China Daily* reports by Jiao Wu, "Ecology Damage Reports Refuted," October 19, 2007, 1; Xinhua, "Dam 'No Threat' to Environment," November 16, 2007, 3; Jiao Wu, "Three Gorges Project 'Under Control'," November 28, 2007, 3; and Jiao Wu, "Dam Impact Less than Predicted," *China Daily*, November 22, 2007, 1.

In late 2009, the backlog of problems for this controversial project was expected to cost an additional $25 billion. Most of the additional funds would be used to support poor farmers displaced by the dam.[153]

The most recent large dam controversy concerns construction of a 13-station dam along the Nu River (*Nujiang*) in Yunnan Province. After at least a decade of construction, the dam would become the world's largest. It would produce more electricity than the Three Gorges Dam, helping address energy shortages while bringing jobs to poor residents of southwestern China and revenues to the provincial and local governments.[154]

This project aroused opposition from residents (at least 50,000 would be displaced, and cropland and fish species would be depleted), environmentalists who questioned the need for yet another expansive hydropower project that would put so many environmental values at risk, as well as international forces such as the International Rivers Network and the World Bank. In this case, however, environmental protest led Premier Wen Jiabao in 2004 to halt (temporarily) construction that, in his words, had "aroused a high level of concern in society, and with which the environmental protection side disagrees."[155] Some parts of the construction will proceed, but this was the first recorded instance of delaying a major project because of environmental protest.

3.4.5 The South–North Water Diversion Project

Another very large-scale project, which if continued would be one of the world's largest, is the plan of the central government to address water scarcity in northern and western regions by transferring water there from the South. Mao Zedong proposed the grand plan in 1952, and it has been in the discussion stage for decades.

The South–North Water Diversion Project (SNWDP) includes three water transfer routes – east, central, and west – which will link the Yangtze River of central China to the Huai, Yellow, and Hai Rivers of North China. The project plan includes "four latitude and three longitudinal water courses regulating and distributing water not only from south to north but also from east to west."[156] Construction began on the first (eastern) phase in late 2002 and was scheduled for completion in 2008. This section is the easiest to construct, as it can take advantage of existing rivers and lakes, including the Grand Canal and its parallel rivers. However, it has required construction of nearly two dozen new, upgraded and expanded pumping stations,

---

153 Chris Buckley, "China Three Gorges Dam Faces New $25 Billion Bill," *Reuters AlertNet*, November 19, 2009.

154 See Gerald A. McBeath and Tse-Kang Leng, *Governance of Biodiversity Conservation in China and Taiwan*. Cheltenham, UK: Edward Elgar, 2006, 1–2, 208–15.

155 Ibid., 2.

156 State Council, Construction Committee of the South-North Water Diversion Project, *China South-to-North Water Diversion Project*. Beijing, 2004, 15.

many reservoirs, and extensive water treatment facilities for polluted water.[157] The cost for just the first, eastern section is estimated to be $6 billion (with the cost of the total project running to $72 billion in 2009 dollars).

The second phase is not expected to be completed so quickly, as the route is longer and more new construction will be required. The western route project is still in the design stage, and many observers believe it will not be built. The plan to divert Qinghai–Tibet plateau water resources away from the Mekong and other international rivers is extremely controversial.[158] The size of the projects is gargantuan. Objections come from provinces losing water to the North and the West, particularly Hunan and Hubei; from scientists who question whether at the time of greatest need for water (in winter months of the North), there will be sufficient water to transfer; from those fearing the disruptive displacement effects on people along the proposed routing; from environmentalists; and from China's neighbors in the southwest who object to loss of water resources from their rivers.[159]

3.4.6 Large-Scale Afforestation and Reforestation Projects

We discussed one of the largest afforestation programs above, in the context of the Yangtze flooding of 1998. This was one of seven different afforestation programs since the 1970s. The others include:

1. The "Three Norths" Shelterbelt program involves establishing plantations in north, northeast, and northwest China (from 1978 to 2050) with the objective of afforestation of 35 million hectares.
2. Protective afforestation in the upper and middle reaches of the Yangtze (1989–2000) involved planting and restoration of 6.8 million hectares.
3. Coastal shelterbelt (1991–2000) led to planting trees on 3.6 million hectares.
4. Cropland protection and agro-forestry in the plains (1988–2000) covered nearly 1,000 counties in four provinces.
5. Afforestation of the Taihang Mountain (1990–2010) involves planting trees on 4 million hectares
6. Combating desertification (1991–2000) had as its objective control of desertification in over 7.2 million hectares[160]

[157] Ibid., 21. For example, pollution in the Huai River poses a threat to diversion of water from the south to the north. See Xiaohua Sun, "Polluters Face Stiff Penalties," *China Daily*, August 27, 2007, 1.

[158] Personal communication to the authors from TJ Cheng, October 8, 2007.

[159] See Economy, 2004, 126.

[160] Runsheng Yin, Jintao Xu, Zhou Li, and Can Liu, "China's Ecological Rehabilitation: The Unprecedented Efforts and Dramatic Impacts of Reforestation and Slope Protection in Western China," Woodrow Wilson Center, *China Environment Series*, Issue 7, 2005, 19. Also see Tianjie Ma, "Interconnected Forests: Global and Domestic Impacts of China's Forestry Conservation," Woodrow Wilson Center, China Environmental Forum, 2008.

In 1998, *China Daily* proudly announced "China now ranks first in the world in both the speed and scale of afforestation."[161] A 2007 report proclaimed that 20% of China would be forested by 2020.[162] Nevertheless, Harkness commented that "(I)ncreases in forest cover have coincided with decreases in the actual amount of wood available for harvesting," which has pushed Chinese logging firms abroad.[163]

Several problems have been identified in the afforestation programs, particularly the development of monocultural plantations, which limit species diversification. A forestry management official said that this needed to be kept in perspective:

> We are now looking at species diversity. We are emphasizing hardwood species, and this is a positive sign. Yes, there are problems, but the scale is unparalleled. We are converting sloping farmlands into forests, and enlarging wetlands. There is massive afforestation, which is good for carbon sequestration.[164]

Yin et al. point out additional difficulties. They object to the top-down nature of campaigns, and insufficient attention paid to local interests and conditions. Often those who have lost access to forests and logging have been inadequately compensated. Finally, the lack of long-range planning and development of good practices may increase other problems, such as erosion and introduction of invasive species.[165] Notwithstanding the criticism, the afforestation and reforestation programs have brought about a significant reduction in erosion, which benefits agricultural productivity.

Forests occupy 280 million hectares and about half of China's rural population depends on them to some extent for a living. In August 2007, the State Forestry Administration (SFA) announced plans to reform the management structure of forests – now owned exclusively and managed either by the state or village collectives (*jiti*). Similar to reforms in management of agricultural lands, the SFA proposal would extend management rights to individual farmers, contractors and in some cases overseas investors, for periods as long as 60–70 years.[166]

3.4.7 Restoration of Forests and Grasslands

The final state program is also the most recent, launched just in 1999–2000. With an overall budget of about $48 billion, the sloping lands conversion program is

[161] *China Daily*, "Afforestation Tops Priority List among Former Loggers," June 10, 1998, 3.

[162] China has planted 53.3 million hectares of forests since 1949, more than any other country in the world, with forest coverage rising from 8.6 percent to 18.2 percent. (Again, percentage of forest coverage varies by SFA definitions.) See Xiaohua Sun, "More Nature Reserves to Protect Forests," *China* Daily, December 5, 2007, 4. and Chao Liang, "China 2020: A Greener and Leaner Landscape," *China Daily*, August 1, 2005, 2.

[163] James Harkness, "Recent Trends in Forestry and Conservation of Biodiversity in China," *China Quarterly*, 156 (December 1998), n. 51, 924.

[164] Personal interview with forestry manager, State Forestry Administration, Beijing, May 18, 2004.

[165] Yin et al., 2005, 28–30.

[166] Jing Fu, "Forest Reform on Horizon," *China Daily*, August 16, 2007, 3.

perhaps China's most ambitious environmental initiative; without a doubt it is one of the world's largest land conservation programs.[167] It was designed after the Yangtze flooding to deal directly with erosion, which is particularly serious on sloped lands. Many of these lands in the Yangtze and Yellow River basins were originally forested, but in previous campaigns of the Maoist era were converted to farmlands. With slopes of 25° or greater, they were especially subject to erosion, and for this reason the original plan was to convert 5.3 million hectares of croplands on steep slopes to forest or grass coverage.

The program began with trials in Sichuan, Gansu, and Shaanxi provinces in 1999, and was then formally inaugurated in 2000 as the Sloping Land Conversion Program (SLCP, also known as "Grain to Green"; in Chinese *Tuigeng huanlin (cao)*). The goals of SLCP expanded to convert about 14.7 million hectares of fragile cropland to forests (or grasslands) by the completion date of 2010. Under the SLCP plan, the state provided extensive benefits to participating farmers. They received 1.5–2.55 t of grain per year (depending on location) for retiring 1 ha of cropland. Also, they received a one-time cash subsidy of $750 yuan (about US$109) per hectare to purchase seedlings or seeds, and $300 yuan ($43) per year for miscellaneous expenses for the duration of the program.[168] By 2004, some 7.2 million hectares of land had been converted, the result of a very rapid expansion of the program from 2001 to 2003.[169]

These incentives made the project quite popular with participating farmers, some of whom received more in food subsidies than they would have through their own productive work. The costs to the state, however, have been in the neighborhood of from $1 to $5 billion per year, making this one of the most expensive major environmental programs in China. The program has been effective in increasing the value of marginal lands, but its sustainability is in question because of the high cost, continuing questions about its effects on rural household income, and its exact environmental consequences, particularly with respect to changes in run-off.[170]

The most penetrating (and obvious) criticism of the Sloping Land Conversion Program has been that it has reduced arable land and led to lowered grain productivity and increased grain prices. In late 2003 and 2004, grain prices rose sharply in response to falling grain production, and the Ministry of Land and Resources and several

[167] Zhigang Xu, M. T. Bennett, Ran Tao and Jintao Xu, "China's Sloping Land Conversion Programme Four Years on: Current Situation and Pending Issues," *International Forestry Review*, Vol. 6, nos. 3–4 (2004), 317.

[168] J. T. Xu and Y. Y. Cao, "Converting Steep Cropland to Forest and Grassland: Efficiency and Prospects of Sustainability" (in Chinese), *International Economic Review*, no. 2, 56–60; also see Yin et al., 2005, 22–23. The importance of economic compensation for farmers' support of the SLCP is presented in Ling Zhi, Nuyun Li, Juan Wang, and Fanbin Kong, "A Discussion on the Economic Compensation System for Conversion of Cropland to Forestland in Western China" (in Chinese), *Scientia Silvae Sinicae*, Vol. 40, no. 2 (March 2004), 1–8.

[169] Xu et al., 2004, 117.

[170] See Jintao Xu, Ran Tao, and Zhigang Xu, "Sloping Land Conversion Program: Cost-effectiveness, Structural Effect and Economic Sustainability" (in Chinese), *China Economic Quarterly*, Vol. 4, no. 1 (October 2004), 160.

researchers hypothesized that the SLCP program was responsible for this price rise.[171] The ministry successfully argued for a slowing of the conversion program.

Researchers have found a co-variance between the SLCP reductions and a reduction in land area sown with grain. However, most findings indicate that SLCP has had a relatively small effect, particularly given that most of the converted land was sloping and of poor quality. Xu et al. attribute just about 1% of the price increase in cereals to SLCP, also noting the large impact it has had in reduction of the buildup of silt in irrigation networks and reservoirs and reduction in downstream flooding.[172] In other words, there does not appear to have been a trade-off between land conversion and agricultural productivity. Nevertheless, the policy is an expensive one and has created new dependencies. Said a policy analyst with the Chinese Academy of Sciences:

> Investment in *Tuigeng huanlin* means a reduction of investment elsewhere. As to people's income, what happens after the policy ends? Now, about 50% of the people in affected areas are reliant on government subsidies.[173]

By 2007, the SLCP had returned 24 million hectares to forest and grass cover, accounting for about 60% of China's new forest area and benefiting 124 million farmers. However, as arable land neared the 120 million hectare threshold, official attention focused again on SLCP, and the government suspended a plan to convert 1.07 million hectares into forest.[174] Agriculture ministry officials said the program curtailment was solely for the purpose of making adjustments in it, and the project seems likely to continue, but at a reduced level.[175] The State Council did fine-tune the program, by limiting the period during which subsidies would be provided, based on category of land, and by making special provisions for "ecological immigrants," especially in minority areas.[176]

An Australian–Chinese research team recently conducted a cost-benefit analysis of the SLCP in northern China, with a specific focus on Qinghai and Shaanxi provinces.

---

[171] Ministry of Land and Resources, *2003 China National Report on Land and Resources.* Beijing, 2004.

[172] Zhigang Xu, Jintao Xu, Xiangzheng Deng, Jikun Huang, Emi Uchida and Scott Rozelle, "Grain for Green versus Grain: Conflict between Food Security and Conservation Set-Aside in China," *World Development*, Vol. 34, no. 1 (2006), 130–48; see also, Xiangzheng Deng, Jikun Huang, Scott Rozelle, and Emi Uchida, "Cultivated Land Conversion and Potential Agricultural Productivity in China," *Land Use Policy*, Vol. 23 (2006), 372–84. For a study of the impact of converting cropland to grassland, see Yongzhong Su and Peixi Su, "Ecological Effects of Converting Crops to Grass in the Marginal Lands of Linze Oasis in the Middle of Heihe River Basin," in Liu, ed., 2006, 445–50. Also, personal interview with former staff member, Chinese Academy of Forestry, Beijing, May 21, 2007.

[173] Personal interview with policy analyst, CAS, Beijing, May 25, 2007.

[174] Yong Wu, "Reforestation Plan Put on Hold," *China Daily*, September 12, 2007, 2.

[175] See Xiaohua Sun, "Tree Plan Being Retooled," *China Daily*, September 20, 2007, 3; and Xiaohua Sun, "Returning Farmland to Forests to Protect Income, Farmers Told," *China Daily*, October 11, 2007, 2.

[176] State Council, "Notification on the Completion of the *Tuigeng Huanlin* Policy," August 9, 2007.

The team found evidence that the social welfare of people across northern China benefitted from the program. Several factors made it difficult for the research group to make clear recommendations, for example the incomplete property rights regime, imperfect credit markets, and high transaction costs. It found that the program was not cost-effective in achieving environmental goals, partly because of poor targeting of resources to areas of greatest ecological risk, and because it did not use market-based instruments. For these reasons, the researchers questioned whether the program would be sustainable after government funding expired.[177]

Most of China's provinces have had SLCP funding, and it is difficult to generalize based on research from two areas only. The SLCP suffers from the defects of all large-scale projects: the inability to focus incentives on areas of greatest need, given the immense variation of conditions in China's diverse terrestrial system.

3.5 Conclusions: Overall Impacts on Current Food Security

During China's reform era (from 1978 to the present), arable land has declined, due to pressures of population growth, urbanization, and exceedingly rapid economic development. These pressures have increased erosion, deforestation, desertification, pollution to land, air, fresh water, and China's marine coastal environment. Nevertheless, through improvements in agricultural technology and practices, China has been able to feed her 1.3 billion residents. (In 2008, China produced 502 million tons of staples,[178] an amount sufficient to provide food nationwide.) Yet the regime pays close attention to the amount of arable land and particularly to grain sufficiency. At the 2007 meetings of the national party congress, Premier Wen Jiabao announced that China must maintain 120 million hectares of arable land, and this statement has been etched into state policy.[179]

---

[177] Jeff Bennett, Xuehong Wang, and Lei Zhang, *Environmental Protection in China: Land-Use Management.* Cheltenham, UK: Edward Elgar, 2008. See especially chapters. 8–9, 196–220.

[178] Xinhua, "Grain Yield Exceeds 500m Ton," *China Daily*, December 24, 2007, 3. This was the fourth consecutive year of increase in grain output, and it allowed China to meet 95 percent of domestic demand. A shortage in production was not expected until 2010. See Ying Wang, "Country is Able to Meet Grain Needs," *China Daily*, November 22, 2007, 3. However, rising global grain prices and shortages of corn and soybeans spurred grain officials to eliminate the export tax rebates on major grains. Also, the State Council enacted an export tax on grains both to insure adequacy of domestic supply and curb food price inflation. See Diao Ying, "Grain Export Tax Rebate Dumped," *China Daily*, December 18, 2007, 13; Diao Ying, "Gov't Sets Export Tariff on Grains," *China Daily*, January 2, 2008, 14; and Jiao Wu, "Tight Supply May Hit Grain Stability," *China Daily*, January 4, 2008, 4.

[179] An official of the Ministry of Land and Resources indicated that Wen's statement did not reflect policy of the State Council, which did not believe that the cited amount of land needed to be retained in the arable land category (personal interview, Beijing, May 27, 2007). A land resources researcher at a university land management institute opined that Wen's statement was "a slogan," designed to outline a conservative approach (personal interview, Beijing, May 23, 2007).

Some Chinese scientists and policy elites are worried about the reduction of arable land; a number of foreign observers, such as Lester Brown, have been pessimistic, suggesting China will not be able to feed itself in the future. Our sense from a reading of the large literature on this topic and interviewing agricultural scientists, land resources experts, and policy-makers is that the amount of arable land is sufficient for the present, and with appropriate changes in policy,[180] grain security can be assured in the near-term. However, without significant changes, land resources will not sustain food production if population increases to 1.6 billion in 2030 and when demand for food increases among China's increasingly well-off population. Also, climate change as well as plant and animal diseases may adversely affect food production too.

We discussed several strategies adopted by the regime to counter environmental stressors and their impact on food security. Another focus of attention in China has been on increasing the efficiency in use and productivity of available arable land. This entails the improvement of cultural practices of farmers, use of improved seeds, more efficient utilization of crop enhancements such as fertilizers, and the like.[181] Indeed, in mid-2007, a coordination group of four ministries – Science and Technology, Agriculture, Finance, and the State Administration of Grain – signed responsibility contracts with 12 major grain producing provinces pledging to make greater efforts to increase crop yields through science and technology.[182] In Chapter 7 we look at biotechnological responses of the regime, primarily through the development of genetically-modified foods, which are targeted directly at increasing agricultural productivity.

In the next chapter, we turn to near-term environmental stressors, with a focus on climate change.

---

[180] See Jikun Huang, "Is China's Grain Security Facing Great Challenge?" (in Chinese), *Science & Technology Review*, (2004), 17. See also Jiao Wu, "Ministry Forecasts Bumper Harvest," *China Daily*, July 12, 2007, 3; and Jing Fu, "Agriculture Chief Has Thought for Food and Plenty on His Plate," *China Daily*, October 15, 2007, supplement 1.

[181] See, for example, Chao Zheng, Zongwen Liao, Kexing Liu, and Xiaoyun Mao, "Effects of Fertilizer on Agriculture and Environment" (in Chinese) *Ecology and Environment*, Vol. 13, no. 1 (2004), 134–38; Shukui Tan and Buzhuo Peng, "Examination of Cultivated Land Utilization to Guarantee our Grain Security" (in Chinese) *Economic Geography*, Vol. 23, no. 3 (May 2003), 371–74.

[182] Ying Wang, "Hi-tech Measures to Increase Grain Yield," *China Daily*, July 17, 2007, 3.

Chapter 4
Near-Term Environmental Stressors: Climate Change

Abstract This chapter begins with an introduction to the climate warming phenomenon globally and briefly chronicles China's involvement in the United Nations Framework Convention on Climate Change (UNFCCC). It treats China's energy policy, its heavy reliance on coal, low energy efficiency, and development of alternate energy sources. Then, China's traditional agriculture is discussed, with an emphasis on methane formation through wet rice cultivation. The chapter examines natural climate variability, for example the East Asian monsoon, which influences climate variability, and then presents information on observed climate change effects: changes in temperature, precipitation, surface evaporation, and sunshine duration. A special section examines extreme weather events in recent years, covering floods, drought, heat waves, sea level rises, and typhoons. The last section of the chapter introduces models used by Chinese scientists and preliminary estimates on ways in which climate change will affect production of rice, wheat, maize, and cotton. It also discusses the limitation of climate change models in predicting outcomes for China's food security.

Keywords United Nations Framework Convention on Climate Change (UNFCCC) • Kyoto Protocol • National Climate Change Program • Energy policy • Energy effi ciency • Alternate energy • Wet rice cultivation • Natural climate cycles • Temperature • Precipitation • Surface evaporation • Sunshine • Floods • Drought • Heat waves • Sea level rises • Typhoons • Climate change models • Mitigation strategies.

4.1 Introduction

In Chapter 3 we considered a variety of immediate environmental stressors, which have influenced food production in China. In this chapter we turn to mid-term and longer range stressors, and focus on climate change. Although scientists already have observed environmental changes in China and elsewhere attributable to climate change, it is the near (generational) and long-term effects that occasion greatest concern.

J.H. McBeath and J. McBeath, *Environmental Change and Food Security in China*, Advances in Global Change Research 35, DOI 10.1007/978-1-4020-9180-3_4,

Since the late 1980s, climate change has been the subject of increased research. Careful scientific measurements establish that the greenhouse gas (primarily, carbon dioxide but also methane, sulfur dioxide, and nitrous oxide) content of the atmosphere has increased nearly 20% since the start of the industrial revolution, a trend which follows the upward growth in global population. Increases in greenhouse gases are largely a consequence of the burning of wood and fossil fuels such as coal, oil and gas. Greenhouse gases trap solar radiation and increase temperatures of the earth. Observations from meteorological stations across the globe show increases in annual mean temperatures of 0.74°C over the last 100 years.[1]

Future consequences of climate change are expected to be more dramatic and disruptive to human life than the near-term observed results. For instance, rising sea levels, erosion, and storm surges may necessitate extensive relocation of coastal communities, huge expenses for reconstruction of infrastructure, and increased costs for fire and pest control as well as greater investment in health services.[2]

Scientists have developed consensus on the buildup of greenhouse gases and threats (and promises also) of climate change, but as we note below, sorting out the anthropogenic causes from the natural ones is quite difficult. A small but influential group of "greenhouse skeptics"[3] recommends that policy-makers defer action on the climate issue, pending the results of further research. They argue that even if climate risks are serious, the penalty for a few decades of inaction will be small.[4]

Chinese scientists have studied climate change causes and effects for more than 15 years.[5] Too, China has participated in the UNFCCC, and in 2002, China ratified the Kyoto Protocol. Critics of China's role in the negotiations suggest that officials

[1] See IPCC, Contribution of Working Group I to the Fourth Assessment Report of the Intergovernmental Panel on Climate Change. Cambridge: Cambridge University Press, 2007; also see, National Science and Technology Council, *Our Changing Planet: The FY2000 U.S. Global Change Research Program.* Washington, DC: A report by the Subcommittee on Global Change Research, Committee on Environment and Natural Resources, 2000; and Sharon L. Spray & Karen L. McGlothlin, *Global Climate Change.* Lanham, MD: Rowman & Littlefield, 2002.

[2] Jerry McBeath and Jonathan Rosenberg, *Comparative Environmental Politics.* Dordrecht, The Netherlands: Springer, 2006, 140.

[3] See the Marshall Institute, *Scientific Perspectives on the Greenhouse Problem.* Washington, DC: George C. Marshall Institute, 1989; S. Fred Singer, ed., *Global Climate Change: Natural and Human Influences.* New York: Paragon House, 1989; Richard S. Lindzen, "A Skeptic Speaks Out," *EPA Journal*, Vol. 16 (March–April, 1990); and for the most influential, Bjorn Lomborg, *The Skeptical Environmentalist.* Cambridge: Cambridge University Press, 2001.

[4] M. E. Schlesinger and X. Jiang, "Revised Projection of Future Greenhouse Warming," *Nature*, Vol. 350 (1991).

[5] A core climate change expert team has been established in China with expertise in economics, the social sciences, energy, meteorology, ecology, environment and other disciplines. More than 1,000 researchers and experts now work in scientific and applied fields of climate change. See Ministry of Science and Technology et al., *China's Scientific & Technological Actions on Climate Change.* Beijing, 2007.

participated "purely to use the political and economic opportunities to increase the inflow of advanced environmental techniques and foreign capital while blocking all aspects of external pressure by asserting political autonomy."[6]

There is merit to this perspective; however, as a non-Annex I country, and not required to reduce greenhouse gas emissions by target dates, China nevertheless has participated in major international climate change activities; and it has developed plans for near-term reduction in greenhouse gases. The National Climate Change Program released by China in mid-2007 pledged a 20% cut in energy consumption per unit of GDP by 2010; also, it raised the proportion of renewable energy, including large-scale hydropower, in primary energy supply up to 10% by 2010.[7] But Ma Kai, Minister of the National Development and Reform Commission (NDRC) told reporters that it was "too early, too abrupt, and too blunt" for the international community to impose emission caps on China, whose historic emissions were lower than those of developed countries.[8]

China, like India, has subscribed to the "common but differentiated responsibilities" that distinguish developing from economically developed countries.[9] At the 2007 meeting of the G-8, China's president, Hu Jintao reiterated this point, while remarking:

> Developed countries should, acting in accordance with this principle, meet the emissions reduction targets set in the Kyoto Protocol, provide assistance to developing countries and continue to take the lead in undertaking obligations to reduce emissions after 2012.[10]

At the 17th Congress of the Communist Party of China, General Secretary Hu Jintao said that all countries "should assist and cooperate with each other in conservation efforts to take good care of the Earth, our only home."[11] This was the first time that global environmental issues entered a political report to the party congress.

---

[6] Hyung-Kwon Jeon and Seong-Suk Yoon, "From International Linkages to Internal Divisions in China: The Political Response to Climate Change Negotiations," *Asian Survey*, Vol. XLVI, no. 6 (November/December 2006), 865.

[7] New China News Agency, "UN Green Body Praises Fight Against Climate Change," *China Daily*, August 3, 2007, 2. Also see Yinan Hu, "Initiative Will Allow for Sustainable Development," *China Daily*, June 5, 2007, 4, and National Development and Reform Commission, *China's National Climate Change Programme*. Beijing: 2007.

[8] Chong Wu, "Action Plan," *China Daily*, June 5, 2007, 1.

[9] National Climate Change Coordination Committee, *China's National Assessment Report on Climate Change* (in Chinese). Beijing: Science Publishing Co., 2007. Also see Xiaohua Sun, "China has a 'Differentiated Responsibility' to Climate Change," *China Daily*, June 9, 2007, 2.

[10] Tian Le, "China Still Committed, Hu Tells G8," *China Daily*, June 9, 2007, 1. The Chinese delegation made similar remarks at the UN Climate Change Conference meeting in Bali in December 2007. See Xinhua, "Common Interest, Separate Onus 'Must Still Combat' Global Warming," *China Daily*, December 4, 2007, 2.

[11] Jiao Wu, "Nation Attaches Importance to Global Warming," *China Daily*, October 22, 2007, 6.

China attended the Bali Climate Change Conference in December 2007 and agreed to the Bali Plan Action 1 (b) (ii), which provided for:

> Nationally appropriate mitigation actions by developing country parties in the context of sustainable development, supported and enabled by technology, financing and capacity-building, in a *measurable, reportable, and verifiable manner* (emphasis added).[12]

This represented a significant change in China's international position, and it led the Group of 77 nations in this more progressive orientation. To fulfill these commitments would not be easy for China, which has not yet developed a national system of statistics to measure its mitigation efforts.[13]

Following its practice of environmental diplomacy,[14] China aggressively has taken advantage of assistance from international NGOs and formed bilateral connections to advance its developing climate change goals. For example, in 2007 and 2008 the government signed agreements with Norway to help Chinese provincial governments assess potential risks caused by climate change and develop ways to respond.[15] China has indicated that technology transfers from developed nations, as provided for in the Kyoto Protocol, have been very slow. Science and Technology Minister Gang Wen commented: "Actually, there has been little progress in negotiations about technology transfers."[16]

Seeming to depart from the practice of developing nations to make no pledges of hard targets on emissions reductions, China announced in late 2009 that by 2020 it would reduce carbon intensity by 40–45%, as compared to 2005. Critics suggested this target could be achieved with little effort, because of the ongoing planning (discussed below) to increase energy efficiency. Nevertheless, the pledge is evidence of continuing Chinese engagement in climate change negotiations.[17]

In Chapters 8 and 9, we indicate that civil society in China has been slow to develop, thereby reducing domestic pressure on government to curb greenhouse gas emissions; nevertheless several NGOs have focused on climate change issues. Greenpeace has stood out in its call for lessened reliance on fossil fuels, but older

---

[12] Personal interview with official, Ministry of Foreign Affairs, December 18, 2007. See also Central People's Government of the People's Republic of China, "News Conference and Related Positions of the Chinese Climate Change Special Delegation." Beijing: December 20, 2007.

[13] Xiaohua Sun, "Now, the Numbers," *China Daily*, January 7, 2008, 4; see also Xiaohua Sun, "Joint Effort Urged on Climate Change," *China Daily*, December 13, 2007, 4.

[14] See Gerald Chan, *China's Compliance in Global Affairs: Trade, Arms Control, Environmental Protection, Human Rights*. Hackensack, NJ: World Scientific Publishing Co., 2006, 143–72 and Jerry McBeath and Bo Wang, "China's Environmental Diplomacy," *American Journal of Chinese Studies*, Vol. 15, no. 1 (April 2008), 1–16.

[15] Tian Le, "China, Norway in New Climate Pact," *China Daily*, March 27, 2007, 3; also see Xiaohua Sun, "New Program Will Take Climate Fight to Provinces," *China Daily*, April 18, 2007, 3, and Xaizhou Zhang, "Provinces to Draft Own Green Plans," *China Daily*, July 1, 2008, 2.

[16] Shanshan Wang, "Rise in Funding Pledged to Tackle Climate Change," *China Daily*, April 24, 2008, 1.

[17] Edward Wong and Keith Bradsher, "China Joins U.S. in Pledge of Hard Targets on Emissions," *New York Times*, November 27, 2009, A18.

and more established organizations, such as Friends of Nature and WWF, have entered the discussion nationally,[18] as have the National Defense Research Council and the Global Environmental Institute.

In this chapter we treat five topics. First, we review China's energy policy and its impact on climate warming. Then we consider traditional agriculture in China, and the extent to which it contributes to climate warming. Next, we examine natural causes of climate variation, such as atmospheric oscillations. This brings us to the state of knowledge on climate change in 2009, which allows us to interpret correlations of climate change and natural disasters. The chapter concludes with an analysis of the impacts of climate change on agricultural production.

4.2 China's Energy Policy

In the early twenty-first century, China remains reliant on coal for more than two-thirds of its energy needs. Although by the end of the twentieth century, China was entering international oil and gas markets, its consumption of petroleum was less than a quarter of overall energy consumption. The third largest source of energy was hydropower, produced through the large dam projects discussed in Chapter 3. Nuclear power contributes little to China's energy needs, and renewables (for example, solar and wind power) are just becoming competitive commercially; they are localized in Inner Mongolia, Xinjiang, and northern Hebei province, among other regions.

China's energy utilization is less efficient than that of most developed countries. In 2006, China used 15% of the world's energy to produce 5.5% of global GDP.[19] For this reason national planning efforts have emphasized improved efficiency. The Eleventh Five Year Plan (2006–2010) on national economic and social development set a target of 20% energy consumption reduction per unit of Gross Domestic Product (GDP). Although the rate of growth in energy consumption declined somewhat in 2006, it remained higher than GDP growth, as it has for nearly each of the last 30 years.[20]

An important factor in the cost of energy in China is the uneven distribution of resources. Coal is located in the North and Northwest, hydropower in the Southwest, and oil and natural gas in the eastern, central and western regions and along the coast. However, the largest consumers of energy are in the eastern and southeastern

---

[18] See Friends of Nature, *A Warming China: Thoughts and Actions for the Chinese Civil Society*. Beijing, 2007; and WWF, *International Climate Change Regime: A Study on Key Issues in China*. Beijing: Environmental Science Press, 2007.

[19] Reasons for China's energy inefficiency include very rapid industrialization and urbanization, energy and resource intensive production during the course of economic globalization, and the extensive growth pattern of the national economy. See Qiwen Zhu, "Pattern of Growth 'Has to Change'," *China Daily*, March 19, 2007, 1. Also, energy resources were not priced by the market for most of this period.

[20] Xiaohua Sun, "Energy Consumption per Unit of GDP Drops," *China Daily*, March 1, 2007, 1.

coastal areas, where the economy is most highly developed. As a result, coal and oil must be transported long distances from the North to the South, while natural gas and electricity are transmitted from the West to the East.[21]

4.2.1 *Reliance on Coal*

China's coal supplies are plentiful, among the largest in the world; and China is the world's largest consumer of coal. The market price of coal is the lowest of any energy source; this explains its prevalent use. Zhang comments that China's recoverable coal reserves are nearly 13 times greater than its recoverable oil and natural gas reserves combined.[22] By the early twenty-first century, China turned from an oil exporter to a net importer (with production expected to peak in the next decade), and currently buys about 6% of the world traded volume, second only to U.S. purchases.[23]

China's coal reserves are in the northern section of the nation. Most of China's thermal power plants are relatively small in size, and for this reason tend to be inefficient. About 400,000 industrial boilers are small, and their thermal efficiency is considerably less than larger boilers used in developed nations.[24] Although government plans call for increasing the size of thermal plants, these changes seem likely to materialize slowly. Zhang observes:

> (I)n the short to medium term (before 2010), China has little alternative but to rely on coal for power generation, given the fact that long lead times and high capital costs pose difficulties for the expansion of both hydropower and nuclear power as well as renewable power to meet the projected rapidly-increasing electricity demand.[25]

Coal not only is used for industrial production, but also fuels cooking stoves in most of China's homes, creating much indoor air pollution.

4.2.2 *Toxic Emissions*

China's power plants largely burn soft coal, which contributes smoke and dust to the atmosphere and spews both carbon dioxide and sulfur dioxide to the upper atmosphere. In 2007 China released about 5 billion tons of carbon dioxide into

---

[21] State Council, *China's Energy Conditions and Policies*. Beijing: Information Office of the State Council, 2007.

[22] Zhongxiang Zhang, *The Economics of Energy Policy in China: Implications for Global Climate Change*. Cheltenham, UK: Edward Elgar, 1998, 237.

[23] Tatsu Kambara and Christopher Howe, *China and the Global Energy Crisis*. Cheltenham, UK: Edward Elgar, 2007, 114.

[24] Zhang, 1998, 67.

[25] Zhang, 1998, 245.

the atmosphere. Because of its very rapid economic growth and use of inefficient thermal plants, China had become the world's second largest contributor of carbon dioxide emissions by the late 1990s (following the United States).[26] Although the rankings game is not an exact science, by late 2007 China had surpassed the U.S. in production of carbon dioxide. Yet on a per capita basis, China's carbon dioxide emissions are relatively low – just about half the world average.

China has the world's highest level of sulfur dioxide emissions. This organic pollutant is the cause of acid rain, and it affects about one-fourth of China's territory, as well as her neighbors in East Asia.[27] Acid deposition has deleterious effects on China's freshwater and ocean fisheries, and it damages productive agricultural land, as noted in Chapter 3.

Clearly, China's reliance on coal and the structure of both coal production and power transmission industries exacerbates the effects of climate change. In the first 2 years of the Eleventh Five Year Plan, indices of sulfur dioxide and chemical oxygen emissions dropped only 2.14% and 3.16% respectively from the 2005 level (less than the 4% plan target). However, in 2008, the two pollutants dropped 7% and 5% respectively from the 2005 level. Deputy MEP Minister Zhang Lijun attributed this to ministry disapproval or delay of 156 polluting projects.[28]

4.2.3 Energy Efficiency

Premier Wen Jiabao connected China's energy consumption pattern with climate change when he addressed party leaders in April 2007, saying "More work on energy conservation and emissions reduction is urgently required to deal with global climate change."[29] Conservation goals as mentioned included cutting 20% of energy use for every unit of GDP from 2006 to 2010, or 4% a year. Leaders have pledged to phase out tax breaks and discounts on land and electricity given to highly polluting industries. In the first year of the conservation program (2006), energy use as a proportion of GDP declined only by 1.2%; in 2007 it improved to 3.7%, but still below the target.[30] However, in the third year, 2008, energy consumption per unit of GDP dropped 4.59%, making officials optimistic that the plan goal might be met by 2010.[31]

---

[26] Zhang, 1998, 47.

[27] Elizabeth Economy, *The River Runs Black: The Environmental Challenge to China's Future*. Ithaca, NY: Cornell University Press, 2004, 72.

[28] Yu Xiao, "Pollution Levels, Energy Use Drop," *China Daily*, March 9, 2009, 6.

[29] Associated Press, "China Looks to End Pollutors' Tax Break," *Los Angeles Times*, April 28, 2007.

[30] Jing Fu and Xiaohua Sun, "Energy Use per Unit of GDP Decreases 3%," *China Daily*, November 30, 2007, 1; also see Xiaohua Sun, "Meeting Energy Goal a Tough Job," *China Daily*, June 23–24, 2007, 1 and Jing Fu, "Performance Figures 'Coming Soon'," *China Daily*, March 12, 2008, 5.

[31] Xiao, "Pollution Levels," 2009.

Observers were skeptical that the goal could be reached given the emphasis of the regime on continued high growth rates, and the efforts of subnational officials to bend the rules to benefit local economic development. Premier Wen expressed frustration in these words: "Understanding is not adequate, responsibilities are unclear, measures are not complementary, policies are incomplete, investment doesn't arrive, and coordination is ineffective." He indicated that "If these problems are not turned around, it will be difficult to achieve any obvious progress."[32]

State policy contradicts the goal of energy efficiency with respect to pricing. China's oil producing, marketing, and refining companies are state-owned enterprises, and the government sets the price of domestic oil and gasoline. During the oil price surge of 2004–2008, China's prices for gasoline rose far slower than those in western nations. In early 2008, the director of the NDRC's Energy Institute said "The government is worried that the higher price of oil could push up the already very high rate of inflation."[33] Yet before mid-year, the government did raise energy prices sharply (to \$3.83 a gallon for gasoline, a 16% increase) as inflation lowered somewhat. The price increases cut losses for China's oil refineries, and also responded to criticism of foreign governments that its fuel subsidies distorted global markets, encouraged greater consumption, and raised oil prices higher for other nations.[34] The state did not increase prices on natural gas, but it did allow modest increases in on-grid electricity rates.[35] Still, continued price controls distorted the market in these sectors. Moreover, the Ministry of Finance distributed \$2.8 billion in subsidies for local transportation, fisheries and grain production.[36]

4.2.4 Alternate Energy Strategies

A recent report by Greenpeace, *Energy Revolution: A Sustainable China Energy Outlook*, urges China to increase its planned development of renewables, increasing target production of wind power from 30 to 118 GW and solar power (through development of a solar photovoltaic system) from 1.8 to 25 GW. With these changes and sharp reductions in use of coal, China would be able to stabilize its carbon dioxide emissions by 2050.[37]

---

[32] Quoted in Howard French, "Far from Beijing's Reach, Officials Bend Energy Rules," *New York Times*, November 24, 2007, A24.

[33] Xinhua, "To Raise Oil Prices or Not," *China Daily*, May 13, 2008, 7.

[34] Zhihong Wan, "Gas, Diesel Cost More from Today," *China* Daily, June 20, 2008, 1, and also, Keith Bradsher, "China Sharply Raises Energy Prices," *New York Times*, June 20, 2008, A7.

[35] Increases in coal prices had squeezed power firms tightly, and the electricity rate hike passed these on to consumers. See Zhihong Wan, "Price Hike to Take Heat Off Electricity Rates," *China Daily*, August 20, 2008, 2; also see Xu Wang, "Call for Reform of Energy Pricing," *China Daily*, August 21, 2008, 18.

[36] Zhihong Wen, "Govt Offers Subsidies to Sectors Hit by Price Hikes," *China Daily*, June 21–22, 2008, 10.

[37] Xiaohua Sun, "Renewable Power Key to 'Green Growth': Report," *China Daily*, April 26, 2007, 3.

China has invested $600 million since 2006 on research into alternate energy. The Ministry of Science and Technology has developed China's Scientific and Technological (S&T) Actions on Climate Change. Early projects include development of renewable and clean energy, clean coal, carbon capture and sequestration, biological means of absorbing carbon, and improved farming practices.[38]

China's National Climate Change Program projects a reduction of CO_2 equivalent to 200 t through coal bed methane (CBM) and coal-mine methane industry upgrades. Other CO_2 reductions include 500 t through hydropower exploitation, 50 t through nuclear power, 110 t through thermal power, 30 t through development of bio-energy sources such as marsh grass, biomass briquettes, and biomass liquid fuel, and 60 t through wind, solar, geothermal, and tidal energy.[39] A 2008 report from the Climate Group affirmed China's progress on renewables, calling it a "low-carbon dragon which will power its future growth, development and energy security objectives." The group, dedicated to advancing business and government leadership on climate change, pointed to China's investment in renewable energy (at 12 billion, comparable to Germany's on a per capita basis). It noted China's leadership in manufacturing solar photovoltaic (PV) technology, export of wind turbines, and competitive advances in manufacture of solar water heaters, energy efficient home appliances and rechargeable batteries.[40]

In Chapter 2 we noted the impact of increased ethanol production on corn prices, and the regime's decision to phase out the use of this crop for fuel production. Instead, the central government plans a number of new crop bases – growing sugarcane, sweet sorghum, cassava, and rapeseed for fuel.[41] Altogether, this suggests that alternative energy strategies are in the experimental stage.

As discussed briefly in Chapter 3, China's huge afforestation and reforestation programs provide the opportunity to create new carbon sinks to absorb carbon dioxide emissions. In 2004, the State Forestry Administration established a Carbon Sink Office dedicated to this issue. Too, China has encouraged international NGOs, such as Conservation International and the World Wide Fund for Nature (WWF) to provide assistance in different aspects of building sustainable power systems and mitigating climate warming.[42]

China stands to gain from the clean development mechanism (CDM) negotiated in the 1997 Kyoto Protocol. It has a large share of certified emissions reductions (CERs), carbon credits permitting countries to emit carbon above their quotas. The CDM allows developing nations such as China to sell their CERs to developed

---

[38] Jing Fu, "Potential of Greenhouse Gases Tapped," *China Daily*, October 31, 2007, 1.

[39] Yinan Hu, "Alternate Power Source Will Help Cut Emissions," *China Daily*, June 5, 2007, 4.

[40] Xinhua, "Nation World Leader in Renewable Energy, Says Eco-Group," *China Daily*, August 4, 2008, 1.

[41] Jiao Wu, "Food before Biofuels," *China Daily*, May 8, 2008, 3.

[42] See Conservation International-China, "Promoting the Use of Natural Regeneration in China to Protect Biodiversity and Mitigate Global Climate Change." Beijing: CI, 2005; and WWF International, *The Asia-Pacific Climate and Energy Programme*. Giand, Switzerland: 2004.

nations, using the proceeds to develop clean coal technology. In late 2007, China's portion of the world supply in carbon trading was 73%.[43]

A white paper issued by the State Council's Information Office on October 29, 2008 said that by conserving and using renewables, China had reduced 835 million tons of CO_2 equivalent in 2006 and 2007. This reduction nearly equaled the combined greenhouse gas emission volume of Belgium and the UK. The paper urged continuation of the CDM after the expiration of the Kyoto Protocol in 2012, but argued that it needed to be amended to promote a higher level of technology transfer to developing countries.[44]

4.3 China's Traditional Agriculture and Climate Change

More than 50% of China's grain production comes from irrigated fields, and they produce about 80% of China's national crop output. We emphasize wet rice (paddy) cultivation, and its contribution to the production of greenhouse gases. We then consider livestock production, another contributor to greenhouse gases, before considering the general impacts of agriculture in China on climate warming.

4.3.1 Wet Rice Cultivation

Rice is the most important crop in China. Rice crops occupy over 290,000 km^2 of harvested area, and yield more than 190 million tons of product. Most of China's rice is irrigated, and about 40% of human-caused methane originates in the decomposition of organic matter in flooded rice paddies. A technical explanation of the process of methane production is:

> CH_4 (methane) emission from rice fields to the atmosphere is the net result of opposing bacterial processes, production in anaerobic microenvironments, and consumption and oxidation in aerobic microenvironments, both of which can be found side by side in flooded rice soils.[45]

Simply stated, this means that the flooding of rice fields cuts off oxygen supply to the soils. This causes fermentation of the organic matter, which results in the formation of methane gas. The methane moves to the atmosphere by bubbling from the soil through the conduit system of the rice plants.

Methane formation from irrigated rice is not considered entirely anthropogenic in origin, because rice originally was grown on soils that were naturally flooded.

---

[43] Huayu Li, "Green Market," *China Daily*, November 19–25, 2007, 3.

[44] Jing Li, "Nation Issues Post-Kyoto Plans," *China Daily*, October 30, 2008, 2.

[45] H. U. Neue and J. Boonjawat, "Methane Emissions from Rice Fields," in James N. Galloway and Jerry M. Melillo, eds., *Asian Change in the Context of Global Climate Change*. Cambridge: Cambridge University Press, International Geosphere–Biosphere Programme, 1998, 187.

However, setting up embankments (bunding) in rice fields along with water harvesting and artificial flooding are human changes to the environment. And of course, increased irrigation and production of wetland rice have expanded methane production.[46]

There is a seasonal pattern to methane emissions into the atmosphere. Within a month of flooding, emission rates rise abruptly, especially if organic inputs are used to fertilize crops. About a month later, during the heading and flowering stages, emissions rise again. Emissions lessen during the ripening and maturing stages, particularly if the plant root system ages and deteriorates (which changes the anaerobic nature of the soil).[47] Emission rates are influenced by temperature changes, the nature of the flooding system, and cultural practices as well.

However, a change in farming practices may be influencing methane formation. In the early 1980s, farmers began to drain rice paddies in the middle of the growing season as they learned that this increased yields and reduced water use. This change reduced methane emissions.[48]

4.3.2 Livestock Production

Domestic farm animals provide important foods in China – especially pigs and chickens – and grazing animals, particularly goats and sheep, are raised and fed in China's grasslands. These animals also produce methane, contributing what is estimated to be nearly one-quarter of greenhouse gases generated through agricultural practices.

Expansion of dairy herds in China likely will increase this problem. In 1980 China had just 640,000 dairy cows; then herds grew at about 11% annually, reaching 4.9 million cows in 2000. In the next 4 years, the rate of increase topped 20% annually, matching the growth in milk production.[49]

4.3.3 Other Agricultural Factors

In general terms, agriculture is thought to be a driving factor for about a third of the global greenhouse gas emissions, and China is close to this average. Wetland rice cultivation is the major source of emissions in China, with livestock production a

---

[46] Ibid.; also see, among others, M. W. Wang, A. Dai, R. Shen, H. Schutz, H. Rennenberg, W. Seiler & H. Wu, "Methane Emission from a Chinese Rice Paddy Field" (in Chinese), *Acta Meteorological Sinica*, Vol. 4 (1990), 265–75.

[47] Neue and Bonjawat, 1998, 189.

[48] Goddard Space Flight Center Top Story, "Shifts in Rice Farming Practices in China Reduce Greenhouse Gas Methane." See http://www.gsfc.nasa.gov/topstory/2002/1204paddies.html (retrieved February 12, 2008).

[49] Frank Fuller, Jikun Huang, Hengyun Ma, and Scott Rozelle, "Got Milk? The Rapid Rise of China's Dairy Sector and Its Future Prospects," *Food Policy*, Vol. 31 (2006), 207.

secondary factor. There are other factors as well, and the most important are land conversion and fertilizer use in farming.

Forests and soil humus are sinks of fossil fuel and land use emissions of carbon dioxide to the atmosphere. By the turn of the 1990s, China had 0.8–1.2% of the world's vegetation terrestrial carbon pool and 5.9–11.9% of the soil carbon pool.[50] The extensive deforestation occurring in China until recently greatly reduced the country's carbon sinks, partly corrected by the regrowth of forests. Ploughing land releases large amounts of carbon dioxide into the atmosphere, as does the burning of biomass in fields and forests.[51] Agriculture also is responsible for much of the nitrous oxide emissions produced in the breakdown of inorganic fertilizers and both manure and urine from livestock.[52]

As mentioned in Chapter 3, changes in state policy (such as the massive afforestation and reforestation programs of the 1990s), mitigated the impacts of human-induced changes to the ecosystem. Below we note some of the ways in which changes of agricultural practice can mitigate both anthropogenic and natural changes of weather and climate.

4.4 Natural Climate Cycles

In Chapter 1 we commented that through its long history, China has experienced a high degree of natural climatic variability. Climate scientists point to two factors that primarily influence this variation: (1) change in precipitation as one moves from south to north, and (2) monsoons and atmospheric oscillations.[53]

China's southern regions are humid, extending to a subtropical zone (Hainan Island), while its northern and western regions are semi-arid and arid. In the North, variability of precipitation is especially high because of rain patterns typical of more arid areas.[54] The arid conditions of the North and West lead to the collection of large

[50] J. Y. Fang, S. L. Su, and G. H. Liu, "Carbon Pools on Terrestrial Ecosystems of China and their Global Significance," in International Geosphere-Biosphere Programme, *Natural and Anthropogenic Changes: Impacts on Global Biogeochemical Cycles*. Beijing: 1995.

[51] A large boreal forest fire in China in 1987 burned more than 1.3 million hectares. Cahoon et al. estimate that this fire, along with a simultaneous fire in Siberia, contributed 20% of the total carbon dioxide, 36% of the total carbon monoxide, and 69% of the total methane emitted by savanna burning in that year. See D. R. Cahoon, B. J. Stocks, J. S. Levine, W. R. Cofer, and K. P. O'Neill, "Seasonal Distribution of African Savanna Fires," *Nature*, no. 359 (1992), 812–15.

[52] Parviz Koohafkan, Ana Rey, and Jacques Antoine, "Soil Carbon Sequestration in Dryland Farming Systems," in R. Lal, N. Uphoff, B. A. Stewart and D. O. Hansen, eds., *Climate Change and Global Food Security*. London: Taylor & Francis, 2005, 516.

[53] C. Vorosmarty, C. Li, J. Sun, and Z. Dai, "Drainage Basins, River Systems, and Anthropogenic Change: The Chinese Example," in James N. Galloway and Jerry M. Melillo, eds., *Asian Change in the Context of Global Climate Change*. Cambridge: Cambridge University Press, 1998, 212.

[54] D. L. Hartmann, *Global Physical Climatology*. San Diego, CA: Academic, 1994.

dust loadings in the atmosphere; this facilitates the movement of particulates, including toxic ones, both in-country as well as to other countries in the East Asia region.

Second and of greater importance in explaining climate variability, China is subject to the East Asian monsoon and interannual variations associated with it. In the summer, strong monsoons carry heavy precipitation to the North China Plain, while drought conditions often afflict southern China.[55] Diaz and Fu have drawn a linkage between droughts in both southern and northern China and the onset of the El Niño Southern Oscillation.[56] Further, Chang and King have linked precipitation patterns of the eastern Yangtze River to large-scale atmosphere–ocean interactions. They believe that flooding of the Yangtze and droughts in both north and south China are associated with the Quasi-biennial Oscillation, which in turn is correlated to the East Asia Circulation Index.[57]

Below we consider impacts of both anthropocentric and natural climate variability on extreme weather events, such as floods and droughts. Suffice it to say at this point that it is quite difficult to tease out the near-term climate changes from regional and temporal oscillations between cold and dry periods, as well as cool and humid periods associated with monsoon and other atmospheric oscillations. In fact, the literature tends to examine the changing Asian monsoon in relation to global climate change. For example, Fu's analysis shows a close relationship between the aridity index of East China (non-periodic oscillations) and the annual mean temperature change (rising) over the Northern Hemisphere land mass since 1880.[58] The aridity pattern is associated with the weakening of the summer monsoon, under the global warming trend.

4.5 Observed Climate Change Effects

Chinese climate scientists increasingly have collected detailed observations on changes in weather and climate.[59] China's National Climate Observation Network was established in 1997 and completed by late 2007. Seven ministries cooperated in the formation of the network: meteorology, water resources, agriculture, environmental

---

[55] Q. Guo, "The East Asia Monsoon and the Southern Oscillation, 1871–1980," in D. Ye, C. Fu, I. Chao, and M. Yoshino, eds., *The Climate of China and Global Climate*. Berlin: Springer-Verlag, 1987.

[56] H. F. Diaz and C. Fu, "Regional Precipitation and Temperature Variability and its Relationship to the Southern Oscillation," in D. Ye, C. Fu, I. Chao, and M. Yoshino, eds., *The Climate of China and Global Climate*. Berlin: Springer-Verlag, 1987.

[57] W. Y. B. Chang and B. King, "Centennial Climate Changes and their Global Associations in the Yangtze River," *Climate Research*, Vol. 4 (1994), 95–103. Also see Vorosmarty et al., 1998, 212.

[58] C. B. Fu, "An Aridity Trend in China in Association with Global Warming," in R. G. Ztpp, ed., *Climate Biosphere Interaction: Biogenic Emissions and Environmental Effects of Climate Change*. Chichester: Wiley, 1994, pp 1–17.

[59] A good introduction to online sources is the paper by Elisa Chihi-Yin Lai, "Climate Change Impacts on the China Environment: Biophysical Impacts," Woodrow Wilson Center, China Environmental Forum, 2009.

protection, forestry, oceans, and scientific research. Sixteen observation areas are the focus of attention:

1. Atmosphere and land systems in the Qinghai-Tibetan plateau
2. Glacier water and ecological systems in the Tianshan Mountain area
3. Xilingol pastures in the Inner Mongolia Autonomous Region
4. Dunhuang desert in Gansu Province
5. Forests in Northeast China
6. Water circulation systems in Sichuan and Yunnan provinces
7. Agriculture in the Yellow and Huaihe river basins
8. The lakes of Dongting and Poyang
9. The atmosphere around Mount Waliguan in Qinhai Province
10. Ecological systems in source regions of the Yangtze, Yellow and Lancang rivers
11. The economic belt around Beijing
12. Economic development zones in the Yangtze and Pearl river deltas
13. Sichuan Basin
14. The land-ocean-atmosphere system around Bohai Sea
15. Air–sea interaction in the South China Sea
16. Comprehensive oceanic observations[60]

Much of the research is based on reports from about 750 weather stations[61] across China, providing information on temperature, aerosols, greenhouse gases, and ozone. Too, observation data are collected on glaciers, frozen soils, accumulated snow, plants, and soils. In more recent years, climate modeling has expanded analytical capabilities. As of the middle of the first decade in the twenty-first century, scientists have noted changes in temperature, precipitation, surface evaporation, sunshine duration, and wind speed. We present some of the latest research findings, and where data are available, we comment on important regional variations.

4.5.1 Temperature Changes

The most significant observed change is that of mean annual air temperatures, which have increased by 1.3°C from 1951 to 2005, with a warming rate of approximately 0.22°C each decade.[62] Warming has been most evident in winter and spring.

---

[60] Xiaohua Sun, "Climate Change to be Better Monitored," *China Daily*, September 12, 2007, 2.

[61] Scientists have corrected for the impact of urban warming on weather stations, influencing rate of temperature increase. See Yaqing Zhou and Guoyu Ren, "Identifying and Correcting Urban Bias for Regional Surface Air Temperature Series of North China over Period of 1961–2000," *Climatic and Environmental Research*, Vol. 10, no. 4 (December 2005), 743–53.

[62] Guoyu Ren, Jun Guo, and Mingzhi Xu, "Climate Changes of China's Mainland over the Past Half Century" (in Chinese), *Acta Meteorologica Sinica*, Vol. 63, no. 6 (2005c), 945–46; and Guoyu Ren, Mingzhi Xu, Ziying Chu, Jun Guo, Qingxiang Li, Xiaoning Liu, and Ying Wang, "Changes of Surface Air Temperature in China during 1951–2004," *Climatic and Environmental Research*, Vol. 10, no. 4 (December 2005b), 718–20, in which the authors revise the mean temperature increase over the period from 1.1°C to 1.3°C.

In the last half of the twentieth century, warming was more rapid than the average values of the world and the Northern Hemisphere. Regionally, warming was greater in the Northeast, North, Northwest,[63] and the Tibetan Plateau[64] than in other regions, yet a cooling trend continued: spring and summer temperatures of southern China, specifically in the middle and lower stretches of the Yangtze River, decreased over the 50-year period. The frequency of temperature-related extreme events (discussed further below) such as cold waves and cool summers, also has not changed in China as a whole, notwithstanding important regional variations. In a 100-year re-analysis, Tang and Ren identify two especially warm periods, the 1930s–1940s and the 1980s–1990s, with 1946 and 1998 as the warmest years in the period.[65]

Effects of temperature increases also vary by altitude. Milder temperatures at high elevations have caused tree lines to move higher, replacing what previously were alpine ecosystems in some cases. Researchers have not noted much change in the temperature of the upper troposphere; however, average annual temperatures of the mid- to lower- troposphere have increased at a rate of 0.05°C each decade, increasing the difference between surface temperatures and those of the mid- to lower troposphere.[66]

Ren et al. compared surface temperature changes of the last 50–55 years to a temperature series of the past 1,000 years, based on tree ring data from western China and the historical record in the East. From this analysis, they believe that the high recorded temperatures of the Medieval Warm Period did not extend to all parts of China; but that temperatures from the end of the eleventh century to the middle of the thirteenth centuries probably were higher than those in any decade of the twentieth century.[67] Based on this study, they hypothesize that surface warming in China during the past 100 years might have been induced by the increased concentration of greenhouse gases, but that other factors – such as solar activities and low-frequency oscillation of the ocean-atmosphere system – might have been important contributing elements.[68]

---

[63] North China temperatures climbed 1.4°C Centigrade in the last years, above the China and global average. See Xiaohua Sun, "Life Will Change, Scientists Warn," *China Daily*, April 24, 2007, 2. Also see Kejia Zhang, "Charting the Pace of Change," *China Daily*, July 25, 2007, 18.

[64] The national weather bureau found temperatures in Tibet rising about ten times the speed of the national average, making the region a barometer for climate change. In 2007, studies centered on problems such as receding snow lines, shrinking glaciers, drying grasslands and desertification, as well as an increase in extreme weather events. See Xinhua, "Tibet Hard Hit by Global Warming," *China Daily*, November 22, 2007, 5.

[65] Guoli Tang and Guoyu Ren, "Reanalysis of Surface Air Temperature Change of the Last 100 Years over China," (in Chinese) *Climatic and Environmental Research*, Vol. 10, no. 4 (December 2005b), 794–97.

[66] Guoyu Ren, Ziying Chu, Yaqing Zhou, Mingzhi Xu, Ying Wang, Guoli Tang, Panmao Zhai, Xuemei Shao, Aiying Zhang, Zhenghong Chen, Jun Guo, Hongbin Liu, Jiangxing Zhou, Zongci Zhao, Li Zhang, Huzhi Bai, Xuefeng Liu, and Hongyu Tang, "Recent Progress in Studies of Regional Temperature Changes in China" (in Chinese), *Climatic and Environmental Research*, Vol. 10, no. 4 (December 2005a), 703–05.

[67] Ibid., 710.

[68] Ibid, 710–12.

4.5.2 Precipitation Changes

Averaged across China as a whole, no significant long-term changes have been observed in annual precipitation. However, significant declines in precipitation have been noted in the Yellow River Basin and the North China Plain; the largest decline in rainfall occurred in Shandong Province.[69] Scientists have detected a slight moistening trend in the Yangtze River Basin and in most parts of western China. In the Yangtze River region, summer rainfall produced most of the increase in annual precipitation, although winters were somewhat wetter as well.

4.5.3 Surface Evaporation Changes

Since the mid-1950s, the amount of pan-evaporation has declined, at a rate of 34.5 mm per decade. The two regions experiencing the largest decrease in pan-evaporation were the North China Plain and the lower reaches of the Yangtze River. In absolute terms, the largest decreases in pan-evaporation have been in the Northeast, areas of Inner Mongolia, and in northwestern China.[70]

Although western China has experienced a slight increase in precipitation, more rapid actual evaporation attributed by Chinese scientists to global warming has contracted wetlands at the source of the Yangtze and Yellow Rivers. Studies in the last 4 decades of the Qinghai–Tibet Plateau indicate a shrinkage in wetlands of greater than 10%.[71] In addition, this area accounting for nearly one-quarter of China's land mass (and called the "water tower" of China[72]) has experienced a receding snow line and more serious desertification than previously.[73]

Glaciers in China, as elsewhere, have melted at an increasing rate in the last half-century. China has 46,342 glaciers in its western regions, with a total area of 59,414 km^2 (about 0.6% of China's total land area, and the fourth largest number of glaciers in the world after Canada, Russia, and the United States). Scientists with the Cold and Dry Areas Environment and Engineering Research Institute of the Chinese Academy of Sciences opine that glaciers in western and northwestern regions have melted up to one-third in certain areas in the last 4 decades.[74]

---

[69] Ren et al., 2005c, 946–47.

[70] Ren et al., 2005c, 947.

[71] New China News Agency, "Scientists Say Climate Change Reducing Flow of Rivers," *China Daily*, July 16, 2007, 3.

[72] The region provides 25% of the water flowing down the Yangtze, 49% of the water in the Yellow River, and 15% of the Lancang River water. See Juan Shan, "Key Water Source Threatened," *China Daily*, August 13, 2007, 2.

[73] Fangchao Li, "Global Warming Threatens Plateau," *China Daily*, December 30–31, 2006, 1.

[74] Lie Ma, "Climate Change Taking Toll on Glaciers," *China Daily*, July 17, 2007, 4.

Qin Dahe of the Chinese Academy of Sciences remarks that "Due to global warming, glaciers on the Qinhai–Tibet Plateau are retreating extensively at a speed faster than in any other part of the world."[75]

China's Tibetan and Xinjiang glaciers are the major source of Asia's largest rivers – the Yangtze, Mekong, Yellow, Indus and Ganges. Although increased meltdown of glaciers has temporarily increased water in the rivers, the mid-term prospects are dire with declining river water and ultimately drought (which has already affected the Yellow River). By 2050, estimates are that one-third the area of China's glaciers will have disappeared.[76]

4.5.4 Sunshine Duration

Scientific observers have noted a large drop in the duration of sunshine in most regions of China since the mid-1950s. In parts of the North China Plain, the annual duration of sunshine is 500 h less than in the mid-1950s.[77] The decline in hours of sunshine is correlated with the decline in pan-evaporation.

Although in the last 40 years, sunny days have been fewer in most parts of China, the temperature-defined growing season has lengthened. From 1961 to 2000, the growing season in China as a whole increased by 6.6 days. Regionally, the largest increase occurred in the Qinghai-Xizang (Tibetan) Plateau (18.2 days), compared to 10.2 days in northern China, and 4.1 days in southern China.[78] The most rapid increase of growing seasons occurred in the 1990s, with the longest year being 1998, also the hottest year in recent Chinese history.

4.5.5 Wind Speed Changes

The average wind speed also has declined in most parts of China since the 1950s.[79] This change too is correlated with the reduction of pan-evaporation.

---

[75] *China Daily*, September 14, 2009, 13.

[76] Shanshan Wang, "Glaciers Melting at Alarming Speed," *China Daily*, July 24, 2007, 1; and "Shrinking Glacier Threat to Rivers," *China Daily*, November 2, 2007, 5; also see Jia Chen, "Targeting Climate Change in Himalayas," *China Daily*, October 10, 2008, 2.

[77] Ren et al., 2005c, 947, 950.

[78] Mingzhi Xu and Guoyu Ren, "Change in Growing Season over China: 1961–2000," *Journal of Applied Meteorological Science*, Vol. 15, no. 3 (June 2004), 306–12.

[79] Ren et al., 2005c, 948, 950.

4.6 Correlation of Climate Change and Extreme Weather Events

About 85% of Chinese food is composed of cereal crops, and as in other countries, grain production is especially susceptible to natural disasters. Indeed, more than half of the arable land in China is potentially affected by natural disasters. Moreover, China's population of impoverished people disproportionately lives in remote areas, under harsh ecological conditions, and is particularly susceptible to rising disasters caused by climate change.[80]

Our question is the extent to which natural variation of climate has been influenced by anthropogenic factors, with the effect of exacerbating extreme weather events. We consider in turn floods, drought, heat waves, sea level rises, typhoons, and other events.

4.6.1 *Floods*

Nearly two-thirds of the grain producing regions of China are below the flood stage of its largest rivers, the Yellow and Yangtze. The Yellow River is estimated to have flooded more than 2,500 times and changed its course 26 times in the last three millennia.[81] In what likely was the world's costliest natural disaster to human life, the 1931 flood of the Yellow River erased four million lives.[82] A flood in 1887 killed nearly one million, making it the fourth most costly natural disaster in world history.[83] Vorosmarty et al. comment that the Yellow River is a good case study of "how land cover change in tandem with climate and water engineering works can potentially alter discharge regime, in particular the occurrence of extreme events."[84] Fang and Xie constructed a time series of flooding on the Yellow River, finding that it varied by size and stability of the population basis and political/military stability. They hypothesized that flooding frequency increased with massive deforestation of the loess plateau for fuelwood, charcoal production, agriculture, and brick-making.[85]

---

[80] Zhuoqiong Wang, "Poverty Reduction 'Still a Challenge'," *China Daily*, October 18, 2007, 3.

[81] R. R. Tregear, *China: A Geographical Survey*. New York: John Wiley, 1980. Erosion has deposited silt in the river bed.

[82] Vorosmarty et al., 1998, 212.

[83] P. Matthews, ed., *The Guinness Book of Records, 1993*. New York: Facts on File, 1993.

[84] Vorosmarty et al., 1998, 220.

[85] J.Q. Fang & Z. Xie, "Deforestation in Preindustrial China: The Loess Plateau Region as an Example," *Chemosphere*, Vol. 29 (1994), 983–99.

Similar events plagued the Yangtze River. The flooding in 1954 of the Yangtze affected 17 million living downstream of the Three Gorges region; it submerged more than 1 million hectares of farmland and caused hundreds of millions in losses.[86] The 1998 flooding of the Yangtze killed more than 3,000, destroyed five million homes, and inundated 21 million hectares of land. Economic losses from this disaster alone were estimated at greater than $20 billion.[87] This flooding was attributed largely to land cover changes, particularly massive deforestation and subsequent erosion.

Records on China's major river systems show the presence of several significant floods, as well as high interannual variety. Climate effects are present as well, but there remains uncertainty concerning the exact impact of greenhouse gas effects. As noted in Chapter 3, the response of the state – from the dynastic period through to the present – has been construction of large-scale water engineering works, which have brought about unintended consequences. They are now a major factor causing disturbance to the food system, along with climate change and variability, land cover changes and human settlement.

In recent years, floods increasingly have been attributed to climate change and variability. A subtitle in *China Daily* announced: "Experts blame freaky weather on global climate change," as the story related that by mid-summer 2007, more than 700 people had been killed in floods, landslides, mud slides and storms across 24 provinces, and 82 million were affected.[88]

Many dikes along the swollen Huaihe River, China's third largest, were washed away. Flooding displaced about one-half million in Henan, Anhui, and Jiangsu provinces. Chongqing (which in the previous year suffered the worst drought in 100 years, leaving 17 million short of water) encountered the worst rainstorms in more than a century.[89] Floods submerged about 9.76 million hectares.

Southern provinces of Guangdong, Guangxi, and Hunan were particularly hard hit by floods too. In 2008, some 12 provinces and regions in the South were lashed by heavy rains, and some areas experienced the heaviest rainfall of 100 years.[90] Increased flooding likely will affect human health in other ways. Experts now predict a 30% increase in malaria-infected land areas in China.[91]

---

[86] C. Liu and D. Zuo, "Environmental Issues of the Three Gorges Project, China," *Regulated Rivers: Research and Management*, Vol. 1 (1997), 267–73.

[87] Economy, 2004, 9.

[88] Dingding Xin, Weifeng Liu, and Fangchao Li, "No Let-up in Floods, Drought," *China Daily*, July 30, 2007, 1; see also Chuanjiao Xie, "66 Killed as Floods Sweep S. China," *China Daily*, June 11, 2007, 1.

[89] Xinhua, "Loose Dykes Spur Flood Fears," *China Daily*, July 25, 2007, 4.

[90] Qiwen Liang, "Floods Take a Toll on Life," *China Daily*, June 17, 2008, 1.

[91] Anna Brettel, "Security, Energy, and the Environment," in In-Taek Hyun and Miranda A. Schreurs, *The Environmental Dimensions of Asian Security*. Washington, DC: United States Institute of Peace Press, 2007, 108.

4.6.2 Drought

An inventory of drought frequency, extent, and severity shows patterns similar to that of flooding. Areas affected by drought on several occasions have exceeded 15 million hectares, which is more than 10% of the arable land in China. Each year, nearly 20 million hectares of land are at risk from drought; crop losses would be greater without the extensive irrigation employed.[92]

Records of moisture conditions in China in the last two millennia display alternating wet and dry periods of relatively long duration.[93] As reported with respect to floods, the incidence of droughts, especially in the oldest area of Chinese civilization (the Yellow River Basin) corresponded with increases in anthropogenic impacts (increased population pressure, removal of land cover) as well as with natural climate variability.

In 2007–2008, widespread and prolonged drought plagued the northern, northeastern, and southern regions of China. The Office of the State Flood Control and Drought Relief Headquarters reported that the sustained drought affected 11 million hectares of arable land (1.7 million more than during the same period in the previous year), 89% of which was crop land. Nearly eight million people and five million head of livestock were left short of drinking water. This drought, in some regions reaching the mark of "once-in-50-years," was attributed to continuous hot weather and insufficient rainfall, some 20–70% lower than normal.[94] Drought conditions continued into 2008, affecting some 14% of China's farmland.[95]

Dry weather brought drought conditions to 22 of China's 31 provinces, among them major grain production bases. Jiangxi, Heilongjiang, Hunan, Jilin provinces and the autonomous regions of Inner Mongolia and Guangxi Zhuang were hardest hit, affecting one-third of arable land.[96]

A report on natural disaster casualties listed deaths of 1,279 and 239 missing as a result of natural disasters, just in the first 7 months of 2007. (Improved emergency response systems reduced fatalities from disasters.[97]) Abnormal weather conditions causing floods, landslides and drought affected about 300 million people, made 6.15 million homeless, and caused $14.5 billion in economic losses.[98]

---

[92] Vorosmarty et al., 1998, 213.

[93] G. F. Gong and S. Hameed, "The Variation on Moisture Conditions in China during the last 2000 Years," *International Journal of Climatology*, Vol. 11 (1991), 271–83.

[94] Zhe Zhu, "Drought Affects 7.5 m People," *China Daily*, August 6, 2007, 3; see also, Quanlin Qiu, "Crops at Risk as Droughts Worsen," *China Daily*, August 3, 2007, 5.

[95] See Ying Wang, "Weather Takes Toll on Farmers," *China Daily*, February 25, 2008, 3; and Zhiming Xin, "Price Surge Becomes Real Food for Thought," *China Daily*, April 16, 2008, 7.

[96] Fangchao Li, "Autumn Grain Harvest Under Severe Threat," *China Daily*, August 10, 2007, 3. See also Hong Chen, "1.25m Hit by Drought in Guangxi, Guangdong," *China Daily*, December 18, 2007, 5.

[97] Xinhua, "Fewer Killed in Disasters: Minister," *China Daily*, August 23, 2007, 3. Also see Xinhua, "Nation Scrambles to Mitigate Impact of Climate Calamities," *China Daily*, January 2, 2008, 2.

[98] New China News Agency, "Natural Disasters Kill 1,279 in Seven Months," *China Daily*, August 8, 2007, 2.

Devastating floods and drought of recent years may have only a marginal impact on grain production in China. In 2006, China's grain output totaled 497.5 million tons, while estimated demand was 507.5 million tons, a gap of 10 million tons or just 2% of annual output. Drought and heavy rains in 2007–2008 were not expected to significantly influence targets for grain production, in the estimation of relevant government agencies (Ministry of Agriculture, the State Grain Administration, the State Flood Control and Drought Relief Headquarters, the Rural Development Institute of the Chinese Academy of Social Sciences, and the Center for Chinese Agricultural Policy of the Chinese Academy of Sciences).[99]

The National Grain and Oil Information Center expected normal to increased production despite extreme weather events. For example, drought conditions in Northeast China affected corn production areas, but the corn harvest increased 2.42% over the previous year. Expansion of sowing areas and remedial measures (such as post-flood re-sowing) accounted for the increases.[100] Other factors – decline in amount of arable land, diminished output of much farmland – appeared to be more troublesome concerns.

A State Council circular issued in late 2007 remarked that "Drought and water scarcity have become increasingly severe in recent years as a result of amplified socioeconomic development and global warming." The circular maintained that droughts had not previously been as "frequent, widespread or destructive," nor had they previously extended to all parts of the country.[101] Meteorologists, on the other hand, attributed the drought to the La Niña phenomenon and abnormal atmospheric circulation.[102]

During the 2008–2009 winter season, many sections of China encountered the worst drought conditions since 1951. The first region affected by drought was North China, and the provinces of Hebei, Shanxi, Anhui, Jiangsu, Henan, Shandong, Shaanxi and Gansu. Drinking water was insufficient for seven million people and five million head of livestock. About 10 million hectares of cereal crops suffered from drought (about 43% of China's winter wheat supply). Similar conditions of low rainfall and lingering high temperatures caused drought in the summer of 2009 and affected Hunan, Hubei, Guizhou and the Guangxi Zhuang autonomous zone, and then the northeast provinces of Jilin, Liaoning and grasslands of Inner Mongolia.[103]

The government responded by supplying drinking water and providing water for irrigation in about half of the wheat fields in the hardest hit provinces. It increased

---

[99] Zhiming Xin, "Floods, Drought 'Won't Hit' Grain Output," *China Daily*, August 8, 2007, 2. Also see Ying Wang, "Drought is Now Nationwide Problem, Officials Say," *China Daily*, December 21, 2007, 3.

[100] Jiao Wu, "Bad Weather Could Eat into Autumn Harvest," *China Daily*, August 22, 2007, 3.

[101] Yinan Hu, "Water Conservation Efforts Might Not be Sufficient," *China Daily*, December 15–16, 2007, 3.

[102] See Jing Li, "North to Suffer Severe Drought in Spring," *China Daily*, March 4, 2008, 4; and Jia Chen, "North Suffering Worst Drought in 5 Yrs," *China Daily*, March 28, 2008, 4.

[103] Yinan Hu, "Regions Reel from Severe Drought, *China Daily*, February 5, 2009, 1 and Dingding Xin and Ce Liu, "Drought Scorches Much of China," *China Daily*, August 31, 2009, 3.

funding for relief work and increased subsidies to farmers. Also, provincial meteorological agencies fired cannons and rockets to seed rain.[104] Because China maintains sizable grain reserves, the Vice Minister of Agriculture, Wei Chao'an said "There has been no major fluctuation in grain prices despite the severity of the drought."[105]

This severe drought prompted some meteorologists to comment that drought would continue to affect northern China into the near term. Leading climatologist Lin Erda reiterated observations of the National Report on Assessment of Climate Change to the effect that interregional differences have increased in the last 100 years, with a decline of rainfall in the North and increase in the South: "Previous research found that climate change intensified atmospheric circulation in southern China," while air in the North has been stagnant recently.[106] Although most analysts believed the immediate threat to food security was limited, they worried about near-term effects. Acceleration in growth of crops because of global warming shortens the time it takes them to mature, and grains do not ripen by the time they are harvested, reducing productivity.[107]

4.6.3 Heat Waves

In the last decade, China has experienced two of the hottest years in recent memory. The China Meteorological Administration reported that the winter season from December 2006 to February 2007 had a national average temperature of −2.4°C, following the warmest winter between 1998 and 1999, with an average temperature of −2.3°C. The average temperature in 2007 was 1.2°C higher than normal, and a CMA spokesman said: "It is the 11th year for the country to experience an abnormally high temperature against the global backdrop of climate change."[108]

High temperatures brought on water shortages and drought as well as increased heavy fog and sandstorm activity. The municipality of Chongqing suffered from scorching temperatures in excess of 40°C in the summer of 2006. Wind gusts from a sandstorm derailed a train in Xinjiang, leaving three dead and more than 30 injured. Qinghai experienced its worst heat wave since 1951. An official with the state drought and flood relief office remarked: "The hot weather has meant a lot less

[104] Chuanjiao Xie, "Drought Relief Effort Underway," *China Daily*, February 3, 2009, 3 and Xiaokun Li and Yinan Hu, "Fighting Drought Top Priority: Wen," *China Daily*, February 9, 2009, 1.

[105] Yingzi Tan, "Grain Costs Stable Despite Dry Fields," *China Daily*, February 13, 2009, 2.

[106] Jing Li, "Experts Warn Droughts to Dry Up North for Decades," *China Daily*, February 14, 2009, 4.

[107] Yu Xiao, "Facing Up to Climate Change," *China Daily*, February 14, 2008, 4.

[108] Xiaohua Sun, "Nation's Average Temperatures at Highest Level since 1951," *China Daily*, December 1–2, 2007, 1.

rain since the early part of the year, while insufficient reserves have made it difficult to provide enough water to feed crops."[109]

Analyses by climate scientists suggest a differentiated response of cereal crops to higher temperatures. Xu et al. surveyed impacts of high temperature events on rice yields in the middle- and lower-reaches of the Yangtze, and found a larger percentage of undeveloped grain when temperatures were hot, reducing yields.[110] Qian and Wang found that yield response varied by crop and region. Higher temperatures increased maize production in all regions, but reduced rice and wheat production only in the south. In the Northeast and Northwest, rice production increased when temperatures rose; in the Northwest and central-north regions, wheat production increased when temperatures became hotter.[111]

Climate scientists blamed global warming for higher temperatures in the first decade of the twenty-first century. They pointed to effects beyond water shortages and drought, reducing cereal crop yields. Forest fires have increased in frequency and severity. Some wheat varieties in the North have become less resistant to cold, and thus more susceptible to spring freezes. Also, noted Wei Xiong, a specialist in the Chinese Academy of Agricultural Sciences, "Warm winters create an environment in which plant diseases and pests thrive, and these pose a serious threat to crops."[112] We discuss climate change effects on plant diseases and pests further in Chapter 5.

4.6.4 Rising Sea Level

Analyses by the State Oceanic Administration (SOA) point to a rise in sea levels off China's coasts, attributable to both global warming and surface subsidence in several areas. The rise has been especially pronounced in China's most populous city, Shanghai, and in Tianjin, which is the closest harbor to Beijing.

The SOA reported that in the previous 30 years, the average sea level rise for China was 90 mm, about 30% higher than the global average during this period. However, in Shanghai, the sea level rose 115 mm (half the length of a chopstick);

---

[109] Quanlin Qiu, "Crops at Risk as Droughts Worsen," *China Daily*, August 3, 2007, 5; also see Xiaohua Sun, "Global Warming Takes Toll," *China Daily*, March 2, 2007, 2; also see Quanlin Qiu, "5 Killed in Guangdong," *China Daily*, September 25, 2008, 2.

[110] Yinlong Xu, Haiyan Zhao, Fengmei Yao, Yong Zhang, Bin Xu, Yanan Hu, and Jing Yuan, "Statistical Analysis of Impacts of High Temperature Events on Rice Yield Components over Middle and Lower Reaches of the Yangtze River" (in Chinese), *Chinese Journal of Agrometeorology*, Vol. 26 (December 2005), 20–23.

[111] Xiaoping Qian and Xiuqing Wang, "Influence Analysis of Climate Change on Grain Supply Function and Yield Response in China" (in Chinese), *Chinese Journal of Agrometeorology*, Vol. 26 (December 2005), 14–19.

[112] Jing Li, "Warm Winter 'Major Threat' to Crops," *China Daily*, November 26, 2008, 3; see also Qian Wang and Hongyi Wang, "'Blame Global Warming' for Higher Temps," *China Daily*, February 14, 2009, 1.

in Tianjin, sea levels rose 196 mm (the length of a pencil). SOA officials suggested that the greater level of sea level rise in the two cities was explained by "indiscriminate exploitation of groundwater resources," causing subsidence.[113]

4.6.5 *Typhoons*

China's climate is part of the East Asia monsoon regime, and historically it has suffered severe impacts from tropical storms. Coastal provinces, especially Guangdong and Fujian, annually are affected by typhoons, with much coastal flooding. Rising sea water levels, attributed to climate warming, have inundated a number of coastal villages.

Although typhoon damage was less in 2007–2009 than previous years, more destructive storms hit the mainland. For example, Typhoon Pabuk brought the heaviest showers of two centuries to Guangdong. One township received 739 mm of rainfall in 24 h. Heavy rains elsewhere washed out roads, railroad beds, and overflowed dams and reservoirs.[114]

4.6.6 *Other Extreme Weather Events*

In early 2008, immediately before Chinese New Years, severe winter storms and blizzards affected 20 provinces and autonomous regions in China, the worst in an half century. The storms affected 19 million hectares of forests and 11 million hectares of crops (about 8% of total planted acreage); blizzards killed four million pigs, 400,000 cows, 1.4 million sheep, and 63 million chickens and ducks, leading to price increases for some cereals and livestock shortages.[115] This extreme weather event was a test of China's forecasting system, and it appeared to be deficient. A spokesperson of the China Meteorological Association said "We underestimated the duration and severity of the weather and failed to pre-evaluate its impact on transport and the power sector."[116] CMA did forecast all the storms 2–5 days in advance, but it failed to alert the public to the extreme danger of the storms. CMA officials attributed this failure to a lack of meteorological monitors in mountainous areas and to forecast models 10 years behind those of advanced western countries.[117]

---

[113] Xiaohua Sun, "Rising Seas Pose Danger to Big Cities," *China Daily*, December 13, 2007, 1.

[114] Quanlin Qiu, "Zhanjiang Hit by Worst Downpour in 200 Years," *China Daily*, August 13, 2007, 1.

[115] Jing Fu, "Measures Taken to Protect Farmers," *China Daily*, February 20, 2008, 2.

[116] Qiwen Zhu, "Meteorologists Ill-equipped to Forecast Bad Weather," *China Daily*, February 15, 2008, 1, and Xiaohua Sun, "Green Cost of Sleet: $8b of Forest Wealth," *China Daily*, February 20, 2008, 1.

[117] Ibid.

Although the May 12, 2008 earthquake in Sichuan Province was unrelated to climate change, the huge death toll (more than 70,000) stemmed partly from failure to heed clear warnings of the devastating quakes. Scientists knew about the risk of a potentially catastrophic earthquake along what is called the Longmenshan belt (where the Wenchuan earthquake struck) and gently raised concerns with government authorities. Yet they did not press the government to impose stricter building codes in the area. Lack of monitoring and enforcement by local government officials, however, is doubtless a more important reason for the high death toll.[118]

4.7 Impacts of Climate Change on Agricultural Production

The impact of climate change on agricultural production in China is expected to be considerable, with both positive and negative results. Government and media reports have emphasized primarily the negative impacts of climate warming. For example, in late December 2006, the government issued its inaugural *National Assessment Report on Climate Change*, which pointed out these challenges of climate change (for example, a projected temperature increase of 2°C by 2050 as compared to 2000) to food security:

- Output of major cereal crops – rice, wheat, corn – could fall by as much as 37% in the second half of the twenty-first century if greenhouse gas emissions are not curbed (plants would grow faster but yields would decline).
- Northern areas of China will become drier, despite projected annual rainfall increases of from 7% to 10%.
- Central and western regions will face annual water shortages of 20 billion cubic meters from 2010 to 2030.
- Floods and drought[119] will become more common as water evaporates more rapidly from rivers.[120]

In late 2008, Greenpeace warned that a continuation of the current climate change trend would significantly reduce China's food supply by 2030. Greenpeace's global campaign against high greenhouse gas emissions was echoed in the report prepared by a number of China's prominent climate scientists. This report predicted that without radical cuts to emissions, China's overall food production would fall 23%

[118] See Howard W. French, "Experts Warned of Quake Risk in China," *New York Times*, June 5, 2008, A1.

[119] Global warming would cause more frequent drought in currently dry, low-lying and mid-altitude areas, reducing rainfall by 10–30% by 2030. Contrariwise, wet high altitude areas would experience more drastic floods. See Jiao Wu, "More Arable Land 'Needed' by 2030," *China Daily*, August 23, 2007, 2.

[120] Xinhua, "Food Security at Risk: Report," *China Daily*, January 4, 2007, 3.

by 2050. It urged "ecologically friendly" agriculture, such as reduction in pesticides and fertilizers and increased use of biofuels.[121]

In January 2007, a group of six agricultural policy experts issued a policy research working paper, printed by the World Bank Development Research Group, and entitled "Can China Continue Feeding Itself?" Wang et al. conducted an econometric analysis of 8,405 farm households in 28 provinces, asking whether near-term expected changes in climate would affect food sufficiency adversely. Their general conclusion was that climate change would likely be harmful, but impacts would vary by region. They conclude:

> The average impact of higher temperatures is negative and the average impact of more precipitation is positive. However, marginal increases in temperature and rainfall have very different effects on different farm types in different regions ... rain fed farmers are more vulnerable than irrigated farmers. Warming is likely helpful to rain fed farmers in very cold places, but it will likely harm rain fed farmers in most of China and especially the far South. More rain is likely to be harmful to rain fed farmers in the wet Southeast but will benefit farmers in the remaining regions. Irrigated farmers are less sensitive to temperature. However, irrigated farmers, like rain fed farmers, will gain from increased rainfall.[122]

The authors stated an important caveat to their analysis is that it assumed no change in water supply; nor were they able to analyze impacts on specific crops.

A third study, the November 2007 report by the Intergovernmental Panel on Climate Change (IPCC), predicted rising sea levels of 18–51 cm leading to land subsidence along China's heavily developed coast. The decline of water volume in China's six largest rivers noticed in the past 50 years would continue, and ground water levels also would continue to drop, with adverse impacts on grain production.[123]

Thus, in recent years, potential impacts have been studied systematically, and we now have the benefit of good preliminary analyses. This section is based on research conducted under the aegis of the China–UK Collaboration Project: *Investigating the Impacts of Climate Change on Chinese Agriculture*.[124] Because it treats different regions, water conditions, and impacts on specific crops, it is most useful to our analysis.

The collaboration involved development of a regional climate change model for China (PRECIS), created by the UK's Hadley Center for Climate Prediction and Research. The PRECIS model simulated China's climate to predict changes in yields of four key Chinese agricultural crops: rice, wheat, maize, and cotton. Rice, wheat, and

---

[121] Jing Li, "Climate Change 'Threatens' Food Safety," *China Daily*, October 16, 2008, 2.

[122] Jinxia Wang, Robert Mendelsohn, Ariel Dinar, Jikun Huang, Scott Rozelle, and Lijuan Zhang, *Can China Continue Feeding Itself? The Impact of Climate Change on Agriculture*. Washington, DC: The World Bank, Policy Research Working Paper 4470, 2007, 15.

[123] Xiaohua Sun, "Rising Sea Levels to Take Toll on Nation," *China Daily*, November 23, 2007, 1.

[124] For a similar appraisal, see F. Tao, M. Yokozawa, J. Liu and Z. Zhang, "Climate Change, Land Use Change, and China's Food Security in the Twenty-first Century: An Integrated Perspective," *Climatic Change*, Vol. 81 (2008), 138–51.

Table 4.1 Average climate change scenarios from PRECIS relative to baseline simulation (1961–1990) (China–UK Collaboration Project: Investigating the Impacts of Climate Change on Chinese Agriculture, 2004, 5)

	A2 (medium-high emissions)			B2 (medium-low emissions)		
Time period	Temp. increase (°C)	Rainfall increase (%)	CO_2 (parts per mil.)	Temp. increase (°C)	Rainfall increase (%)	CO_2 (parts per mil.)
2011–2020	1.0	3.3	440	1.16	3.7	429
2041–2050	2.11	7.0	559	2.20	7.0	492
2071–2079	3.89	12.9	721	3.20	10.2	561

maize account for 85% of total food supply in China.[125] The climate change scenarios are based on a consistent set of changes in meteorological variables (such as temperature and rainfall) and generally accepted projections of future levels of carbon dioxide and other greenhouse gases. The scenarios are derived from projections of climate change from global climate models (GCMs).[126]

The PRECIS model has two emissions scenarios, as noted in Table 4.1:

This table projects an average temperature increase in China by the end of the twenty-first century of between 3°C and 4°C. As we shall see, modeling suggests that climate change could reduce yields of cereal crops up to 37% in the next 20–80 years. However, carbon dioxide fertilization likely would increase yields, reducing the adverse effects of climate change. (Higher levels of CO_2 mean that plants absorb more CO_2, which spurs plant growth.[127]) To optimize carbon dioxide fertilization, however, would require both water and nutrient availability, and these were factors going beyond the modeling study.

As we indicate below with respect to specific crops, the PRECIS model was validated by comparing historical temperature and rainfall data across China for 1961–1990 with modeled data for this baseline period. Team members found "generally good agreement" between observed and simulated data for this period, increasing their confidence for projections into the twenty-first century.[128] We turn now to impacts on specific crops.

[125] C. L. Tong, C. A. S. Hall, and H. Q. Wang, "Land Use Change in Rice, Wheat and Maize Production in China (1961–1998)," *Agricultural Ecosystem and Environment*, Vol. 95 (2003), 523–36.

[126] China–UK Collaboration Project, *Investigating the Impacts of Climate Change on Chinese Agriculture*. London: Department for Environment, Food and Rural Affairs (Defra), 2004.

[127] Erda Lin, Wei Xiong, Hui Ju, Yinlong Xu, Yue Li, Liping Bai, and Liyong Xie, "Climate Change Impacts on Crop Yield and Quality with CO2 Fertilization in China," *Philosophical Transactions of the Royal Society*, Vol. (2005, in press); also see Haiyan Zhao, Qiaojuan Cui, Fengmei Yao, Zhan Tian, Changli Chen, and Yinlong Xu, "Response of Rice and Wheat Net Photosynthetic Rate to CO2 Concentrations and PAR," *Chinese Journal of Agrometeorology* (in Chinese), Vol. 26 (December 2005), 65.

[128] China–UK Collaboration Project, 2004, 4.

4.7.1 Impacts on Rice Production

Rice is China's most important cereal crop. It covers over 290,000 km^2 of land, and in 2000 yielded 190–195 million tons of product. Rice is double (and often triple) cropped in southern areas, and mostly single cropped in the North. A rice crop simulation model called Crop Environment Resource Synthesis (CERES) was used to simulate changes in rice production under the two emissions scenarios.

When averaged across China, rice yields generally increased under the A2 emissions scenario (for both rain fed and irrigated crops), while decreasing under the B2 scenario, when the CO_2 direct effect was included in the simulation. More simply stated, the CO_2 fertilization effect offset decreases in yield caused by shorter growth duration because of higher temperatures. The carbon dioxide effect was more pronounced in the A2 than the B2 scenario. Were the CO_2 effect removed from the simulation, average yields are projected to fall under both emissions scenarios.[129]

Regional differences explain these general findings. Under the A2 emissions scenario, yields increased in most areas but particularly in central China. Areas with the highest increases lay between Hubei, Sichuan, Hunan, and Jiangxi; areas with declining yields include the North China Plain and Shanxi Province. Although yields decrease over most parts of China under the B2 scenario, in some areas such as between Hubei and Sichuan and in hilly areas of south China, yields would increase.

The area suitable for rice growing likely would expand in the future. Single-cropped rice would expand to northern parts of Heilongjiang Province in China's Northeast.[130] Double-cropped rice would expand north of the Yangtze River in central China and to other elevated regions. The main obstacle to this predicted expansion is likely to be lack of sufficient water for irrigation.

4.7.2 Wheat Production

Wheat is China's second most important grain crop, producing more than 20% of total cereals in 2000. Within recent years, wheat production has been adversely influenced by water shortages due to higher temperatures and increased drought. As noted for wheat and other cereal crops, soil degradation too affects wheat yields. These adverse impacts are likely to be exacerbated by climate change.

A CERES-Wheat model was used to examine projected changes in wheat production under the two emissions scenarios.[131] Including the effect of CO_2 fertilization,

[129] Ibid., 6.

[130] See Fengmei Yao, Yinlong Xu, and Bin Xu, "Analysis of the Temporal and Spatial Distribution and the Character of the Period of Rice Climate Yields in the Northeast Rice Areas" (in Chinese), *Chinese Journal of Agrometeorology*, Vol. 26 (December 2005), 31–36.

[131] See Yinlong Xu and Zhan Tian, "Exploring the Methodology to Assess the Impacts of Climate Change on Wheat Production in China Using RCM Scenarios" (in Chinese), *Chinese Journal of Agrometeorology*, Vol. 26 (December 2005), 46–51.

average wheat yields would increase in most areas of China by 2080 under the A2 scenario for both rainfed and irrigated wheat. Yet, for irrigated wheat to benefit from the effects of CO_2 fertilization, there would need to be sufficient water and nutrients. Without CO_2 fertilization, wheat yields would be from 10% to 20% lower for the A2 and B2 emissions scenarios, as compared to current yields.[132]

Yields would vary by wheat-growing region as well as by wheat type. Under the A2 scenario, yields are expected to increase all over China, but with greater increases in north China and some areas of the Yangtze River Valley.[133] Correspondingly, yields would fall in the northeast and southwest under the B2 scenario, especially for rainfed wheat. Spring wheat, mainly grown in northeast and northwest China (where climate changes are expected to be larger than other regions) likely would encounter more adverse impacts than winter wheat. Yields of the former are expected to decline by about 30% as compared to 14% for the latter, by 2080. Because of higher projected temperatures in the North, spring wheat would experience more heat stress, whereas winter wheat would benefit from less risk of frosts and more frost-free days during the growing season.

4.7.3 Maize Production

Maize (corn) is grown in most parts of China, as it adapts well to a wide range of temperature and rainfall conditions. The harvested area of maize in 2000 was more than 220,000 km^2, when it produced 106 million tons of product. Because maize assimilates carbon dioxide more efficiently than rice and wheat, it responds less positively to CO_2 fertilization. The CERES-Maize crop simulation model was used to prepare climate change projections.[134]

Incorporating the direct effect of CO_2 fertilization, average yields of maize would increase for rain fed crops while decreasing for irrigated maize – under both the A2 and B2 emissions scenarios, for the 2080s. The reason that rain fed maize yields would increase under the A2 scenario is probably because higher CO_2 concentrations would boost production under present water-limited conditions found in North China (the largest maize cultivation area). Although maize does assimilate CO_2 well, it still responds to CO_2 fertilization; without that effect, the average yield for both rain fed and irrigated maize likely would fall for both emissions

[132] China–UK Collaboration Project, 2004, 8–9.

[133] Yanling Song, Deliang Chen, and Wenjie Dong, using a different model (the World Food Studies crop model) report similar findings: an increase of winter wheat yield in northern China and a reduction in southern China. See their "Influence of Climate on Winter Wheat Productivity in Different Climate Regions of China, 1961–2000," *Climate Research*, Vol. 32 (2006), 219–27.

[134] See Qiaojuan Cui and Yinlong Xu, "Validating CERES-Maize Model in China Using Field Observation Data Sets" (in Chinese), *Chinese Journal of Agrometeorology*, 37–41.

scenarios, because higher temperatures would shorten the growing period by around a week. Good irrigation would counter declines in yields.[135]

Regionally, yields likely would decline in the major production areas of the southeast plain, North China and the Sichuan basis. They would increase in northern parts of the Northeast and South China. Without irrigation in North China, the maize grown there – a summer variety – likely would be adversely affected by higher temperatures, lower rainfall, and evaporation.

4.7.4 Cotton Production

China is one of the world's largest producers of cotton, and it is an important cash crop. In the case of cotton, a Cotton Production Regional Assessment System (COPRAS) was developed in a collaborative project between China's Agrometeorology Institute and Agricultural University. The simulation accounted for solar energy utilization, photosynthesis, yield formation, and the water balance when evaluating effects of weather, soil, variety and management. This model was validated by comparison with field data from the Xinjiang region.[136]

The preliminary analysis indicates that yields of cotton likely would increase somewhat in northern regions of the North China Plain, remain the same in cotton-growing areas of the Yangtze River Basin, and decline in the northern edge and eastern part of the southern Xinjiang region.[137]

4.7.5 Mitigation Difficulties

Chinese climate scientists Xiong, Lin, Ju, and Xu summarized research on application of the PRECIS model to future carbon dioxide emissions scenarios. They did not find a critical threshold of temperature increase for food production: "Overall, the impact of climate change on China's three main crops' production would have more positive than negative impacts if the direct effects of CO_2 are included."[138] Although their summative analysis includes some adaptation strategies, such as the use of fertilizers and Genetically Modified (GM) crops, it does not include future adaptation strategies. Also, it leaves out of consideration choices made by

[135] China–UK Collaboration Project, 2004, 7–8.

[136] Ibid., 9.

[137] Ibid., 9.

[138] Wei Xiong, Erda Lin, Hui Ju, and Yinlong Xu, "Climate Change and Critical Thresholds in China's Food Security," *Climatic Change*, Vol. 81 (2007), 218.

farmers and changes in government policy respecting agricultural production. Climate scientists, for these reasons, urge caution in the interpretation of their findings.[139] We discuss six areas of uncertainty that should cushion the foregoing projections.

First, the different climate models assume that water sufficiency will not limit production of cereals, but certainly this is a questionable assumption. In Chapter 3 we indicated that in 2009, China's national water supply is barely sufficient, but it suffers from two problems: uneven distribution and pollution. Some of the areas where increases in cereal crop yields are anticipated – for example North China – are short of water with rapidly depleting ground water supplies. The second and third stages of the South-North Water Diversion Project are still on the drawing boards, and serious environmental and fiscal objections have been raised to their construction. Already in most parts of China, use of water for irrigation has dropped in priority behind industrial and personal consumption use. Water pollution is now regarded as China's most serious water problem. Many of China's rivers and lakes have been polluted to such an extent that the water is of sewage quality and cannot be used for irrigation purposes. Similar observations have been made of groundwater because of extensive contamination.

Second, the models assume no changes in land use, and indeed it has become a priority of the regime to insure that China's arable land is not reduced below the 120 million hectare figure. However, pressures of population increase, urbanization, and rapid economic development seem likely to cross this line, reducing the flexibility of the state to shift crops and to open new areas for cultivation as older agricultural regions become less hospitable to cultivation of cereal crops.

Third, the models do not incorporate the impact of climate change on insect pests, plant diseases, and invasive species. As we indicate in Chapter 5, by most accounts, problems of disease and pest infestation are likely to increase significantly as temperatures warm. Historically, China's crop yields have been greatly affected by infestations of insects, such as locusts. Plant diseases such as rice blast too have hugely reduced harvests. Just in 2007, the Ministry of Water Resources head said China's crops had been hit by "serious" plant diseases and insect pests. The Ministry of Agriculture predicted that insects would affect 69 million hectares of rice fields.[140]

Fourth, the models are based on changes in average temperatures. They cannot account for extreme climate events, which by definition are unpredictable. As we discussed above, the recent spate of floods, drought, heat waves, and tropical storms have wreaked enormous destruction on major regions of China, affecting hundreds of millions of people, and destroying cultivated crops and harvests.

Fifth, the models do not account for the wide range of agronomic adaptation strategies, which are of great potential significance in agricultural productivity.

---

[139] See C. Rosenzweig, M. L. Parry, G. Fischer, and K. Frohberg, *Climate Change and World Food Supply*. Oxford: University of Oxford, Environmental Change Unit, Research Report No. 3, 1993.

[140] Zhe Zhu, "Autumn Harvest Under Severe Threat," *China Daily*, August 30, 2007, 3.

Some of these adaptations are dependent on technological advancements, such as development of plant cultivars that are resistant to drought, diseases and pests, that mature earlier or later in the growing season, and that produce enhanced yields. Other potential adaptations depend on decisions of farmers under China's relatively new household responsibility system, which increasingly responds to market incentives. Here we refer to what varieties of crops to plant and at what density, how much and what type of plant fertilizers, disease, weed, and pest controls farmers elect to use, their willingness to adjust planting dates, their flexibility in planting substitute crops, and the extent to which they conserve water in irrigation systems and control soil erosion.

Sixth and finally, the climate change models do not factor in costs of investment in agriculture, by farm families, and by governments at the local, provincial, and national level.[141] The investments required to mitigate adverse impacts of climate change are expected to be huge. For example, to take full advantage of the CO_2 fertilization effect would require adding nitrogen and other fertilizers to compensate for yield losses caused by climate change, and this will be costly. (Increased fertilization, however, would exacerbate contaminated runoff into rivers and aquifers.) Government agricultural subsidies already are an expensive part of the state budget, and they are likely to increase to insure farmers against climate change-related crop failures and stabilize crop production. Disaster assistance too, as well as agricultural research sponsored by governments, will require increased investments. Overall, then, costs and investment needs will increase greatly because of the changes in agricultural conditions.

The central government has paid attention to developing research on climate change, and particularly its effects on agriculture, a concern also necessitated by global negotiations in the last 2 decades. In November 2008, the State Council Information Office issued a white paper "China's Policies and Actions for Addressing Climate Change."[142] The paper highlights impacts on agriculture, forestry, water resources and coastal zones. Proposed strategies include a continuation of policies to increase energy efficiency and foster technological innovations, among others. The lengthy section on adaptation chronicles past efforts in building agriculture infrastructure, improved land conservation, and cultivation practices. The report concludes in aspiration terms that

[141] Government planning documents display additional difficulties – general and vague policies related to climate change mitigation without clear guidance from officials charged with their implementation. See, for example, People's Republic of China, *Initial National Communication on Climate Change*, and especially, chapter 4 "Politics and Measures Related to Climate Change Mitigation." Beijing: 2003.

[142] State Council, Beijing, 2008. Also see excerpts, "Comprehensive Plan to Fight Climate Change," *China Daily*, October 30, 2008, 4–6, and Jing Li, "Sweeping Change," *China Daily*, December 8, 2008, 4.

> China has onerous tasks to develop the economy and improve the people's livelihood, and faces a more severe challenge of climate change than developed nations do. China will continue to follow the guidance of the Scientific Outlook on Development,[143] unswervingly stick to the road of sustainable development, and adopt more powerful policies and measures to strengthen the ability to deal with climate change in an all-round way.[144]

4.8 Conclusions

China is one of the world's largest producers and consumers of agricultural products. Given its large population relative to the amount of arable land, it probably has operated closer to the margin than any other country. Historically and certainly since 1949 under the People's Republic, China has pursued a policy of self-sufficiency in grains, intending to produce at least 95% of the food requirements of its population. It is a testament to the effectiveness of the food system that in most years, under conditions of increased domestic demand, farmers have been able to provide most of the cereals needed. Only in exceptional times has importation of cereals exceeded 5% (and in most recent years it has hovered at the 1–3% range).

In this chapter we have explored the impact of climate change on China's food security. We began by considering energy utilization, as China has become the world's largest emitters of greenhouse gases, which affects regional as well as global warming. Heavy reliance on coal for industrial and household use continues to describe policy in China, notwithstanding efforts to diversify energy sources and emphasize those producing fewer greenhouse gases. We then examined the contributions of traditional agriculture and livestock husbandry to climate change. Agriculture in its different dimensions contributes about the same amount to climate warming in China as in other developing nations.

Distinguishing natural sources of variability in climatic conditions from anthropogenic-induced changes is difficult in all countries, and we noted the problems in teasing out near- and mid-term climate changes from regional and temporal oscillations between cold and dry periods and cool and humid periods associated with monsoon cycles. In fact, human-induced changes have interacted with natural cycles.

Based on the most recent research findings, we illustrated observed climate change effects in China: the rise of temperatures and changes of precipitation patterns, surface evaporation, and sunshine duration. Following this, we considered a series of extreme climate events – floods, drought, heat waves, and storm surges. This discussion led to the central issue of the chapter: the nature of the impact that climate changes will have on agricultural production. Our assessment based

---

143 For the linkage between climate change and development in China, see Jun Pang and Ji Zou, China's Policy-Making Process on Climate Change," in Y. Kamayama, A. P. Sari, M. H. Soejachmoen, and N. Kanie, eds., *Climate Change in Asia: Perspectives on the Future Climate Regime.* New York: United Nations University Press, 2008, 66–82.

144 Ibid., *China Daily*, October 30, 2008, 6.

on climate projections under different carbon dioxide emissions scenarios is optimistic. Increased temperatures across China will not reduce yields of cereal crops, *if* carbon dioxide fertilization effects are included. If they are not, then future yields of major cereals – rice, wheat, and maize – will be reduced.

We concluded our analysis by pointing out the need to exercise caution in interpreting projections generated by climate change models. They do not (because in most cases they cannot) factor in important variables, such as sufficiency of water for irrigation uses, land use changes, effects of pests and diseases, extreme climate events, agronomic adaptations, and the adequacy of private and government investment to mitigate climate change effects.

Nevertheless, we remain optimistic about the ability of China to mitigate adverse impacts of climate change, because of the quick learning curve of the scientific community in China and the attention that the central government has given this critical issue.

Chapter 5
Plant Diseases, Pests and Food Security

Abstract Chapter 5 discusses the plant diseases and pests that affect agricultural production in China, and thus food security. Because diseases of plants and insect pests are anthropocentric concepts, we begin the chapter by defining the functions they play in ecological renewal and the environmental conditions favoring pathogens and pests. Then we explore the broad range of economic damage they do to food crops, with data from recent years. Our examination then turns in some detail to migratory locusts, rice blast disease and wheat rust, which afflicted food crops throughout history and remain of great significance to the Chinese economy today. Other diseases and insect pests have reached epidemic scales, and we treat them by the major food crops they threaten: rice, wheat, maize (corn), potatoes and soybeans. Some plant pathogens not only affect crop yield but also endanger human and animal health, and we consider rice false smut, wheat scab, and other post-harvest diseases. Too, we consider farm cultural practices having an impact on diseases and pests as well as changes in government policy, including improvement of domestic transportation. Climate change is the broadest environmental change that may affect incidence and severity of disease and pest infestation. Finally, we describe the primary mitigation (or control) measures China has adopted. These include regulations and standards, cultural controls, biological control measures, and physical and chemical controls.

Keywords Plant pathogens • Insect pests • Ecological services • Economic impacts • Migratory locust • Rice blast disease • Wheat rust • Other diseases/pests of rice • Wheat • Corn • Potatoes • Soybeans • Post-harvest diseases • Farm cultural practices • Biological • Physical and chemical controls

5.1 Introduction

In Chapter 4, we discussed the adverse impact of floods, drought, heat waves and other natural disasters on crop production. In this chapter we turn to biological factors, such as plant diseases and pests and the role they play in food security.

J.H. McBeath and J. McBeath, *Environmental Change and Food Security in China*, Advances in Global Change Research 35, DOI 10.1007/978-1-4020-9180-3_5,

Although many varieties of weeds also reduce crop yield, their impacts are less dramatic because they are subjected to control in a labor intensive agricultural system such as China's. The subject of this chapter is different from that of Chapter 6 on invasive species. The microorganisms and insects discussed here are considered indigenous today, even though they may have accompanied food crops on their introduction to China many centuries ago.

Our discussion begins with a definition of primary concepts, and then we provide examples of important diseases and insect pests on major crops such as rice, wheat, corn, soybeans, and potatoes. Many of the diseases and pests also are significant historically. Next we consider their impact on cereal production. We consider the correlation of specific diseases, insect pests, and environmental conditions and treat the impact of climate and policy changes on their occurrence, dissemination and severity. Because of the adverse impact of these diseases and insect pests on the Chinese food supply, we also address mitigation methods used and the problems they cause.

5.2 Definition of Primary Concepts

Diseases of plants and insect pests are anthropocentric concepts. A microorganism or an insect is viewed as a pathogen or a pest when it reduces the quality and/or the yield of plants which humans grow or harvest as food or timber. In nature, plants, microorganisms and insects are all part of the ecosystem. Many microorganisms (bacteria, fungi, phytoplasmas, protozoa, viruses, viroids, etc.), nematodes (tiny worms), insects and even other plants co-evolve with plant hosts and develop a close parasitic relationship with them. Pathogens and pests rely on host plants for their own sustenance and shelter. There are also other microorganisms and insects that prey on plant pathogens and insect pests, using them as their food sources. These microorganisms and insects are considered beneficial to food and timber production. In yet other cases, microorganisms, nematodes and insects work together, sometimes synergistically, to cause diseases on plants. These different relationships are all part of a dynamic and very complicated web of food chains in nature.

A microorganism or insect ceases to be regarded as a pathogen or insect pest when it can produce products of high economic value. The silkworm is a very good example. China was the first and for a long period the only country in the world where silkworms were raised for the purpose of silk production. Originally, the wild silkworm was a pest on mulberry trees. Ancient people living in the Yellow River Basin discovered the high quality silk of the cocoons when harvesting pupa for food. No less than 4,000 years ago, the technology had evolved from gathering, to rearing by domestication both silkworms and mulberry trees, and silk production became a major pivot of the Chinese economy. The annals of the Zhou Dynasty (1,000 BC) called mulberry trees and grains (sorghum, millet, soybean, wheat, barley, rice, flex) the "foundation of the nation."[1]

[1] Genpan Li, *Ancient Agriculture in China* (in Chinese): Taipei: Taiwan Commerce Publishing House, 1994, 30–31.

It should also be noted that many microorganisms and insects which humans consider pathogens and pests in an economic sense actually serve important functions in ecological renewal. Wood rotting fungi, for instance, cause decay in trees and reduce the value of timber, but they also play a significant role in the degradation of wood fibers and plant tissues, thereby facilitating the recycling of nutrients. Other fungi and bacteria serve similar functions, broadly called ecological services.

To cause diseases, virulent pathogens must come in close contact with susceptible host plants under environmental conditions favoring the pathogens. Close contact with host plants is also necessary for parasitism by nematodes and insects to occur.

Thousands of pathogens and insect pests affect cultivated as well as wild plants. Generally, each plant can be affected by a hundred or more pathogens and nematodes. With few exceptions, a single pathogen can affect dozens or even hundreds of species of plants.[2] This is also the case with respect to insect pests.

Among the microorganisms that can cause diseases on plants, fungi compose the largest group. For instance, more than 250 species of fungi can cause diseases on rice, compared to 12 bacteria and 14 viruses and phytoplasmas. Interactions among microorganisms may have synergistic or antagonistic effects. For instance, Potato Early Dying is a disease complex, caused by synergism between a pathogenic fungus (*Vreticillum dahliae*) and a nematode (*Pratylenchus* spp.), and each by itself also causes significant economic losses.

Environmental factors play a significant role in the health of plants, microbes, nematodes and insects, and affect their interactions. The primary environmental factors are temperature, moisture, oxygen, light, soil structure and soil pH; other factors such as nutrients, air pollutants, and harmful chemicals are important too.[3] For example, epiphytic microbes living on the plant surface respond strongly to UV radiation and leaf surface water, whereas soil inhabiting microbes and nematodes are sensitive to soil structures and pH. Wind and splashing rain also are important for the dissemination and dispersal of microbes and insects. Plants, microbes, nematodes, and insects growing under environmental conditions that exceed their normal range of requirements (either excessive or deficient) frequently sustain severe damage to their health or suffer injuries, which may lead to their death.[4]

Microbes, nematodes, and insects vary greatly with respect to their normal range of environmental requirements. In a community of mixed populations, as different species (as well as strains or races) of microbes, nematodes and insects compete for nutrients, for space, and for other resources, those species (strains or races) that can adapt to a wider range of environmental conditions compete better than those displaying less tolerance. When environmental conditions become very favorable to a specific species, strain, or race of a pathogen or pest in the presence of susceptible hosts, the endemic population escalates and expands rapidly, which results in an epidemic or pandemic.

---

[2] G.N. Agrios. *Plant Pathology*. 5th edition. Burlington, MA: Elsevier Academic, 2005, 7.

[3] Ibid, 249–63.

[4] G. L. Schumann and C. J. D'Arcy. *Essential Plant Pathology*. St. Paul, MN: APS Press, 2006, 8.

Plants grown under favorable environmental conditions with balanced nutrient regimens often can successfully fend off the attack of pathogens, using their natural defense mechanisms, unless the pathogens are especially virulent. However, plants grown under unfavorable conditions are more susceptible to diseases when coming into close contact with pathogens, particularly if the environmental conditions are conducive to development of diseases. Simply put, plants in good health show more tolerance toward pathogens, nematodes and insect pests.

5.3 Economic Impact of Diseases and Insect Pests on Food Production

In China, there are thousands of known microbes, viruses, nematodes, and insects that can adversely affect production of agricultural crops, fruits, flowers, and forest trees. Indeed, the list continues to lengthen as more information is gained through research. Although the number of pathogens and pests is very large, most losses on crops, trees and grasslands are caused by a relatively small number of plant diseases and pests. For instance, Xie Jianjun, director of the Tianjin Plant Protection and Inspection Station, reports that in the Tianjin area, of the more than 500 pathogens and insect pests causing diseases and damage to various plants each year, fewer than 40 cause most of the damage.[5] They affect an average of 1.4 million hectares of agricultural land and reduce grain harvest by more than 100,000 t.[6]

In the *2006 China Agriculture Yearbook*, the Ministry of Agriculture reported that damages caused by economic insects, and escalation of disease and insect pest infestation had reached 342.8 million hectares (a figure repeating the count of lands which were double- or triple-cropped), an increase of 16.5 million hectares. Losses of grains, cotton and oil seeds were 59.1, 1.4, and 1.8 million tons, respectively. With 484 million metric tons of grains harvested in 2005, the loss was about 12% of total grain production. The major diseases listed were rice blast, wheat rust, and soybean *sclerotinia* stem rot. The most important insect pests were the brown rice planthopper, rice borer, cotton boll weevil, corn borer, and migratory locust.[7] In 2005, an explosion in the brown planthopper population occurred. Planthoppers infected 25.8 million hectares of agricultural land – an increase of 6.2 million hectares over the previous year – an all time record. Yield losses of 1.9 million metric tons were three times higher than in the previous year. The brown planthopper has developed high resistance to several commonly used chemical insecticides and is becoming increasingly difficult to control. In this year, 20.3 million hectares of rice were

[5] Yianhua Huo. Northern Net News (in Chinese): http://news.enorth.com.cn/system/2007/02/28/001561574.shtml, February 28, 2007.

[6] Ibid.

[7] Ministry of Agriculture, China Agriculture Yearbook Editorial committee, Ed. *2006 China Agriculture Yearbook* (in Chinese). Beijing: Chinese Agriculture Press, 2006, 376.

affected by the rice leaf folder, an increase of 634,000 ha from the previous year. The damage to the mid-rice crop was more severe than to the early-rice crop. The loss of rice grains was 670,000 t. In the major rice growing areas of the middle and lower reaches of the Yangtze River, south of the Yangtze River and in southwest China, leaf folder damage reached epidemic proportions. Damage caused by the leaf folder in other rice regions also was severe. This pest was especially active in Guizhou and caused the death of rice plants in many fields.
The report made the following observations:

1. The overall damage caused by agricultural diseases and insect pests was more severe (than in previous years).
2. Both plant diseases and insect pests were equally serious.
3. Outbreaks of migratory insect attacks were more frequent.
4. There was an increased severity of disease epidemics.
5. There were substantial increases in damages caused by major diseases and insect pests.
6. The severity of minor diseases and insect pests also had increased.
7. New diseases and insect pests increased acutely in certain areas.

The Ministry of Agriculture attributed the escalation in severity to abnormal weather conditions, the cultivation system and methods in use, and limited resistance of plant varieties to diseases and pests.[8]

In December 2006, the China Agricultural Extension and Service Center revealed that in recent years, despise redoubled efforts by the government to improve disease and pest monitoring and control measures, the damage caused by diseases and pests on agricultural crops and grasslands still reached 25 million metric tons per year. The increase in scope and severity of diseases and insect pests resulted in economic losses of 57 billion yuan (about US$8.4 billion) per annum.[9] The report provided several examples of the seriousness of the situation:

- Eight consecutive years of migratory locust epidemics in agricultural regions
- Six consecutive years of stem borer outbreaks
- Six consecutive years of rice brown planthopper epidemics at detrimental scales in rice growing regions of the south
- Six consecutive years of wheat stripe rust epidemics
- Three consecutive years of rice blast outbreaks

Furthermore, the occurrence and severity of what previously were thought to be minor diseases (rice sheath blight, wheat smut) and insect pests (rice borer, wheat midge) were also on the rise.[10]

---

[8] Ibid.

[9] Jinhuai Wu. "China Harmful Biologicals Cause Loss of 57 Billion Yuan" (in Chinese), *China Chemical Engineering News*. December 6, 2006.

[10] Ibid.

Jiangxi Province demonstrated the severity of these problems. In 2006, it reported a 10-year record of rice blast disease in major rice areas; early rice (the first of two crops) sustained the heaviest losses. Infestation of rice leaf folder was also the most severe in recent years. Yield losses of rice due to diseases and insect pests in Jiangxi alone were 1.3 million metric tons.[11]

These diseases and insect pests continue to plague food production in China. The Ministry of Agriculture reported in 2008 that epidemics of wheat stripe rust, rice brown planthopper, and small rice brown planthoppers occurred not only earlier than in previous years, but also were more severe and affected larger areas. Heavy infestations of insect pests (rice leaf folder) and diseases (rice blast, rice sheath blight, bacterial blight, bacterial leaf streak) occurred in all rice growing areas of the Southwest, the Yangtze River and other major river valleys. There were outbreaks of the Oriental migratory locust, the Tibetan migratory locust, and other locust species too. Invasions of Ciaodimin in ranches and grasslands of the Northeast, Huabei (North China) and Xinjiang reached epidemic proportions.[12] *China Daily* reported that approximately 2 million hectares of grassland in central and eastern Inner Mongolia were severely affected by migratory locusts. The regional government mobilized more than 33,000 people, including 1,100 technical staff, to combat the plague. In their arsenal were 200 t of chemical pesticides and 100,000 large and small spray guns. The government booked four planes for locust eradication. Herdsmen aided the operation by urging their chickens and ducks to feed on locusts as a means of control.[13]

5.4 Diseases and Insect Pests of Historical and Contemporary Importance

Migratory locusts, rice blast disease and wheat rust are insect pest and plant diseases of great significance not only to the Chinese economy today, but also throughout Chinese history. We discuss each in turn.

5.4.1 Migratory Locusts (Locusta migratoria)

Among all the pestilences in Chinese historical records, those of migratory locusts were by far the most spectacular. The magnitude of devastation and the close

[11] Anonymous. "Experts Predict 2007 Natural Disasters in Jiangxi Severer than Other Years" (in Chinese), *Jiangxi Daily*, January 19, 2007.

[12] Ministry of Agriculture, China Agriculture Yearbook Editorial Committee, Ed. *2008 China Agriculture Yearbook* (in Chinese). Beijing: Chinese Agriculture Press, 2008.

[13] Jing Li, "Locust Epidemic in Inner Mongolia," *China Daily*, July 2, 2008, 3.

association of locust plagues with floods and drought resulted in labeling them the "three natural disasters" of ancient China.[14]

The earliest Chinese recording of a migratory locust epidemic was 707 BC, in the *Shandong Tongzhi* (*The General Record of Shandong*) of the *ChunQiu* (*The Spring and Autumn Annals*), written by Confucius.[15] Throughout Chinese history, archives of regional and national offices and reports of each dynasty were peppered with accounts of locust epidemics. For instance, 60 locust epidemics were recorded in the Han Dynasty[16] and 42 in the Tang Dynasty,[17] an average of approximately once every 7 years. After AD 960, the frequency of locust epidemics increased dramatically to an average of once every 3.5 years in the Song Dynasty, every 1.6 years in the Yuan Dynasty, and every 2.8 years in the Ming and Qing dynasties.[18]

The increase in locust epidemic frequency was due to changes of climate (drought, floods), shift of river channels, and redirection of agricultural policies, as well as incompetence of government officials. For example, in the Song Dynasty (eighth century), locust epidemics in regions south of the Yangtze River increased dramatically. Prior to the Song, there were nine locust pandemics in regions south of the Yangtze River for every one in the North, a ratio of 9:1; afterwards, the ratio increased to 2:1. This is explained by a change of the agricultural production structure following the shift of the state's economic center, and exodus of the Han people from northern China (then occupied by the Jin, a government established by a nomadic minority) to south of the Yangtze River, still under Song control.

Because wheat is the major staple of the northern Chinese, the Song government encouraged cultivation of wheat in the South by giving peasants tax exemptions for growing wheat but not for rice. This policy resulted in the rapid expansion of wheat cultivation into the paddy rice-growing area. Warm temperatures in the South also favored locust propagation (the optimal temperature for migratory locusts is 33°C); expansion of crop areas on dry land provided ample food to support the locust population. This created optimal conditions for the development of migratory locust epidemics. The introduction of corn and sweet potatoes in the mid-eighteenth century made agricultural cultivation in the foothills and on other arid land possible,[19] which further increased the vulnerability of the South to locust plagues.

Locusts are part of the grasshopper family. In fact grasshoppers, with more than 10,000 species, are one of the most diverse insects on earth. Scientists have identified

---

[14] Yuanzheng Suan and Yonglian Yuan, Eds. *Shandong Locust* (in Chinese). Beijing: China Agriculture Press, 1999, 3.

[15] Ibid, 3. 220.

[16] Wenhua Zhang, "Plagues of Locusts in the Han Dynasty" (in Chinese), *Tangdu Journal*, Vol. 19, no. 75 (2003), 45–47.

[17] Shoucheng Yan, "Plague of Locusts in the Tang Dynasty" (in Chinese), *Journal of Capital Normal University*, no. 151 (2003), 12–18.

[18] Yente Deng. *Zhongguo Jiou Huan Shi* (*Disaster Relief in China* [in Chinese]): Taipei: Taiwan Commercial Press, 1987, 348–357.

[19] Faiyuan Zheng, "Analysis of Locust Plagues in Chinese History" (in Chinese), *Journal of Chinese Agriculture History*, no. 4 (1990), 38–51.

at least 800 species in China alone.[20] Among the some 60 species of grasshoppers that cause serious damage to agricultural crops, forests, pastures and grasslands, the most devastating is the migratory locust *(Locusta migratoria)*, including three subspecies, namely the Oriental migratory locust (*Locusta migratoria manilensis*), the Asian migratory locust (*Locusta migratoria migratoria*), and the Tibetan migratory locust. These three migratory locust subspecies are distributed across China. Among them, the Oriental migratory locust is by far the most damaging, because its habitat covers the most productive agricultural areas in the Yellow, Huai, Yangtze, and Pearl River basins, the richest agricultural areas of China.

The migratory locust usually lays eggs in the banks of rivers and lakes. After over-wintering in the egg stage, development and generation are influenced by temperature, humidity, rainfall, and sunlight. Customarily, there are two generations of migratory locusts annually in the flood plains of the Hai, Yellow, and Huai rivers; three generations per year along the tributaries of the Huai and Yangtze rivers; and four generations per year along tributaries of the Pearl River.

The occurrence of a migratory locust epidemic is determined largely by climatic conditions. Under normal conditions, it takes 70–80 days for the completion of one generation. However, under drought conditions, it requires just 35 days, compared to from 95 to 100 days in rainy years. The first and second highest probabilities for migratory locust epidemics are in a year of drought and the subsequent year. Epidemics can also occur in years when drought follows flood.[21]

Migratory locusts have several unique characteristics which contribute to their destructiveness. They prefer to live close to each other in an aggregate or a pack. They are extremely productive. Females deposit capsules of 50–90 eggs into the damp sandy soils of river banks and lake shores. On average, a female locust is capable of producing 300–400 eggs in her lifetime, and from 700 to more than 1,000 eggs in extreme cases.[22] Migratory locusts also are capable of parthenogenesis (self-reproduction) and can lay fertile eggs without the benefit of a male.

Locusts have chewing mouthpieces and ravenous appetites. They usually feed during daylight hours, but may eat through the night in populous communities. Too, they eat greater amounts and more frequently when temperatures are high (in the range of 15–38°C). Unlike many herbivorous grasshoppers, migratory locusts are omnivorous. They prefer reeds and grasses; however, when starved, they will enlarge their menu to include wheat, rice, corn, sorghum, soybeans, cabbages, cotton, sunflowers and other dicotyledonous plants (except for potatoes, sweet potatoes, hemp and a few others dicots). In extreme conditions as noted in historical records, they will devour everything in their path, such as trees and animals, paper, clothes, and human flesh. They also cannibalize each other.

---

[20] Chinese Academy of Agricultural Sciences (CAAS), Plant Protection Institute. Ed. *China Diseases and Insects on Agricultural Crops*, Second Edition (in Chinese). Beijing: China Agriculture Press, 1995. Vol. I, 653–69.

[21] Ibid., 654–61.

[22] Ibid., 660.

For this reason, Chinese use the phrase "a thousand leagues of red earth" to describe the aftermath of a locust epidemic. Migratory locusts can fly long distances. When they are in flight, the massive number of locusts is described as a huge patch of dark cloud, obstructing the sunlight and darkening the sky.

Four chronicles from the Tang, Yuan, and Qing dynasties describe the magnitude of locust epidemics and the damage they caused. The *Wuxingzhi* (*Record of Wuxing*) from the late Tang Dynasty reports:

> In the first year of Zhenguan (785 AD), a locust plague occurred in the summer. Masses of locusts flew continuously for more than ten days, starting on the eastern seaboard and ending in Helong (the Shanxi region), and they blotted out the sun. Along their path, they left not one leaf on trees, one blade of grass, or one hair on animals. Roads were littered with corpses of humans who had died of starvation.[23]

In ancient China, the climate in Shandong, Hebei, and Henan provinces was fairly mild; landscapes were crisscrossed with rivers and streams, and dotted with reed-filled swamps. These regions were ideal habitat for migratory locusts. Archives of the Yuan dynasty described a locust epidemic in these terms:

> In the fifth month of the nineteenth year of Zhizheng (1359 AD), locust outbreaks occurred in Shandong, Yunzhou, Jiaozhou, Boxingzhou.... All crops were devoured. The flying locusts obscured the sun. People and horses could not gain footholds on the road. Locusts filled ditches and canals to the brim. Starving people trapped locusts for food or stored them. When the dried locusts were exhausted, people ate one another.[24]

The record of Wei County in Shandong Province, the *Weixianzhi*, reported that in the 15th year of Emperor Qianlong (1785), "A locust plague occurred in the seventh month of this year. Men who lost their way were eaten by locusts."[25] A different chronicle, the Shandong *Anchiu Xinzhi* (*New Record of Anchiu County*), made this observation in the same year:

> Severe drought occurred in the spring and summer, and there was a locust plague. Masses of locusts covered the sky. On their descent, the locusts piled several feet deep. Many large trees broke from the weight of the locusts. People who could not escape after falling into ditches filled with locusts were devoured."[26]

The misery of famine caused by migratory locusts is unfathomable. For the people who sought to survive under these circumstances, cannibalism was the last resort. County records of Shandong Province[27] note the areas where cannibalism occurred during famines caused by migratory locust epidemics:

- One county in AD 194 (the last emperor of the Han Dynasty)
- Five counties in AD 1359 (the last emperor of the Yuan dynasty)

---

[23] Ibid., 654.

[24] Suan and Yuan, *Shandong Locust*, 1999, 236.

[25] Ibid., 262.

[26] Ibid.

[27] Ibid., 220–78.

- Two counties in AD 1457, one county in AD 1465, and two counties in AD 1485 (third emperor of the Ming Dynasty)
- Two counties in AD 1528 (sixth emperor of the Ming Dynasty)
- Two counties in AD 1616 (eighth emperor of the Ming Dynasty),
- Two counties in AD 1640 (the 11th and last emperor of the Ming Dynasty)
- One county in AD 1785 (fourth emperor of Qing Dynasty)

The occurrence of locust plagues and famine was correlated with official incompetence and corruption of governments. Officials did not effectively implement policies to control migratory locust populations; nor did they provide rapid relief measures. The scope and speed of measures to alleviate famine caused by locust epidemics (and other natural disasters as well) were insufficient.

5.4.2 Rice Blast Disease

Rice blast, caused by *Magnaporthe grisea* (a fungus) is found in approximately 85 countries where rice is grown commercially on a large scale, and is one of the most widely distributed plant diseases in the world. Rice blast is highly destructive to lowland rice, but it also affects upland rice. It is generally considered to be the principal disease of rice because of its wide distribution and its destructiveness.

Rice blast can afflict all parts of the rice plant. It is a seed borne disease, and contaminated seeds are important sources of primary infection in the following year. The pathogen produces diamond-shaped lesions, and as the disease progresses, many lesions coalesce. The leaves appear grayish-white, giving the appearance of scald. A collar infection can kill the entire leaf blade, while a nodal infection weakens the culm (rice stem) considerably and makes it susceptible to breaking. In the case of severe infestation, plants shrivel and die of root rot. If disease breaks out during the late stage of plant development, infection occurring on the panicle (neck) node is the most destructive. This is because the infection may severely interrupt development of the grain (seed), resulting in poor quality rice. If the infection occurs during the middle stage of grain development, white heads (empty kernels) form, which produce no rice.[28]

The development of *M. grisea* is highly dependent on environmental conditions. The temperature range of the rice blast fungal body (mycelia) is 8–37°C; the optimal temperature for spore production is 25–28°C, which is also the optimal temperature for rice growth. *M. grisea* is a prolific spore producer. Spores can be disseminated a large distance by wind, splashing rain, running water, or insects. Sporulation is strongly influenced by humidity and leaf wetness.[29] Inasmuch as sunlight is a deterrent to spore germination, conditions of cloudiness, drizzling, and

[28] R. K. Webster and P. S. Gunnell, Editors. *Compendium of Rice Diseases*. St. Paul, MN: APS Press. 1992, 14–15.

[29] Ibid.

fogginess greatly facilitate the development of the disease. Furthermore, cell walls of rice plants formed under insufficient sunlight (or with high nitrogen fertility) are tender because of a slowing down of metabolites; hence they are more susceptible to rice blast infection.

In China, rice blast is known as *daowenbing* (rice plague disease, a name commonly used in mainland China) or *daorebing* (rice fever disease, commonly used in Taiwan). The earliest recording of rice blast was in 1637 (late in the Ming Dynasty). In his manuscript *Utilization of Natural Resources* (*Tiangong kaiwu*), Song Yingxing provided the first description of plant diseases in the world literature.[30] In the chapter on rice maladies, he described scalding of rice seedlings and its relationship to health of the seeds and to warm and humid climatic conditions. He also observed that seedlings, after transplanting, could suffer severe damage under prolonged cloudy and drizzling conditions. He correctly pointed out that the disease problem was located in the roots. In addition, he described the scalding of leaves and flowering spikes,[31] which fit the symptoms of rice blast diseases on seedlings, leaves, and panicles.[32] Song attributed the cause of scalding of seedlings to "too much heat" stowed in rice kernels from the previous year, after these kernels were dried under a hot sun and put in storage too hastily without sufficient cooling.

The heat or fever considered to be the cause of the disease was perhaps the original basis for calling it "rice fever disease," a name which has been extensively used in the Chinese and Japanese literature.[33] Song also explained the death of seedlings after days of continuous rain by "roots not having yet developed well," and scalding of kernels and leaves of rice by "ghost fire" (phosphate). These misinterpretations are understandable, especially in an era prior to the discovery of microscopes and before the development of germ theory. Significantly, Song observed correlations between the severity of maladies of seedlings and environmental conditions (warm and humid southeast winds). He openly contested the common belief that diseases of rice seedlings were caused by the interference of the supernatural. In his manuscript, he pointed out simply that it was the seed, and not spirits or deities, which was the source of the problem. In an era dominated by superstitions, Song's contribution was remarkable.

Today, rice blast is found in all paddy rice growing areas in China, from Heilongjiang Province in the North to Xinjiang and Tibet in the West, Taiwan and Fujian in the Southeast and Hainan Island in the South. This disease is especially severe in river valleys where the climate is mild, with high humidity and long periods of fog, and relatively limited periods of direct sunlight, as well as in coastal areas with heavy rainfall. Generally, infestation of rice blast is the heaviest on late

[30] Gau Tan Wen Hua, Editor, *Jiaoni Kandong Tiangong Kaiwu* (Instruct You on How to Understand Utilization of Natural Resources) (in Chinese). Beijing: Contemporary World Publisher, 2006, 7–9.

[31] Ibid.

[32] CAAS, *China Diseases and Insects on Agricultural Crops*, 1995, 3–13.

[33] S.H. Ou, *Rice Diseases*. London/Reading, UK: Eastern Press, 1972, 97–98.

rice with the exception of southern China; there, rice blast infestation is heaviest on early rice. In an epidemic year, yield reduction caused by rice blast ranges from 10% to 50%, and it can reach 100%.

In 2005, rice blast was the most widely spread disease in China, causing the most damage among all plant pathogens. Even in Heilongjiang, the magnitude and severity of rice blast epidemics reached a 10-year record. Rice blast affected 5.6 million hectares of rice fields, an increase of 570,000 ha as compared to the previous year. The disease was also more intense in most of the infested areas. In some districts, panicle neck rot and leaf blast were equal in incidence and severity. Many fields suffered from "melt-down." Rice blast was especially severe in Jiangxi Province, where some 80–100% of the rice leaves and 40–80% of the panicles of rice plants in 30 counties were affected; many fields had to be plowed under.[34]

Among rice varieties, short grain rice (*Oryza sativa japonica*) and Africa rice (*O. glaberrima*) are more susceptible to the disease than long grain rice (*O. sativa indica*). In addition to rice, rice blast also infects grasses including fescus (*Festuca arundinacea*), and rice cutgrass (*Leersia oryzoides*).

As mentioned above, rice blast is a prolific producer of spores. Spores are disseminated by air currents, splashing rain, running water, and insects. Most of the spore dispersal reaches to about 10 km; however, long distance dissemination also occurs. The great variability of pathogenicity adds to the difficulty in breeding resistant varieties, and explains the short useful life of resistant varieties.

Generally, a rice blast epidemic requires a prolonged period of relative humidity of 90% to saturation. Rice blast is remarkably adaptable to various environmental conditions. Survivability of rice blast spores and mycelia under a certain temperature is influenced by humidity. Under conditions of high humidity, these spores can survive for only 5–7 min at 52°C. However, under dry conditions, spores of rice blast fungus are extremely resilient and can survive more than 60 min under high heat (100°C) and more than 75 days under −10°C conditions. This ability to survive under sub-freezing conditions makes it possible for the rice blast fungus to overwinter, and it insures that the disease reappears in the following year in northern China.[35]

One of the characteristics of *M. grisea* is its genetic variability and its ease in breaking down the host plant's genetic resistance. For instance, spores produced from the same colony may consist of many pathogenic races, which not only vary in nutrient requirements and cultural characteristics, but also in their pathogenicity. This phenomenon is especially acute in host plants appearing to possess complete resistance to blast. Generally, these rice varieties remain effective for 2–3 years in environments conducive to disease.[36] In 2005, rice blast was especially severe in old rice blast areas and on susceptible cultivars.[37] In Sichuan Province, a total of 108

[34] *2006 China Agriculture Yearbook*, 2006, 377.

[35] CAAS, *China Diseases and Insects on Agricultural Crops*, 1995, 4–7.

[36] Webster and Gunnell, *Compendium of Rice Diseases*, 1992, 15–16.

[37] Ibid.

cultivars grown in 133 counties became susceptible to rice blast. The scope and intensity of the 2005 epidemic and the speed of its spread were 10 year records; furthermore, the infestation arrived 5–35 days earlier than previously.[38]

5.4.3 Wheat Rust Diseases

There are three wheat rust diseases, namely stripe rust (*Puccinia straiiformis*), stem rust (*Puccinia graminis*), and leaf rust (*Puccinia recondita).* In addition to wheat, the stem and stripe rusts also cause diseases on barley, rye and some grasses. The leaf rust however affects wheat primarily. The stem rust fungus requires an alternate host such as barberry to complete its life cycle.[39] Unlike the rust situation in the USA and Canada, stem rust is not a serious problem in China. In China, stripe rust is the most prevalent of the three diseases.

Rust diseases are believed to have co-evolved with wheat and are among the oldest diseases in the world. The earliest recorded rust epidemic was in 700 BC. In ancient Rome, temples were erected to worship the rust god, Rubigo, and rust goddess, Rubigos. The festival Rubiyalia was held every spring to pray for protection and a good harvest. Archeological evidence indicates that wheat cultivation occurred in Gansu more than 5,000 years ago. Rust diseases migrated to China following the spread of wheat. The earliest mentions of wheat rust are in *Mashou nongyan* (1836) and *Chimin yaoshu* (1846) as "jaundice," doubtless because of the coloring. *Chimin yaoshu* states that "(If there is) heavy rainfall in the spring (there will be) more frequent jaundice."

Rusts are prodigious producers of spores. Dissemination relies primarily on the wind, and they have been known to disperse long distances when carried by jet streams. Rusts reduce vigor of the wheat plant, as well as seed yield, forage value and winter-hardiness; they predispose plants to infection by other diseases.[40]

In China, wheat stripe rust occurs primarily on winter wheat in the regions of Huabei, Huaibei, the Northwest, and the Southwest. Occasionally, stripe rust also occurs on late maturing spring wheat in the Northwest, Southwest and Huabei regions.[41] Under favorable conditions (cooler temperatures), stripe rust can cause repeated infections. Although stripe rust does not require high humidity, drought reduces sporulation significantly. It over-summers on high plateau wheat regions, and over-winters in wheat regions of river basins.[42]

---

[38] *2006 China Agriculture Yearbook*, 2006, 377.

[39] M. V. Weise, *Compendium of Wheat Disease,* Second edition. St. Paul, MN: APS Press, 1987, 40–41.

[40] Ibid, 37.

[41] CAAS, *China Diseases and Insects on Agricultural Crops*, 1995, 271.

[42] Zhenqi Li and Shimai Zeng, Editors. *China Wheat Rust Diseases* (in Chinese). Beijing: China Agriculture Press, 2000, 205–07.

Wheat stripe rust is one of the most important diseases on wheat in China, and it often reduces yields by 20–30%. In severe cases, yield loss may reach 50–100%. Yield losses caused by wheat stripe rust in the epidemic years of 1950, 1964, 1990 and 2002 were 60, 30, 25, and 14 million metric tons, respectively.[43] Areas of infestation ranged from 3.3 to 6.7 million hectares.[44] A number of medium- to large-scale epidemics have affected wheat production. For example, in 1975, stripe rust occurred in Shandong, Hebei, Henan, and northern Jiangsu, on more than 4.3 million hectares; it produced wheat production losses of 0.8–1.0 million metric tons. In 1983, stripe rust epidemics influenced wheat production in ten provinces and affected 7.9 million hectares, with a yield loss of more than 1.0 million metric tons. The 1990 epidemic resulted from loss of genetic resistance in main wheat varieties due to emergence of a new stripe rust pathogenic race. This epidemic (which included some wheat leaf rust), occurred in the Yellow and Huai river basins, encompassed 9.8 million hectares and occasioned a loss of 1.8 million metric tons of wheat.[45] In 2005, the area of wheat rust infestation encompassed 14 provinces, autonomous regions, and counties, some 6.1 million hectares.[46]

Wheat leaf rust is also a major rust disease on spring and winter wheat. Three epidemics occurred in the years of 1973, 1975 and 1979 and resulted in serious yield losses.[47] Wheat stem rust in China is relatively insignificant in terms of its impact on crop production.[48]

5.5 Emerging and Re-emerging Diseases and Insect Pests

A number of other diseases and insect pests also have reached epidemic scale and significantly affect Chinese food production today. The discussion is organized by the major food crops the diseases and pests threaten: rice, wheat, maize (corn), potatoes and soybeans.

5.5.1 Rice Diseases and Insect Pests

Rice is China's most important food crop, and it also hosts the largest reported number of plant diseases and pests. Many of these diseases have become wide

[43] G. Chen, H. Wang and Z. Ma, "Forecasting Wheat Stripe Rust by Discrimination Analysis" (in Chinese), *Plant Protection*, Vol. 32 (2006), 24–27.

[44] Li and Zeng, *China Wheat Rust Diseases*, 2000, 3.

[45] Ibid., 3.

[46] *2006 China Agriculture Yearbook*, 2006, 377.

[47] Li and Zeng, *China Wheat Rust Diseases*, 2000, 3.

[48] Ibid.

spread and cause severe losses of rice production due to the emergence of virulent strains/races of pathogens, changes of policy, rice cultivars, cultural practices and environmental conditions. We introduce seven diseases and pests below.

5.5.1.1 Rice Sheath Blight

Rice sheath blight is caused by the fungus *Thanatephorus cucumeris*. This pathogen is soil borne and endemic in most rice production areas.[49] Prior to the 1950s, rice sheath blight was an obscure disease in China. However, following the adoption of new cultivars (semidwarf, high tillering species) and new cultural practices (high fertilizer use, close spacing of plants), rice sheath blight became a major obstacle to increased production. In 1982, the estimated reduction of rice production caused by sheath blight was 50 million metric tons.[50] In the early twenty-first century, this disease has surpassed the destructiveness of rice blast and become the most important disease of paddy rice in China.[51]

Prior to 1990, rice sheath blight rarely occurred in Heilongjiang Province. Then, as farmers successfully cultivated paddy rice in the Northeast, infestation of rice sheath blight followed the expansion of paddy rice fields. Yield losses caused by this disease ranged from 7% to 60% or higher.[52] Typically, intensity of disease is relatively light in early season and escalates in mid and late season.

The pathogen can spread rapidly via irrigation water and the movement of soils during land preparation and cultivation.[53] Under conditions of low sunlight, high humidity (95%) and high temperature (28–32°C), infection of rice plants occurs easily. Young rice plants are more resistant than old ones to the infection of sheath blight. The rice plant is the most susceptible during early heading, and infection often results in poorly filled grains. Furthermore, infection by *T. cucumeris* seriously weakens the sheath, which can result in lodging (the plant falls over). The sheath blight fungus produces sclerotia (a seed-like fungal structure), which is essential for the initiation of infection, dissemination, and survival of the pathogen. Sclerotia can survive several years in the soil; they are resistant to chemical fungicides, which makes control of the disease especially difficult.

5.5.1.2 Bacterial Blight

Bacterial blight is caused by *Xanthomonas oryzae pv.oryzae*, and is one of the most serious rice diseases worldwide. It was first reported in China in the 1930s, in

---

[49] Ibid., 22.

[50] J. M. Hong and X. M. Tung, Editors. *China Rice Diseases and Control* (in Chinese). Shanghai: Shanghai Century Publisher, 2006, 62.

[51] Ibid.

[52] H. P. Xin, Chief Editor, *Northern Rice, Pictorial Guide to Diseases and Insect Pests and Control* (in Chinese). Beijing: China Agriculture Press, 2002, 20.

[53] Webster and Gunnell, *Compendium of Rice Diseases*, 1992, 15.

Jiangsu and Zhejiang. Presently, bacterial blight is found in all rice growing areas of China, except for Xinjiang and Gansu provinces. Rice blight affects yields seriously. For example, in 1974, a rice blight epidemic covered six provinces in southwest, southeast, and central China. The areas of infestation encompassed 1.5 million hectares, and losses of rice were 605,000 t.[54]

Bacterial blight benefits from warm temperatures (25–30°C) and high humidity. Severe winds (which create wounds) and excessive nitrogen also abet spread of this disease.[55] Bacteria infect the host through wounds on the leaves and roots, and cause wilt in seedlings. On older plants, the pathogen produces water-soaked stripes on the leaves and interferes with the transport of water and nutrients to the host. *X. oryzae pv. oryzae* survives primarily in rice stubble and on weed hosts. It is disseminated by irrigation water, by splashing rain or plant-to-plant contact, and by trimming tools.[56] The recent practice of transporting hybrid rice seeds, which have been mass produced in Hainan, to other parts of China has brought bacterial blight to areas previously free of this disease, because the bacteria are able to survive the long distance transportation.[57]

5.5.1.3 Bacterial Leaf Streak

Bacterial leaf streak is caused by *Xanthomonas oryzae pv.oryzicola*. This disease is widely distributed in tropical Asia. First found in Guangdong in the early 1950s, in the following years, outbreaks occurred not only in Guangdong but also in Guangxi and Fujian provinces, with severity greatest in the Pearl River Delta.[58] The bacteria reduces the "thousand seed weight" (a measure of quality) and yield of rice by 20–60%. In the late 1960s, the disease was effectively controlled through the development of "leaf streak-free seeds," seed treatments and selections of resistant cultivars. However, the disease re-emerged in the late 1970s and early 1980s, for two reasons. First, scientists promoted hybrid rice that was not genetically resistant to the bacterium. The second factor was the policy of *Nanfan Beidiao* (multiplying seeds in the South and transporting them to the North). Long distance dissemination facilitated the rapid spread of bacterial leaf streak to Jiangxi, Hunan, Hubei, Zhejiang, Guizhou, Yunnan, Sichuan, and other places.[59] In 2005, bacterial leaf streak infestation was fairly heavy in the Yangtze and Huai river rice regions. Infestation rates ranged from 0.5% to 5%, yet 100% infestation rates were common in some fields.[60]

---

[54] Hong and Tung, *China Rice Diseases and Control*. 2006, 141–42.

[55] Webster and Gunnell, *Compendium of Rice Diseases*, 1992, 10.

[56] Ibid.

[57] Hong and Tung, *China Rice Diseases and Control*, 2006, 62, 148.

[58] Ibid., 156.

[59] Ibid., 157.

[60] *2006 China Agriculture Yearbook*, 2006, 377.

Bacterial leaf streak is a seed borne disease. It survives primarily on infected seeds and straw. Development of this disease is abetted by rain, high humidity, and high temperatures (28–30°C).[61] Although the disease can occur at any growth stage of rice production, in southern China from July through September, this disease is especially severe at the panicle stage of early rice, seedlings, and the tillering stages of rice, when storms and flooding are frequent.[62]

5.5.1.4 Rice Stripe Disease

Rice stripe disease is a viral disease, caused by the rice stripe virus. It is transmitted by the small brown planthopper *Laodelphax striatellus*, which is an insect pest of rice (discussed below). The discovery of this disease in Zhejiang is fairly recent. However, areas of infestation of this disease have expanded rapidly. In 2006, rice stripe disease reached epidemic proportions and infested more than 100,000 ha. In heavily infested rice fields, the incidence was greater than 30%,[63] with corresponding losses of yields.[64]

5.5.1.5 Rice Brown Planthopper and Smaller Brown Planthopper (Gray Leafhopper)

Of the several migratory, plant sap sucking insects in China, the brown planthopper, *Nilaparvata lugens*, is by far the most important rice pest. This insect prefers a warm climate, and it feeds exclusively on rice.[65] Rice plants infested by brown planthoppers sustain damages not only from loss of sap, but also from disruption of translocation of water and nutrients due to plugging of the vascular tissues by saliva secreted by the pest during feeding. The infected rice plants display severe stunting, wilting and lodging. The brown planthopper is also an insect vector and can transmit wilted stunt,[66] grassy stunt, ragged stunt and other rice viruses.[67]

---

[61] Webster and Gunnell, *Compendium of Rice Diseases*, 1992, 11.

[62] Hong and Tung, *China Rice Diseases and Control*, 2006, 62, 159.

[63] H. D. Wang, Z. R. Zhu, J. L. Zhu and Y. Xu, "Occurrence and Seasonal Epidemics of Rice Stripe Disease," *Proceedings of Annual Meeting of Chinese Society for Plant Pathology* (2007), 224–26.

[64] H. D. Wang, X. L. Sun and J. L. Zhu, "Relationship Between Rice Stripe Disease (Rice Stripe Virus) and Rice Yield Losses," *Proceedings of Annual Meeting of Chinese Society for Plant Pathology* (2007), 231.

[65] X. N. Cheng, J. C. Wu and F. Ma, Eds. *Brown Planthopper Studies and Control* (in Chinese). Beijing: China Agriculture Press, 2002, 2.

[66] C. C. Chen and R. J. Chiu, "Transmission of Rice Wilted Stunt Virus by Brown Planthopper (*Nilapavata lugens Stål*)," Taichung Agricultural Improvement Station Research Proceedings, Vol. 23 (1989), 3–10.

[67] X. N. Cheng, J. C. Wu and F. Ma, Eds., *Brown Planthopper Studies and Control* (in Chinese). Beijing: China Agriculture Press, 2002, 7.

The brown planthopper is widely distributed in Asia and the Pacific. It prefers a warm climate. In China, a small population of brown planthoppers over-winters along coastal areas, in Taiwan and Hainan, but most brown planthoppers migrate to China in the spring from the Mekong Delta.[68] From the spring through the fall, brown planthoppers travel northward to the southern region of Heilongjiang, and west to Lanzhou and western regions of Sichuan Province.[69]

The smaller brown planthopper (also known as the gray leafhopper), *Laodelphax striaitellus*, is a sap sucking, migratory insect pest too. It has a relatively higher tolerance to low temperatures. Gray leafhoppers are endemic pests in the rice regions south of the Huai River. The range of these two insect pests overlaps, but gray leafhoppers cover the more northern rice regions. Both insect pests cause severe depletion of nutrients and damage to plant cells.

The smaller brown planthopper is an effective vector, capable of transmitting a large number of virus diseases, including rice stripe virus,[70] rice black streaked dwarf virus, barley yellow dwarf virus, barley yellow striate mosaic virus, Northern cereal mosaic virus, cereal tillering, and wheat chlorotic streak virus.[71] Results of a recent study conducted by Wang Huadi et al. indicate that 9.7% of the adults carried rice stripe virus.[72]

5.5.1.6 Rice Leaf Folder

The rice leaf folder (*Cnaphalocrocis medinalis*), a moth, is also called the rice leaf folder. It is one of the major insect pests on rice crops in central and southern China. This pest is a leaf feeding insect. Larva feed inside the folded leaf removing the chlorophyll content of the leaf (especially flag leaves) leading to a considerable reduction in yield. This pest prefers high humidity and shady areas.[73] It has a very broad host range; in addition to rice, it affects maize, oats, barley, wheat, sorghum, millet, coconut, tobacco, and sugarcane.[74] The broad host range and its migratory nature make control of the rice leaf folder very difficult.

5.5.1.7 Striped Rice Borer

The striped rice borer (*Chilo suppressalis* Walker) is another major insect pest on rice. Every year *C. suppressalis* produces from one to five generations, and it encompasses all the rice growing areas, except Xinjiang and Qinghai.

---

[68] Ibid., 3–5.

[69] CAAS, *China Diseases and Insects on Agricultural Crops*, 1995, 124–26.

[70] Wang et al., "Occurrence and Seasonal Epidemics of Rice Stripe Disease," 2007, 224–26.

[71] Wiese, *Compendium of Wheat Disease*, 1987, 67–68.

[72] Wang, Sun and Zhu, "Relationship Between Rice Stripe Disease . . . ," 2007, 231.

[73] W. B. Rani, R. Amutha, S. Muthulakshmi, K. Indira and P. Mareeswari, "Diversity of Rice Leaf Folders and Their Natural Enemies," *Research Journal of Agriculture and Biological Sciences*, Vol. 3, no. 5 (2007), 394–97.

[74] Ibid.

The rice borer is omnivorous. It will infect not only rice, but also wheat, corn, sorghum, rapeseed, other crops, and graminous weeds. When infestations occur in the early stage of rice development, caterpillars of the insect burrow into the rice stem (or stems of other food plants) and cause "hollow heart"; if infestation occurs in the late stage, white heads (empty husks) are produced. In 2005, the area of infestation was 18.1 million hectares, an increase of 413,000 ha over previous years. Infestation has reached an epidemic scale in the middle reaches of the Yangtze River. It also became increasingly active in southwestern rice regions. Damage caused by the stripe rice borer in the northeastern rice region also was significant. The escalation of damage is explained by increasing numbers of insects surviving the winter, expanding areas of infestation, high numbers of eggs laid and large numbers of insects remaining alive after insecticide treatments. Losses of grain due to this insect pest were 670,000 t. In Jiangxi, epidemics of stripe rice borer occurred in the first and second generations of early rice. They were especially serious in the second generation; hollow heart of the stem spread to more than 70%, which resulted in the lodging of the entire field, a phenomenon rarely seen in history.[75]

5.5.2 Wheat Diseases and Pests

Wheat is China's second most important cereal, and we present information on five diseases and insect pests which have gained increasing importance in wheat production.

5.5.2.1 Wheat Powdery Mildew

Wheat powdery mildew is caused by the fungus *Erysiphe graminis f.sp. tritici*. Powdery mildews, like rusts, smuts, downy mildews and viruses, are obligate parasites (meaning that they require a live host) and have formed close relationships with their hosts. *E. graminis f.sp. tritici* attacks wheat exclusively.[76]

In China, wheat powdery mildew is found in all wheat growing regions.[77] It can affect all parts of the plant. Heavily infected wheat plants not only produce fewer heads and kernels, but also the kernels have a lighter weight. Yield reduction is directly correlated with the severity of the disease. Heavily infected leaves and plants can be killed prematurely, which causes a high yield loss.[78]

---

[75] *2006 China Agriculture Yearbook*, 2006, 377.

[76] M. V. Weise, *Compendium of Wheat Disease*. St. Paul, MN: APS Press, 1987, 30–1.

[77] Zhiping Dong and Jingyu Jiang, Eds, *Wheat, Pictorial Guide to Diseases, Insects and Weeds and Control* (in Chinese). Beijing: China Agriculture Press, 2007, 4.

[78] Ibid.

The powdery mildew fungus prefers cool weather. The optimal temperature for its development is 15–22°C, and growth slows above 25°C. Powdery mildew is prodigious in the production of spores. Dissemination of spores relies primarily on air currents. Spores germinate over a wide temperature range (1–30°C) and do not require a layer of moisture.[79]

In 2005 powdery mildew infestation was heavier than in previous years in Hebei, Shandong, Jiangsu, Gansu, Sichuan, Henan, and Shanxi provinces. The total area of infestation was 7.0 million hectares, an increase of 1.1 million hectares over the previous year. This disease was most severe in situations where wheat was intercropped, soil was not tilled, and in mountainous areas. Yield losses in wheat production due to powdery mildew were 320,000 t.[80]

5.5.2.2 Wheat Sharp Eyespot

Wheat sharp eyespot disease, caused by the fungus *Rhizoctonia cerealis*, is becoming another major disease on wheat. It causes not only yield loss, but also quality reduction. In the past, this disease was most destructive in wheat production areas of southern China. However, in recent years, severity of this disease in wheat growing areas of the North has escalated.[81] Wheat sharp eyespot has become endemic in 20 provinces and districts in China, and it affects all stages of wheat development. In infested wheat fields, yield loss has ranged from 5% to 40%. In heavily infested wheat fields, wheat plants produce white heads and nothing of value.[82] In 2005, sharp eyespot affected areas of 8.1 million hectares, and the severity of the disease increased. Yield loss was 530,000 t, an increase of 80,000 t.[83]

5.5.2.3 Wheat Yellow Mosaic Virus and Barley Yellow Dwarf Virus

Wheat yellow mosaic virus and barley yellow dwarf virus are two virus diseases causing severe damage to wheat production in China. These two viruses are found in southern and northern parts of the wheat-growing areas along the Yangtze and Yellow rivers.[84]

---

[79] M. V. Weise, *Compendium of Wheat Disease*. St. Paul, MN: APS Press, 1987, 31.

[80] *2006 China Agriculture Yearbook*, 2006, 377.

[81] Ibid.

[82] J. F. Yu, T. Y. Zhang and G. X. Zhang. "Genetic and Virulence Variation of Wheat Sharp Eyespot Isolates (*Rhizoctonia cerealis* AG-D) from Shandong Province" (in Chinese), *Acta Phytopathologica Sinica*, Vol. 37, no. 4 (2007), 418–25.

[83] *2006 China Agriculture Yearbook*, 2006, 377.

[84] C. G. Han, D. W. Li, Y. M. Xing, K. Zhu, Z. F. Tian, Z. N. Cai, J. L. Yu and Y. Liu, "Wheat Yellow Mosaic Virus Widely Occurring in Wheat (*Triticum aestivum*) in China," *Plant Disease*, Vol. 84 (2000), 627–30.

Wheat yellow mosaic virus is transmitted by a fungus, *Polymyxa graminis*, which prefers low lying wet areas.[85] It was first found in the 1960s in Sichuan Province and spread gradually into the middle and lower reaches of the Yangtze and Huai rivers in the wheat production belt of China, extending more than 2,000 km. Estimates of yield losses ranged from 20% to 30% in most years, but on occasion increased from 50% to 70%.[86]

Barley yellow dwarf virus (BYDV) is one of the most widely distributed and destructive virus diseases of cereals. BYDV is transmitted by more than 20 aphid species, including the greenbug and corn leaf aphids, both prevalent in China. Many of BYDV's hosts remain symptomless. Damage to wheat varies in accord with the cultivar, virus strain, time of infection and environmental conditions. Severely infected crops often yield no grain. Over large production areas, BYDV is estimated to reduce yield by from 5% to 25%.[87]

5.5.2.4 Greenbug and Other Wheat Aphids

Aphids are sap sucking insects, and several types infest and cause damage to wheat in China. The predominant subspecies are greenbug, *Schizaphis graminum* and the cereal aphid, *Sitobion avenae.*[88] Infestation of aphids on wheat causes yellowing of leaves and spikes; it results in reduction of kernel weight and even premature death. Aphids multiply rapidly and can produce 20–30 generations a year.[89] Greenbug can reproduce without mating (parthenogenesis) in warm, mild climates; it is an insect vector and can transmit barley yellow dwarf virus to wheat, barley, rye, wild oats, millet, corn, grasses and even to some dicots.[90]

In 2005, infestation of aphids on wheat was moderate to heavy. The area of infestation was 15.1 million hectares; yield losses of wheat production were 530,000 t, an increase of 80,000 t from the previous year.[91]

5.5.2.5 Wheat Midge

Wheat midge (*Sitodiplosis mosellana*), a minute fly, is a major insect pest in the wheat belt of northern China. It causes damage primarily in flowering spikes by sucking sap from immature kernels, which results in shriveled grain of poor quality. Midges produce one generation a year. The larva over-winter or over-summer in soils.

---

[85] Weise, *Compendium of Wheat Disease*, 1987, 78–9.

[86] Han et al., "Wheat Yellow Mosaic Virus . . . ," 2000, 627–30.

[87] Weise, *Compendium of Wheat Disease*, 1987, 70–1.

[88] Dong and Jiang, *Wheat Diseases, Insects and Weeds and Their Control*, 2007, 34.

[89] Ibid.

[90] Weise, *Compendium of Wheat Disease*, 1987, 67.

[91] *2006 China Agriculture Yearbook*, 2006, 377.

In the spring when weather is humid and rainy, larvae emerge from the soil, form pupa and reach the adult stage when wheat flowers. Female midges fly toward the wheat spikes and lay 30–90 eggs. After hatching, the larvae burrow into the kernel and feed on the sap. Midge larvae remain dormant in years of severe drought. They rely on air currents for dissemination.[92]

There were three wheat midge epidemics in the 1960s, 1980s and 1990s. During the outbreaks, yields of wheat were seriously affected; in severe cases, nothing was produced.[93] Based on indices of insect numbers, date of emergence from soils, population increases in newly infested areas, and severity of damages, the infestation of wheat midges in China is still in the expansion phase. In 2005, wheat midges infected 1.95 million hectares of wheat fields, which is an increase of 125,000 ha. Yield losses of wheat were 110,000 t.[94]

5.5.3 Diseases and Insect Pests on Corn

Corn is China's third most important cereal. Although many diseases and pests of rice and wheat also affect corn, here we present information on four diseases and insect pests of increasing importance.

5.5.3.1 Northern and Southern Corn Leaf Blight

Northern corn leaf blight, also called large corn leaf blight, is caused by the fungus *Setosphaeria turcica*. This pathogen affects the leaves or outer husks of corn, but not ears.[95] Lesions range from 2 to 15 cm in length and 1–3 cm in width. The same fungus causes similar but smaller and darker lesions on sorghum.[96] Northern corn leaf blight disease development is favored by moderate temperatures (18–27°C) and heavy dews during the growing season. Disease progression is retarded by dry weather. Spores produced on leaf lesions are wind borne over long distances to leaves of maize plants.[97] In 2003 and 2004, northern corn leaf blight epidemics affected both the northern spring corn region and the southwestern corn region. In 2004, epidemics of northern corn leaf blight also occurred in the summer corn regions of the Yellow, Huai and Hai Rivers due to cool, rainy weather.[98]

---

[92] Dong and Jiang, *Wheat Diseases, Insects and Weeds and Their Control*, 2007, 37–40.

[93] Ibid.

[94] Ibid.

[95] M. C. Shurtleff, *Compendium of Corn Diseases*, 2nd edition. St. Paul, MN: APS Press, 1980, 16.

[96] Agrios, 2005, 468.

[97] Shurtleff, 1980, 16.

[98] X. M. Wang, Q. M. Jin, J. Shi, Z. Y. Wang and X. Li, "The Status of Maize Diseases and the Possible Effect of Variety Resistance on Disease Occurrence in the Future" (in Chinese), *Acta Phytopathologica Sinica*, Vol. 36, no. 1 (2006), 2.

Southern corn leaf blight, also called small corn leaf blight, is caused by the fungus *Cochliobolus heterostrophus*. This pathogen causes numerous small (0.6 by 2.5 cm) lesions on leaves. Some races also attack stalks, leaf sheaths, ear husks, shanks, ears, and cobs.[99] A black, velvety mold may cover the affected kernels. Seedlings from infected kernels may wilt and die within 3–4 weeks after planting. Early death of leaves predisposes the plants to stalk rot.[100] Southern corn leaf blight prefers a warm (20–32°C), damp climate. In China, southern corn leaf blight is a disease focused on summer corn growing in the corn-belt of the Yellow, Huai and Hai rivers.

In 2005, epidemics of northern and southern corn leaf blight occurred in the Northeast and Huabei regions. Areas of infestation exceeded 5.7 million hectares, an increase of 1.4 million hectares. Yield losses were 500,000 t.[101]

5.5.3.2 Asian Corn Borer and Yellow Peach Moth

The Asian corn borer (*Ostrinia furnacalis*) is one of the most destructive insect pests on maize in China. In southern China, especially in the Pearl River Delta region, where multiple generations (many overlapping) develop each year, the Asian corn borer is a limiting factor in the production of sweet corn.[102] The corn borer damages corn when larvae tunnel into the corn stalk, disrupting vascular tissues and interfering with the internal transfer of water, sugars, and nutrients. Additionally, some infectious diseases establish themselves after borer damage. The strength of the corn stalk and ear shank can be compromised, resulting in severe lodging and ear drop. The Asian corn borer has a fairly wide host range, including maize, sorghum, sugarcane, ginger, millet, Indian hemp, wild grasses, bell peppers, cereals and wormwood (*Artemesia*). The pest is a major problem where there is continuous cultivation of hosts throughout the year. In temperate regions, the corn borer becomes a problem in reduced tillage situations as larvae are able to over-winter in crop stems and debris.

In China, the Asian corn borer causes a loss in yield from 6 to 9 million tons in an average year. In 2005, corn borer infested areas covered 14.5 million hectares, which was an increase of 1.2 million hectares from the previous year. Infestation by the corn borer was especially severe in the northeast region.

The yellow peach moth (*Dichorcrocis puncrtiferalis*) is an omnivorous insect pest affecting more than 40 plant species, including corn, other grains, beans, fruit, nuts, and trees. In China, this moth produces two to three generations in the North and up to five generations in the western and southwestern regions. It over-winters

---

[99] Agrios, 2005, 466.

[100] Shurtleff, 1980, 15–16.

[101] *2006 China Agriculture Yearbook*, 2006, 378.

[102] D. S. Li, S. H. Huang, B. X. Zhang, Y. J. Wang and H. L. Li, "The Seasonal Occurrences of *Ostrinia furnacalis* and Its Egg Parasitic Wasps on Sweet Corn in Pearl Delta Region" (in Chinese), *Acta Phytopathologica Sinia*, Vol. 34, no. 2 (2007), 173–76.

in the debris of corn, sorghum, sunflowers or in the cracks of trees, soil and rocks. Larvae of yellow peach moths bore into the stalk and ears of corn and cause stalk lodging, rotten ears as well as production of poor quality corn (low kernel weight). For example, in Yibin county of Sichuan province, infestation of fall corn crops ranged from 30% to more than 80%. The overall reduction of yield has been in the range of from 20% to more than 30%.[103] In recent years, the yellow peach moth has become increasingly prevalent on summer and fall corn crops; the damage it causes in some regions surpasses that caused by the Asian corn borer.

5.5.4 Potato Diseases and Insect Pests

China is the largest potato producing country in the world. It is also the world's largest potato consumer. Potatoes are now ranked fourth after rice, wheat and corn. As the areas of potato cultivation increase, the importance of many potato diseases and insect pests emerges as an issue of plant protection. We consider several of these diseases.

5.5.4.1 Potato Late Blight Disease

Late blight, caused by the fungus *Phytophthora infestans*, is one of the most destructive diseases of potatoes (and tomatoes) worldwide. *P. infestans* is an obligate parasite and affects all parts of the host plant; it penetrates into and spreads quickly in plant tissues. The affected tissues quickly turn black and die. Under favorable conditions of high humidity (near 100%) and moderate temperatures (10–24°C), a field of potatoes can be destroyed in 5–7 days. Late blight disease was the cause of potato crop failure in the 1840s, resulting in the Great Irish Famine. Prior to the mid-1980s, late blight was successfully managed in most countries through the use of resistant varieties and effective chemical fungicides. In the 1980s, after the emergence of *P. infestans* isolates resistant to metalaxyl, a commonly used systemic fungicide, and the immigration of the *P. infestans* A2 mating type to other countries from Mexico, late blight again became a serious threat to the potato industry globally. Hybridization of the immigrant A2 with the local A1 resulted in rapid genetic exchange and vastly improved genetic diversity of the *P.infestans* population. Commercial potato varieties have no resistance to A2 and most of the new *P. infestans* isolates. Moreover, A2 and many of the new isolates are also highly resistant to chemical fungicides; they are more resilient than A1 to high and low temperatures, making them more difficult to control.

[103] Z. Y. Wang, K. L. He, H. Shi and S. Y. Ma, "Reasons for the Increasing Damage of Peach Stem Borer on Corn and Control Tactics," (in Chinese), *Plant Protection*, Vol. 32, no. 2 (2006), 67–69.

In China, late blight is a limiting factor in potato production. In the 1950s' epidemic, potato yield reduction in northern China was from 30% to 50%. In the 1960s, as resistant cultivars were developed, the damage caused by late blight on potatoes declined sharply. In the late 1980s, late blight again became a serious disease in all potato growing areas.[104] The presence of the A2 mating type was first noted in 1996, and strains resistant to metalaxyl were first confirmed in 1998. Presently, China has a very wide-spread *P. infestans* population; many isolates from Yunnan, Sichuan and Chongqing in the southwestern regions are closely related to those from Gansu, Inner Mongolia, Heilongjiang and Jilin in the western, northern and northeast regions. The practice of moving seed potato tubers across regions and the exchange of breeding materials among breeders explains the homogenization of *P. infestans* isolates.[105] Most commercial potato cultivars are susceptible to *P. infestans*. Hezuo 88 (Cooperate 88), one of the very few late blight resistant cultivars widely grown in southwest China, has now become susceptible to this disease.

5.5.4.2 Bacterial Ring Rot

Bacterial ring rot, caused by the bacterium *Clavibacter michiganensis* subsp. *sepedonicus*, is one of the most important diseases on potatoes. This bacterial pathogen causes wilt of leaves and stems and rot of tubers. Ring rot bacteria survive in infected tubers. This pathogen is extremely contagious. Cutting and planting of potato tubers provide a vehicle for the spread of this disease in the field. The transportation of potato tubers provides a long distance means of disease dissemination. Bacterial ring rot is quarantined by many countries.[106]

In China, ring rot was first found in the 1950s in Heilongjiang Province. In the 1970s, the disease spread to Qinghai, the Beijing municipality and other provinces. Because of the movement of seed potatoes across different regions and the exchange of breeding materials, bacterial ring rot now is present in all potato growing areas in China.

5.5.4.3 Potato Viruses

Potato degeneration, caused by various viruses, is a major problem on potatoes in all potato production regions. Viruses infections have led to yield reductions of

---

[104] Z. M. Zhang, K. Q. Cao, H. Zhang, X. M. Guai and Z. H. Yang, "Research Progress on China Potato Late Blight Epidemics and Forecasting" (in Chinese). In Y. L. Chen and D. Y. Chiou, eds., *China Potato Research and Industry Development*. Harbin, China: Harbin Engineering University Press, 2003, 31.

[105] X. Q. Zhu, Y. H. Wang and L. Y. Guo, "Genetic Diversity Revealed by RAPD Analysis among Isolates of *Phytophthora infestans* from Different Locations in China," *Acta Phytopathologica Sinica*, Vol. 36, no. 3 (2006), 249–58.

[106] W. R. Stevenson, R. Loria, G. D. Franc and D. P. Weingartner, *Compendium of Potato Diseases* 2nd Edition. St. Paul, MN: APS Press, 2001, 9–10.

from 15% to 90%. Potato virus diseases are especially prevalent in the South, because environmental conditions are more conducive to the transmission of the disease. For example, in Yunnan Province, yield of the Zhongdianhong cultivar declined significantly, due to virus infection.[107]

Viruses cause systemic infection in host plants. Potatoes are asexually propagated by potato tubers. Viruses carried by the mother plant are transmitted to the cells of the progeny directly through contaminated seed potatoes. In addition, many potato viruses can be transmitted by contact as well as by aphids (the green peach aphid, *Myzus persicae*, is the most proficient) and other insect vectors. In China, there are more than ten important potato viruses.[108] Potato virus X, potato virus Y and spindle tuber viroid diseased potatoes were detected in potato growing areas in the 1950s. Since the 1970s, following the introduction of potato germplasms from other countries, and the improvement of detection techniques, a large number of additional potato viruses have been identified: potato virus S, potato virus M, potato aucuba mosaic virus, potato virus A, tobacco rattle virus (transmitted by *Paratrichodorus* and *Trichodorus* spp., stubby root nematodes), tobacco mosaic virus, alfalfa mosaic virus, cucumber mosaic virus, potato yellow dwarf virus and beet curly-top virus.[109] Recently, potato virus A has had a serious impact on potato production.[110]

5.5.5 Important Soybean Diseases and Pests

Soybean aphids, the soybean borer (*Leguminivora glycinivorella*) and stem borer (*Loxostege sticticalis*) can cause serious yield losses. However, soybean rust, together with soybean mosaic virus, sclerotina stem rot and the soybean cyst nematode (*Heterodera glycines*) are the three most important diseases and pests affecting soybean production in China.

5.5.5.1 Soybean Rust

Soybean rust is caused by the fungus *Phakopsora pachyrhizi*. Spores of *P. pachyrhizi* infect the leaves of soybeans and cause severe defoliation. *P. pachyrhizi* prefers warm,

---

[107] Z. K. Zhang and Y. Li, Eds., *Yunnan Plant Viruses* (in Chinese). Beijing: Science Publisher, 2000, 77–117.

[108] CAAS, *China Diseases and Insects on Agricultural Crops*, 1995, 529–34.

[109] Z. F. Li, Ed., *Pictorial Identification of Major Potato Viruses in China* (in Chinese). Beijing: China Agriculture Press, 2004, 7-116. See CAAS, *China Diseases and Insects on Agricultural Crops*, 1995, 529–34.; also see Z. K. Zhang, Q. Fang, Z Q. Wu, Y. K. He, Y. H. Li and H. Q. Huang, "The Occurrence and Distribution of Main Solanaceae Crops Viruses in Yunnan" (in Chinese), *Journal of Yunnan University*, Vol. 20, no. 2 (1998), 128–31.

[110] X. Q. Wu, S. H. Chen, G. B. Wei, Z. J. Wu and L. H. Xie, "Molecular Identification and Detection Method of Potato Virus A" (in Chinese), *Journal of Agricultural Biotechnology*, Vol. 12, no. 1 (2004), 90–95.

humid conditions, and this disease is most severe in the southern soybean production area (south of 27°N). Yield losses caused by soybean rust range from 10% to 30% in moderately infected fields and up to 50% when there is heavy infection.[111]

In 1914, soybean rust was first observed in Taiwan and in 1938 in Guizhou. Today, it is found in 23 provinces and districts. Rust also was reported in the north-east region, but damages were insignificant.[112]

5.5.5.2 Soybean Aphid and Mosaic Virus

The soybean aphid, *Aphis glycines*, is the insect most seriously threatening soybean productivity. It is found in all soybean production areas between 18 and 52°N and 75 and 135°E, except for the northern and Tibetan plateaus, where it is too cold or dry for soybean growth.[113] This insect is especially prevalent and severe in the Northeast, Huabei and Inner Mongolia.[114] Soybean aphids feed on the soybean by sucking sap from the plant, which results in wrinkled and distorted foliage, early defoliation, underdeveloped roots, shortened stems and leaves, reduced branch numbers, lower pod and seed counts, reduced seed weight and plant death.[115] Under especially heavy infestations, reduction of soybean yield has reached from 50% to 70%.[116] In 1998, a heavy aphid infestation occurred in the Suihua prefecture of Heilongjiang Province, causing a 30% yield loss and a total reduction of 112,500 t.[117] In addition to direct infestation, *Aphis glycines* is an important vector of the soybean mosaic virus, a major soybean disease.

Soybean mosaic virus (SMV) is the most important and widespread of the seven viruses infecting soybeans in China. SMV is readily transmitted by sap, aphids and by seed. Diseased plants are short, stunted and have fewer root nodules. Seeds produced in diseased pods are usually smaller and poorer in quality than those from healthy pods.[118]

---

[111] Y. K. Xu, C. J. Li, D. Zhao, X. M. Liu, F. J. Pan and H. Wang, "Current Status and Advances in Soybean Rust" (in Chinese), *Plant Protection*, Vol. 32, no. 4 (2006), 9–10.

[112] Ibid.

[113] Z. S. Wu, D. Schenk-Hamlin, W. Y. Zhan, D. W. Ragsdale and G. E. Heimpel, "The Soybean Aphid in China: A Historical Review," *Annals of the Entomological Society of America,* Vol. 97, no. 2 (2004), 209–11.

[114] H. P. Xin, Ed., *Soybean Diseases and Insects, Their Protection and Pictorial Guide* (in Chinese): Beijing: China Agriculture Press, 2003, 47–49.

[115] Z. S. Wu, D. Schenk-Hamlin, W. Y. Zhan, D. W. Ragsdale and G. E. Heimpel, "The Soybean Aphid in China: A Historical Review," *Annals of the Entomological Society of America,* Vol. 47 (2004), 209–11.

[116] Ibid.

[117] B. Sun, S. B. Liang and W. X. Zhao, "Outbreak of the Soybean Aphid in Suihua Prefecture in 1998 and its Control Methods," (in Chinese), *Soybean Bulletin* , no. 1, (2000), 5.

[118] J. B. Sinclair and P. A. Backman, *Compendium of Soybean Diseases,* 3rd ed. St. Paul, MN: APS Press, 55–57.

Both of the soybean aphids and SMV are sensitive to climatic conditions. The incubation period and symptom expression of SMV are greatly affected by temperature. The optimal temperature for SMV is 20–30°C. At 29.5°C, the incubation period of SMV is 4 days compared to 14 days at 18.5°C.[119] The most favorable temperature range for aphid reproduction is 20–24°C, with an optimal humidity of 78% or less.[120] Furthermore, warm and dry weather also facilitate aphid migration and thus the dissemination of the SMV virus.[121]

5.5.5.3 Sclerotinia Stem Rot

Sclerotina stem rot, caused by the fungus *Sclerotinia sclerotiorum*, is also called white rot disease. It has an extremely broad host range and is found in all regions. In the 1960s, it was found mostly in the eastern portion of Heilongjiang Province. After 1990, the expansion in acreage of sunflowers, rapeseeds, beans and flexes (all of which host *S. sclerotiorum*), the severity of this disease in soybean fields increased. *S.sclerotiorum* produces large numbers of schlerotia, which are highly resistant to chemical fungicides and can remain viable in soils for more than 50 years. In 2002, because weather conditions in Heilongjiang were cool and rainy, the soybean crop was heavily infected by *S. sclerotiorum*, and the infestation rate exceeded 50%.[122] In 2005, outbreaks of Sclerotina stem rot occurred again in Heilongjiang; the area of infestation, 85,300 ha, was the largest and heaviest in history.[123]

5.5.5.4 Soybean Cyst Nematode

Soybean cyst nematode (SCN) is caused by *Heterodera glycines*, which is a destructive pest on soybeans. *H. glycines* infestation severely interrupts the development and functioning of the root and root nodules, resulting in stunting and ultimately death of the plant. This nematode prefers sandy and alkaline soils; its reproduction is suppressed in acid soil (pH 5.0 or lower). The nematode cysts are highly resistant to drought, flood, cold and salinity, and also they are insensitive to microbial activities. The cysts of SCN can survive in soils from 3 to 4 years up to more than 10 years.[124]

The distribution of SCN matches the premier soybean production regions, but the pest is most prevalent in the Northeast and the Yellow and Huai river valleys.

---

[119] Ibid.

[120] Xin, *Soybean Diseases and Insects,* 2003, 47–49.

[121] Ibid., 27.

[122] Ibid. 19–21.

[123] *2006 China Agriculture Yearbook*, 2006, 378.

[124] Xin, *Soybean Diseases and Insects*, 2003, 28–29.

SCN damage is especially severe in the northeast region where large tracts of land are sandy and alkaline in nature. In these soils, SCN produces three to four generations during a growing season.[125]

5.6 Plant Pathogens Affecting Crop Yield and Human/Animal Health

Generally, plant pathogens affect only the health of plants and quality and quantity of their products. In the following cases, plant pathogens at the post-harvest stage may also produce, because of inappropriate storage conditions, large quantities of toxins which can seriously affect the health of humans and animals.

5.6.1 Rice False Smut

False smut is a re-emergent disease in China and worldwide. It is caused by the fungus *Ustilaginoidea viren*, and it affects primarily the grains of rice. After infection occurs at the stage of flowering, the fungus transforms individual grains of the panicle into greenish spore balls.[126] *U. viren* is most commonly found in rice production areas of the South, where it is warm and humid. In the northeastern region, false smut occurs often in Liaoning, yet outbreaks of false smut have occurred in Heilongjiang Province recently. This disease not only severely affects yields, but also affects the quality of the grain. The fungus produces spores that are toxic to humans, other animals, and birds.[127] In 2005, the area of infestation was 2.6 million hectares, encompassing rice regions in the middle and lower reaches of the Yangtze, and the southern portion of the rice region in the Northeast, and northern rice regions of the Southwest. Definitely, it is a serious problem on late rice.[128]

5.6.2 Wheat Scab

Wheat scab is also called head blight, and is caused by a group of fungi. The primary causal agent is *Gibberella zeae*. (Other causal agents are *G. avenacea*, *Fusarium*

[125] L. J. Chen, Y. Zhu, B. Liu and Y. X. Duan, "Influence of Continuous Cropping and Rotation on Soybean Cyst Nematode and Soil Nematode Community Structure" (in Chinese), Acta Phytopathologica Sinica, Vol. 34, no. 4 (2007), 347.

[126] Ou, *Rice Diseases*, 1972, 289–93.

[127] Ibid. See also Xin, 2003.

[128] *2006 China Agriculture Yearbook*, 2006. 377.

culmorum and *Monographella nivalis*.[129]) This disease occurs primarily in the wheat regions of the middle and lower reaches of the Yangtze River; however, recently it has become prevalent in wheat regions of northern China.

In 2003, an epidemic of head blight occurred in the wheat production areas of central and southern Hebei Province and caused white heads in wheat and a large scale yield reduction; it also affected the quality of the grain.[130] Since then, infestation of scab has been less severe. In 2005, wheat scab adversely influenced wheat productions in Hebei, Anhui, Henan, Hubei and other provinces and resulted in yield losses of 200,000 t.[131]

Pathogens causing wheat scab prefer humid and rainy weather. Scabs affect flowers of cereal crops and cause blighting of spikelets, producing "white head" and leading to premature death of the plant. Warm (25–30°C) temperatures and humid (continuous moisture) conditions are nearly ideal for development of scabs in wheat and other cereal crops[132] In addition to yield losses, grains produced from head blighted fields may contain mycotoxins that induce muscle spasms and vomiting (vomitoxin) as well as food poisoning in human and non-human animals.[133] The toxins usually remain stable for years in stored grain. Grains are not edible when diseased kernels exceed 4%.[134]

5.6.3 Maize Ear, Kernel Rot and Post-harvest Diseases

Maize is susceptible to a large number of ear and kernel rots caused by several fungi. These rots cause considerable damage to corn crops in humid areas, especially when rainfall occurs during the period from silking to harvest. Ear and kernel rots can reduce yield, quality and feed value of grain. Since the 1980s, yield loss and grain quality declines due to ear rot have increased, especially in Northern China. More than ten fungi species, including several species of *Fusarium*, *Aspergillus* and *Penicillium*, were identified.[135] The incidence of ear rot ranged from 10% to 20% in most years but from 20% to 40% at the time of outbreaks.[136]

In addition to causing yield losses which affect food security, *Aspergillus*, *Penicillium* and *Fusarium* (*Gibberella*) also have significant impacts on food safety,

---

[129] Weise, *Compendium of Wheat Disease*, 1987, 16–17.

[130] Dong and Jiang, *Wheat Diseases, Insects and Weeds and Their Control*, 2006, 7–9.

[131] *2006 China Agriculture Yearbook*, 2006, 377.

[132] Weise, *Compendium of Wheat Disease*, 1987, 16–17.

[133] Ibid.

[134] Dong and Jiang, *Wheat Diseases, Insects and Weeds and Their Control*, 2006, 7–9.

[135] S. F. Xu, J. Chen, Z. G. Gao, Q. D. Zou, M. S. Ji and H. N. Liu, "Maize Stalk Rot and Ear Rot in China," *Acta Phytopathological Sinica*, Vol. 36, no. 3 (2006), 193–203.

[136] L. S. Kong, and P. C. Luo, "Identification of Pathogen of Corn Ear Rot and Its Pathogenicity" (in Chinese). *Corn Science* (1995), Supplement: 29–31.

because of their ability to produce metabolites under humid conditions that are toxic when consumed by humans, animals, fish, and birds. Mycotoxins can accumulate in maturing corn in the field and in grain during transportation and storage. Among the mycotoxins, aflotoxins produced by *Aspergillus* spp. are the most potent. Aflatoxin B1, present in corn, is among the most potent of known carcinogens and causes fatal liver cancer. Experiments conducted in China by Groopman et al. (1996) showed a high incidence of hepatitis B virus infection when dietary exposure to aflatoxins was prevalent; aflotoxins and hepatitis B virus act synergistically in causing liver cancer.[137] Mycotoxins produced by *Penicillium* spp. can cause hemmorage of the lung, brain, and liver and damage to the kidney.[138] Mycotoxins produced by *Fusarium* spp. cause vomiting (vomitoxin), diarrhea, hyperestrogenism and infertility in animals.[139]

The effects of mycotoxins on humans and animals range from loss of appetite, food refusal and decreased food efficiency to mortality.

5.7 Effects of Social and Environmental Changes on Diseases and Pests

In China, food security has been a primary issue for policy makers in each of the 13 dynasties, in the Republican era, and after 1949 in the communist government. The need of the government to insure survival of the population put enormous stress on the land, at the expense of sustainability of soils and health of ecosystems. Many plant diseases and insect pest problems Chinese encounter today are the consequences of past policies and practices.

5.7.1 Agricultural Cultural Practices

China has one of the most complex cropping and cultivation systems in the world. To feed the world's largest population requires maximum efficiency in the utilization of farm land. Small plots and cheap labor make intercropping possible. The intercropping of rice-lotus, rice-tea, corn-rapeseed, corn-mung bean, corn-peanuts, wheat-saddle bean, wheat-rapeseed, wheat alfalfa, cotton-rapeseed, and potato-garlic

---

[137] A. S. Bommakanti and F. Walliyar, "Aflatoxin: Importance of Aflatoxins in Human and Livestock Health," *International Crops Research Institute for the Semi-Arid Tropics*, 2000, 1–6. See also J. D. Groopman and W. Kensler, "Temporal Patterns of Aflatoxin-ablumin Adducts in Hepatitis B Surface Antigen-positive and Antigen-negative Residents of Daxi, Qidong County, People's Republic of China," *Cancer Epidemiology Biomarkers and Prevention* (1996), 253–61.

[138] Shurtleff, *Compendium of Corn Diseases*, 1980, 57–58.

[139] Ibid.

has brought benefits in disease and insect pest control by reducing the inoculum/population density of plant pathogens; it has encouraged beneficial natural predators of insect pests.

However, the practice of growing generations of one crop in close proximity to one another, multiple cropping, and inter-cropping without careful consideration of diseases, inadvertently provide a fertile ground for many pathogens and insect pests by providing them continuous hosts to sustain their survival. For instance, many potato viruses can also affect tobacco. Inter-cropping of potatoes and tobacco understandably exacerbates virus disease problems for both crops and should not be encouraged.

In order to achieve maximum yield, Chinese farmers often rely on the application of a high rate of nitrogen to promote plant growth. However, excessive nitrogen use weakens plants' natural defense mechanism to pathogens and insect pests because plants become more succulent.

Planting semi-dwarf crops has many advantages. Not only can this reduce use of water and fertilizers, but the short stature of the plants also makes it possible to reduce spacing between rows, and it increases efficiency of land use. However, many semi-dwarf crops have low levels of resistance to plant pathogens which make plants more prone to severe damage. Moreover, close spacing also alters the microclimate by increasing humidity of the plant canopy, making the microenvironment more conducive to plant diseases.

Hybrid rice has been credited for the recent dramatic increase of rice production in China. However, it has also contributed to an increase in the incidence and severity of rice leaf blight and bacterial leaf streak, because the rice varieties selected for hybrid rice lack resistance to bacterial diseases.

5.7.2 Government Policy

In the past 50 years, the Chinese government has instituted many policies in order to expand arable land (see Chapter 3). Many of these policies have had inadvertent consequences. For instance, Luliang county, a major rice and potato producing area in Yunnan Province, suffers a chronic shortage of water even though it is located in an area with sufficient rainfall. The reason is that the farmland today once was a huge lake which was filled by leveling nearby mountains during the Maoist era.

Expansion of paddy rice cultivation to Heilongjiang Province was touted as a great agronomic accomplishment by the central and provincial governments. However, intensive, large-scale paddy rice cultivation exacerbated damages caused by rice sheath blight from previously isolated incidents to epidemic proportions.

The Maoist policy on conversion of forests to farmland by large scale deforestation has had serious impacts on water conservation. The dry climate produced a high incidence and increased severity of migratory locust damage.

5.7.3 Improvement of Domestic Transportation

In the past 20 years, the Chinese government has promoted construction of highways and other transportation infrastructure. Consequent to the improvement of ground transportation, goods and products easily can be transported from one part of China to another. Communities benefited from this improved accessibility, yet the reduction of isolation also facilitated the movement of plant pathogens and pests throughout China (and also the movement of invasive species, as noted in Chapter 6).

Recently, the practice of increasing seed production (of hybrid rice, vegetables, melons) in Hainan and transporting these seeds to the North makes long distance dissemination of plant pathogens possible. The re-emergence of rice false smut and bacterial diseases as major diseases is due largely to transporting hybrid rice from Hainan to other parts of China. The phenomenon of long distance dissemination of plant diseases from Hainan also applies to many vegetable and melon diseases.

5.7.4 Climate Change

It is well known in China that migratory locust epidemics are closely associated with weather patterns. The explosion of the migratory locust population is expected in a year of drought following a year of floods.

Temperature, humidity and sunlight are important environmental factors affecting the growth, development and survival of plant pathogens and insect pests. Under favorable environmental conditions, the increasing incidence and severity of diseases caused by these pathogens and insect pests on susceptible hosts will increase. Changes in temperature can also affect the symptom expression of diseases on host plants. For viruses and potato bacterial ring rot, a diseased plant may not display symptoms if the temperatures are too high or too low. This makes proper disease diagnostics difficult.

Each of the plant pathogens has its favored range of temperature and humidity. For instance, sheath blight, rice bacterial blight, rice bacterial leaf streak and southern corn leaf blight diseases prefer hot and humid conditions; rice blast and northern corn leaf blight also require high humidity but prefer more moderate temperatures. Soybean rust, wheat rust and wheat powdery mildew all prefer low humidity, but wheat rust and powdery mildew prefer cool temperatures.

It is to be expected that the change of weather pattern or climatic condition also affects the dominance of species. In a microbial community of mixed populations, the plant pathogens which have the greatest fitness under that environmental condition will become dominant. The phenomenon of replacement has been observed on rice (rice blast replaced by rice sheath blight) and corn (Asian corn borer replaced by yellow peach moth). Replacement by virulent races/strains also has been observed on rice blast, wheat rust and potato late blight diseases.

In addition, weather facilitates the dissemination of plant pathogens and insect pests. Soybean rust, wheat strip rust, mites, greenbugs, aphids and planthoppers rely on dry wind for dissemination locally and on the jet-stream for long distance dissemination. Scientists recently have produced evidence showing that planthoppers of China came from Vietnam on the jet-stream.

5.8 Control Measures

The plant diseases and pests we have introduced in this chapter present daunting challenges to the Chinese state. No nation of the world has an effective "pathogen/pest control regime," because pests and pathogens, parts of a hugely complex network of ecosystems, are constantly evolving and adapting to changes in the natural and human environment. Nevertheless, China has a plentiful array of mitigation measures, and these have provided food security to the state in the near-term.

We group control measures used in China into six broad categories: regulatory, cultural, breeding for disease resistance, biological, physical and chemical.[140] Regulatory measures refer to attempts to exclude a pathogen or pest from a host or a particular geographic area. Cultural control measures refer to changes in environmental conditions so that they become inhospitable to the pathogen or pest, including the reduction or eradication of pathogens/pests in a plant, a field, or an area. Use of genetic manipulation in plant breeding also is important in producing disease-resistant varieties of seeds. Biological controls attempt to improve the health of plants by applying microorganisms or predators that are antagonistic or predatory to the pathogens and pests. Physical and chemical control measures attempt to protect plants from the pathogen inoculum or pest, or to cure an infection that already has occurred. We provide examples of each control measure.

The use of these control measures depends on knowledge of plant diseases and insect pests, and in this respect agricultural science in China is still catching up with developments in the West. In Chapter 2 we discussed the "10 lost years," when no attention was paid to scientific research; for some diseases, there is a 40 year gap of information. While university professors and agricultural institute researchers today do cutting-edge research, overall knowledge of plant diseases and insect pests is incomplete and dissimination of information to agricultural workers and farmers is disjointed. Nevertheless, surveys of diseases are becoming more common. GPS is widely used in disease surveys; and disease forecasting has made strides. Forecasting, of great potential value to agricultural workers and farmers, has been used on locusts, but in the monitoring of other diseases and pests is still theoretical; it has not yet been broadly disseminated for the practical management of crop diseases. It involves modeling or computer simulations, as for

[140] Agrios, 2005, 293–353.

example forecasting the race dynamics of plant rusts.[141] Researchers developed a model of interaction between multi-cultivars and multi-races of wheat stripe rust and incorporated in the model historical data analysis of cultivar resistance to different races of the pathogen. Continued enhancement of models and simulations holds promise for eventual improvements in farm cultural practices and disease control, as does monitoring weather factors affecting disease development.

As we introduce the different control measures, we cannot emphasize strongly enough that each has strengths and weaknesses, and that no single means is likely to be efficacious in all cases.

5.8.1 Regulatory Controls

Like most nations, China uses phytosanitary standards and quarantines to prevent movement of plant pathogens and insect pests into its borders and onto its fields. Too, China sends plant inspectors abroad to guarantee the quarantined disease- and pest-free status of agricultural imports, and to insure that they are in accordance with bilateral agreements. However, China seems to achieve more success in preventing introduction of unwanted pathogens and insect pests from foreign countries than in curtailing once introduced pathogens and pests from crossing municipal boundaries. For instance, pine wilt nematode (PWN) which is a pathogen quarantined by the Ministry of Agriculture was found in China in 1981; it has spread steadily through PWN-infested wooden crates and packaging materials. However, information on PWN is not freely published in the media. Discussion, research and publication of the status of PWN in China is restricted by the government. This restriction also applies to other diseases and insect pests of a sensitive nature, as in PWN. Concerning diseases and insect pests of a less sensitive nature, agricultural inspection agents from the national to the local levels conduct inspections of plants in farms as well as in storage facilities.

5.8.2 Cultural Control Measures

This disease control category includes a broad and diverse collection of production practices such as crop rotations, soil tillage, changing the time of planting and plant spacing, providing soil drainage and making other irrigation system changes and improving sanitation. Significant advances have been made in controlling plant

---

[141] Yue-yan Xiao, Li-ren Wu, Jia-huai Hu, An-min Wan and Wen Chang, "Forecast of Race Dynamic of Wheat Stripe Rust," *Acta Phytopathologica Sinica*, Vol. 34, no. 3 (June 2007), 257–62.

diseases and insect pests by mixture planting and inter-planting of crops.[142] Also, increasingly, scientists have paid attention to ecological regulation strategies against crop pests, and to biodiversity-rich cropping practices.[143]

A large number of cultural practices have been used to control plant diseases, and a specific example is the mitigation of rice blast. One suggested means is to reduce the use of nitrogen and increase the potassium/nitrogen ratio of the soil. A second approach is to regulate irrigation through the growth stages of the rice plants. Prior to seed germination, the seeds are submerged in water, which inhibits the germination of rice blast spores. During the vegetative stage, the roots of the plant seed are maintained in shallow water that is frequently replenished. At the reproductive stage, draining most of the water from the field stimulates aeration of the soil and regulates nitrogen uptake to promote root development and enhance disease resistance.

Polyethylene mulching is used extensively in China for the purposes of raising soil temperature, conserving water, and eradicating many soilborne pathogens. Tunnels of various heights are used to protect plants from infection by airborne pathogens, and also to conserve heat and water.

5.8.3 Plant Breeding for Disease Resistance

Historically, Chinese farmers developed plant breeding programs based on the selection of plant lines increasing crop production, and through happenstance, some of these had disease-resistance properties. Breakthroughs in the science of genetics in the twentieth century made the breeding of resistant varieties systematic, and agricultural stations disseminated the seeds of these varieties. In the reform period, sales from seeds and chemical pesticides became major income sources supporting operations of many provincial or county agricultural stations. Because of keen competition and incomplete intellectual property protection, new varieties of crops are released almost every year, and some of them are not well tested. Recently, awareness of damage to the environment through intensive use of chemical controls also has increased interest and government support of programs breeding plants for disease resistance.

Given a choice, breeders and farmers usually select varieties possessing traits such as high yield, early maturing and semi-dwarf and not disease resistance,

[142] Jun-hui Shen, Qin Nie, De-run Huang, Guang-jie Liu and Long-xing Tao, "Recent Advances in Controlling Plant Diseases and Insect Pests by Mixture Planting and Inter-Planting of Crops," *Acta Phytopathologica Sinica*, Vol. 34, no. 2 (April 2007).

[143] See also Youyong Zhu, "Cultivating Biodiversity for Disease Control, A Case Study in China," Yunnan Agricultural University, no date; Jia'an Cheng and Zengrong Zhu, "Analysis on the Key Factors Causing the Outbreak of Brown Planthopper in Yangtze Area, China in 2005," *Plant Protection*, Vol. 32, no. 4 (2006), 1–4; and Yuyuan Guo, "Illustrations with Real Examples of Using Ecological Regulation Strategies Against Crop Pests in China," *Plant Protection*, Vol. 32, no. 2 (2006), 1–4.

because diseases are perceived to be manageable by chemical pesticides and other means. Varieties of rice, wheat and maize now often are bred for resistance against important pathogens, such as rice blast, wheat rust, wheat powdery mildew and corn leaf blight, because chemical control is costly and not always effective. Research is advancing rapidly in the analysis of resistance reaction of rice germplasms to rice blast,[144] as well as many other pathogens. When rice blast and other pathogens, which possess great genetic variation, emerge, new virulent strains/races will cause a breakdown of resistance in crops; then new plant varieties resistant to the new virulent strains/races will be needed. In the case of rice blast, the average efficacy of a resistant cultivar is approximately 5 years. Hybrids developed primarily to increase yields also may have disease-resistant qualities.

5.8.4 Biological Controls

Biological control involves using one biological species against another. The concept that everything in nature is vulnerable and has natural enemies was recognized in ancient China; it is a fundamental principle of Taoism. Biological control has been part of the Chinese agricultural system since time immemorial. For instance, the practices of using ducks in controlling weeds in rice paddy fields and also using ducks and chickens to eat grasshoppers and locusts are still part of the pest management regime. The development of biological controls today involves using antagonistic microorganisms and predators to mitigate plant diseases and insect pests by encouraging competition for space and nutrients and by inhibiting, parasitizing and potentially eradicating pathogens and insect pests.

Many biological control agents have been developed and commercialized in China. Some are sold as biopesticides, others as soil amendments and growth enhancers. These products utilize the mycoparasitic and/or growth promotion characteristics of *Trichoderma atroviride*, *T. harzianum* and other species and isolates of *Bacillus cereus* and *Bacillus subtilis*. As for insect pests, the entomopathogenic fungus *Beauveria bassiana* and toxin producing bacterium *Bacillus thuringensis* (which paralyzes insects' digestive systems) are most common. In addition, natural enemies of insect pests such as aphids, Asian corn borer and locusts have been identified and cultivated, including egg, larval and pupal parasitoids.[145]

---

[144] Xiao-ping Yuan, Xing-hua Wei, Han-yong Yu, Yi-ping Wang and Sheng-xiang Tang, "Analysis of the Resistance Reaction of Some Chinese Rice Germplasms (*Oryza sativa L.*) to Rice Blast," *Plant Protection*, Vol. 31, no. 3 (2005), 27–34.

[145] See Rani et al., "Diversity of Rice Leaf Folders …," 2007; M. A. Legrand, H. Colinet, P. Vernon and T. Hance, "Autumn, Winter and Spring Dynamics of Aphid *Sitobion avenae* and Parasitoid *Aphidius rhopalosiphi* Interactions," *Ann. Appl. Biol*, Vol. 145 (2004); Li et al., "The Seasonal Occurrences of *Ostrinia furnacalis* …," 2007; and De-ying Ma, Ian Denholm, Kevin Gorman and Wan-chun Luo, "The Resistance Status and Management Strategies of *Bemisia tabaci* B Biotype in Xinjiang," *Acta Phytopathologica Sinica,* Vol. 34, no. 3 (June 2007).

For example, the egg parasitoid Trichogramma has been found to be effective in the control of the Asian corn borer,[146] and *Nosema asiaticus* is efficacious against the Oriental migratory locust[147] and other subspecies of this pest.[148] Other examples include the antagonistic bacterium *B. subtilis* B-322 strain, which has been shown to be efficacious in controlling rice blast.[149]

The advantage in use of biological controls is that they cause little or no damage to the natural environment or to human health. The disadvantage is that they are slow-acting. Similar to chemical pesticides, the efficacy of biological pesticides is varied. Yet due to environmental concerns and the growing number of organic farms, acceptance of biological control of diseases and insect pests has increased.

China is a leading nation in the development of agricultural biotechnology, and the regime has allocated huge resources to it. Today, about 70% of China's cotton producers use Bt cotton, which has a gene resistant to boll worms. Other transgenic species, such as Bt maize, have not yet been commercialized for reasons discussed at length in Chapter 7.

5.8.5 Physical Control Means

Physical control means such as heat or light conditions are used less frequently than chemical ones. Examples of physical controls are sterilization of soils by heat, hot water and hot air treatments. More prevalent are heat drying of grains and fruit in storage.

5.8.6 Chemical Control Measures

In China, chemical means are preferred by farmers because they produce fast results. However, increasingly highly toxic, systemic pesticides are applied; these not only endanger the environment, but more importantly they are harmful to

---

[146] See Zhenying Wang, Kanglai He and Su Yan, "Large-Scale Augmentative Biological Control of Asian Corn Borer Using *Trichogramma* in China: A Success Story," Second International Symposium on Biological Control of Arthropods, Davos, Switzerland, 2005.

[147] Yin Wang, Jing-zeng Wen, Yi-ping Shi, Bao-feng Zhou and Hong Huang, "Pathogenicity Studies of *Nosema asiaticus* on Migratory Locust." In Proceedings of 3rd National Young Plant Protection Worker Conference and Outlooks of 21th Century. Beijing: China Science and Technology Publisher, 1998, 632–47.

[148] Jing-Jeng Wen, She-ping Li, Bao-feng Zhou and Fu-lai Fan. "Pathogenicity of *Nosema asiaticus* on Two Subspecies of Grassland Locusts." In Proceedings of 3rd National Young Plant Protection Worker Conference and Outlooks of 21th Century. Beijing: China Science and Technology Publisher, 1998, 639–41.

[149] Chang-qing Mu, Xue Liu, Qing-guang Lu, Xi-liang Jiang and Chang-xiong Zhu, "Biological Control of Rice Blast by *Bacillus subtilis* B-332 Strain," *Acta Phytopathologica Sinica*, Vol. 34, no. 2 (April 2007), 123–28;

human and animal health. In China, it is widely believed that farmers do not eat the products they produce for the market.

Chemical pesticides are used to reduce pathogens before the inoculum reaches the plant, for example by fumigating soils. Also, they are sprayed on plant surfaces to protect them from infection. They are toxic to pathogens and inhibit germination, growth, and proliferation of the pathogen. Insecticides are applied to control insects and insect vectors (carriers) of viruses and phytoplasmas.

Fungicide seed treatment and foliar spray on tillers are the most common practices for addressing plant disease; however pathogens such as rice blast, stripe rust, powdery mildew and many others have developed resistance to chemical treatments, after continuous intensive use of chemical fungicides in controlling these diseases. As for pathogens that produce sclerotia such as rice sheath blight and soybean sclerotinia disease, the continuous application of chemical fungicides actually facilitates rapid escalation in population of these pathogens in soils, because sclerotia are resistant to chemical fungicides. Soybean stem rot disease is especially hard to control once it has become established in the environment. It not only produces enormous numbers of chemical fungicide-resistant sclerotia, but it also has an extremely wide host range, from vegetables to flowers, and this makes selection of rotation crops for soybeans very difficult. A similar phenomenon has been noticed in insect pests. Resistance to chemical insecticides has been observed on brown rice planthoppers, smaller brown planthopper, greenbug, wheat aphids, green peach aphids and other insect pests, due to frequent applications of insecticides (eight to ten times or more per crop season).

Chemical pesticides also may have varied impacts on pathogens and insect pests. For instance, triazophos, an organophosphate insecticide which is efficacious against Chinese rice grasshoppers, stimulates the reproduction of the rice brown planthopper and results in more severe damage to plants.[150]

Continued heavy use of pesticides while their impacts on pathogens and insect pests decline (an indication that these species are developing resistance) cannot be sustained. China already has the world's highest application rate per hectare of chemical pesticides. This has prompted interest in integrated pest management (IPM), which in many countries has allowed reduction of 50–80% usage if farmers are appropriately trained.[151]

One large obstacle to improvement of disease management at the local level is the inherent conflict between objectives of China's agricultural stations. These are outreach program offices with few books, little equipment, and no tools to conduct fundamental scientific work. Instead, they have become enterprises relying for revenue on the products they sell farmers. While they sell seeds, including those with some disease-resistance properties, they make more money from selling pesticides, which farmers buy because they produce fast results.

---

[150] Jia-an Cheng and Zengrong Zhu, "Analysis on the Key Factors Causing the Outbreak of Brown Planthopper in Yangtze Area, China in 2005," *Plant Protection,* Vol. 32, no. 4 (2006), 1–4.

[151] "Rice and the Asian Environment," *Pesticide News*, October 10, 2006, 3.

5.9 Conclusions

At the start of the chapter, we emphasized the role plant diseases and pests play in diverse ecosystems. They are major forces of ecological renewal. However, to humans they present serious challenges to agricultural production, and as we noted, they cause damages in the billions of yuan annually. While the exact costs of plant diseases and insect pests to the Chinese economy are unknown, estimates suggest that they reduce harvests at least by 10–15% in most years. It would be more advantageous for the Chinese government to increase funding for research in plant pathology and entomology and in mitigation methods that are environmentally benign, than to open marginal lands for cultivation.

Of the thousands of insect pests and plant diseases, we focused on those producing the most damage historically and in the contemporary period, giving ample space to the epidemics of migratory locusts, rice blast, and wheat rust diseases. Then we described briefly more than two dozen other diseases and pests, presenting them in the order of the staples they threatened: rice, wheat, maize (corn), potatoes, and soybeans. Also, we discussed plant pathogens affecting both crop yield and human/animal health, providing examples of rice false smut, wheat scab and other post-harvest diseases. We examined past cultural practices of farmers as well as government policies (including improvement of domestic transportation) which have impaired the sustainability of soils and health of ecosystems. These problem areas potentially will be compounded by broad-scale effects of climate change.

Then we introduced the six categories of mitigation methods used in China today: regulatory, cultural practices, breeding for resistant varieties, biological control, physical and chemical controls. Of course, these are not used in equal measures. Cultural practices were the primary means of defense in traditional China, and they still are used today, with effect. The tendency during the reform period, however, has been to use chemical controls and especially pesticides. Initially, these means significantly reduced insect pest and disease infestation, but with a very high cost to the environment and human health. Of importance as well is the developing resistance of plant pathogens and insect pests to chemical pesticides. More recently, scientists have promoted biological controls – the identification and cultivation of antagonists/predators to pathogens and insects, which in some cases have produced promising results.

In the next chapter, we treat invasive pests, and note the mostly similar methods used to control them. In Chapter 7, we return to the biological control process through detailed study of genetically-modified organisms.

Chapter 6
Invasive Species and Food Security

Abstract This chapter defines invasive species, noting that they may arrive in new areas either intentionally or unintentionally, and it discusses their economic and non-economic effects. It provides examples of plant, insect, fish, and other invasive species, which have impacts on food production. Immediate environmental stressors, such as socio-economic change, cause the unintentional transmission of invasive species. In recent years, improved domestic transportation systems in China as well as rapidly growing international trade and tourism have increased opportunities for invasive species to enter China, as has climate change. China recently has strengthened regulations to better protect against the spread of invasive plants and insects, and some scientists, influenced by global guidelines issued by the World Conservation Union, recommend using the precautionary principle to eliminate threats. To the present, four methods have been used to prevent or control harmful invasives: physical, biological, chemical, and integrated pest management. Scientists and government officials also are developing research, outreach, and international cooperation strategies to counter the growing invasive species challenge.

Keywords Invasive species • Water hyacinth • American white moth • Mosquitofish • Amazonian snail • World Conservation Union (WCU) • Global Invasive Species Program (GISP) • Precautionary principle • Integrated pest management

6.1 Introduction

In Chapter 5 we considered traditional plant pests and plant diseases, and their impact on China's food security. In this chapter we turn to external threats to food production and food security, those species (including diseases afflicting plants and animals) alien to China, which nevertheless influence the Chinese ecosystem.

Our discussion begins with a definition of terms and then a differentiation between species spread intentionally or unintentionally, and those with economic

J.H. McBeath and J. McBeath, *Environmental Change and Food Security in China*, Advances in Global Change Research 35, DOI 10.1007/978-1-4020-9180-3_6,

as opposed to non-economic effects. Next we consider the impacts of invasive species on food production, with examples of plant, insect, fish, and other species.

We consider those factors responsible for unintentional transmission of invasive species, including population growth, economic development, transportation improvements, and the broad category of changes included within the rubric of globalization (and particularly, international trade). We also treat the impact that global warming is having on the severity and distribution of invasive species problems in China.

Because invasive species have been recognized as a problem in China's environment and food system for little more than a decade, no regulatory "regime" has yet developed to comprehensively address the issue. Nonetheless, policy makers and administrators have adapted existing regulatory systems, and they also have been influenced by global recommendations.

6.2 Nature of Invasive Species

China covers a vast territory with highly diverse ecosystems and a wide range of native species. Its diverse bioregions and species megadiversity (the third most biologically diverse country in the world) make it particularly susceptible to invasive species. We define this critical term and then consider the nature of transmission and broad effects.

6.2.1 Definitions

An invasive species is one which is not native to the area in which it now lives. It may be introduced to the new area either through intentional or unintentional transmission. In its new home, the species has established a sufficient population to reproduce. The final part of the normal definition of an invasive species, which makes invasive species problematical, is that it brings about changes to the native ecosystem. The changes may be adverse either with respect to economic or non-economic values, such as threats to valuable crops or biodiversity.[1] In China, invasive species also are considered to include those native to one region in China, but are introduced to other regions in which they are foreign.

Alien invasive species can be found in each of China's 31 provinces and autonomous regions, in urban as well as rural areas, and in virtually every type of ecosystem. Although China's species catalog is not complete, invasive species are present in

[1] See Secretariat of the Convention on Biological Diversity, *Handbook of the Convention on Biological Diversity*. London: Earthscan Publications, 2001. For an interpretation of invasive species in the Chinese context see, among others, Yan Xie, Zhenu Li, William P. Gregg, and Dianmo Li, "Invasive Species in China – An Overview," *Biodiversity and Conservation*, Vol. 10, no. 8 (2000), 1318.

many taxonomic groups, for example birds, mammals, reptiles, fish, amphibians, crustaceans, algae, plants, fungi, viruses, bacteria, and other micro-organisms.[2]

6.2.2 *Transmission of Invasive Species*

Broadly, there are two classes of transmission: intentional and unintentional. In its long history, a large number of species were brought into China by those who saw their introduction as benefiting the food system, the landscape, economic development or for some other purpose. Well before the unification of China in the third century BC, the tamarind tree (valued for its fruit) was introduced into Yunnan. In the early second century BC, messengers of the Han Dynasty brought grape vines to China from central Asia. Traders and travelers in the Tang Dynasty introduced plants, foods, and medical drugs thought to be of value, a practice which continued in the Song Dynasty as traders carried Barbados aloe from northeast Africa into China. Two crops of large economic value – sweet potatoes and tobacco – arrived in China courtesy of colonizers and traders of the seventeenth century who took these native American species to India and Southeast Asia, and then overseas Chinese brought them to China.[3]

In modern times, institutions of the state such as state ministries and agencies of agriculture, forestry, and fisheries have been actively involved in the adoption into China of economically valued species. Traders have imported dozens of species into China from Taiwan for the purpose of improving food production.[4] A number of marine and freshwater fish species have been introduced, such as prawns, salmon, tilapia, European eels, and bass.[5] Most of the lawn grass seed in China comes from abroad (particularly, from the state of Oregon in the USA.).[6]

It is important to keep these facts in mind when considering the invasive species problem in China. We have not mentioned wheat, corn, and potatoes, which collectively produce nearly half of the carbohydrates in the typical Chinese diet. None of these crops is native to China; all were "invasive" when they entered the Chinese landscape.

Our focus here, however, is on species alien to China, which endanger local species of value and local ecosystems. Most of these came to China unintentionally, as a consequence of travel and trade. Examples are legion, including a large number of invasive weeds (for example, Spanish needles), insects (such as the banana moth), and diseases of plants and animals. As China's connections with the rest of the

[2] Xie et al., 2000, 1321.

[3] Ibid., 1324. See also E. N. Anderson, *The Food of China*. New Haven: Yale University Press, 1988, and Shiu-ying Hu, *Food Plants of China*. Hong Kong: The Chinese University Press, 2005.

[4] Qixian Yi and Li Hua, "Strengthening Quarantine of Plants Introduced from Taiwan," *Plant Quarantine*, Vol. 12 (1997), 185–87.

[5] Linming Qiu and Guosheng Deng, *Developing Technology for Special Freshwater Species Aquaculture*. Beijing: Zhongnan Industry University Press, 1994.

[6] Xiaogang Xu, Xinquan Zhang and Yanqi Wu, "A Discussion of Some Problems in Introducing and Breeding Turfgrass in China," *Grassland of China*, Vol. 1 (1999), 57–61.

world increased, first through imperialism and colonialism of the nineteenth and early twentieth century, and second through economic reforms beginning in the late 1970s, the unintentional transmission of invasive species accelerated.

6.2.3 Valorization of Invasive Species

We study invasive species because of the negative value attached to them in an era of globalization. Regarding values, invasive species customarily are appraised in terms of their economic and non-economic effects. Economic values refer to the extent to which the species advances or retards the economic development and the wealth of the state. Most recent discussion omits consideration of the tremendous benefit in the introduction of alien species that add exponentially to the agricultural productivity of nations, for example as in the introduction of maize. Instead, for perhaps obvious reasons, discussion centers on the damage to food crops and animal husbandry of foreign pests. Indeed, more than half of the alien species listed as dangerous by the World Conservation Union have been found in China. As we note below, economists attempt to measure the actual cost to nations of such dangerous invasive species.

Economic costs attract attention because they can be expressed in numbers, connoting significance in the marketplace. However, invasive species also have non-economic costs and benefits. By definition, invasive species change ecosystems. These effects may be malign when, for example, they crowd out and may exterminate native species, thereby reducing biodiversity with all its attendant benefits in terms of providing ecosystem services, biological resources, and social benefits. Little serious attention has been paid to the benign consequences of invasive species on ecosystems.

The clearly negative consequence of invasive species is their threat to native species as a whole. Invasive species may directly reduce the number of native species, reduce the number of other species relying on local species, lower the ability of native species to resist pests and fire, and alter mechanisms of soil and water conservation.[7] Because most native species and environments do not have established economic values, it is difficult to mobilize support in their defense.

6.3 Invasive Species with Impacts on Food Production

Inasmuch as the invasive species problem is relatively new in China, information about species' numbers and spread is less than exact. In fact, the first inventory was published only in 2004.[8] By 2000, an estimated 500 invasive species had been

[7] Yan Xie, *Alien Invasive Species in China.* ed from http://www.chinabiodiversity.com/shwdyx/ruq/ruq2-4n.htm on October 16, 2007.

[8] Haigen Xu and Sheng Qiang, *Inventory: Invasive Alien Species in China* (in Chinese). Beijing: China Environmental Sciences Publishing Co., 2004.

recorded, based on recent surveys.[9] Understandably, greater attention has been paid to invasive species whose introduction has had economic effects. We provide a number of examples of invasive species from major taxonomic groups, based on the work of Xie Yan and her colleagues.[10] Most of the examples include the country of origin of the species, the approximate time it was introduced to China, areas to which it has spread and damage caused.

6.3.1 Plant Species

- Alligator Weed (*Alternanthera philoxeroides*). This weed originated in Brazil; it entered eastern China in the 1940s and was used for pig forage in areas of south China, from which it spread into the wild. By the mid-1980s, the species covered 13 million hectares and was a major weed in vegetable gardens, citrus orchards, and sweet potato fields.[11]
- Ragweed (*Ambrosia* spp.). Ragweed is native to North America, but varieties have been observed in China since the 1930s. From the late 1980s, both common ragweed and giant ragweed have spread to 12 provinces. They reproduce well in infertile and dry soils, and they affect crop productivity because they block sunlight. In addition, pollen from the species causes allergic reactions (hayfever) in humans.[12]
- Crofton Weed (*Eupatorium adenophorum*). Although this weed is native to Central America, it had spread along roads from Burma to southern parts of Yunnan by the 1940s. Today it covers hundreds of thousands of hectares in Yunnan Province and, because it is poisonous, stunts the growth of plants and may kill plants and domestic animals.
- Water Hyacinth (*Eichhornia crassipes*). This invasive aquatic weed is native to South America. It was introduced to China at the turn of the twentieth century for use as forage for domestic animals, ornamental use, and to purify wastewater. By the end of the century, it had spread widely through south and southwest China. Producing dense mats, the water hyacinth has crowded out native aquatic plants, threatening biodiversity, while adversely affecting fisheries and tourism.[13]

---

[9] Personal interview with invasive species specialist, Chinese Academy of Sciences, Beijing, May 21, 2007.

[10] Xie et al., 2000. Duplicated in Biodiversity Working Group/CCICED, 1997–2001, *Conserving China's Biodiversity (II)*. Beijing: China Environmental Sciences Press, 2001, 101–24. Page numbers for this section follow the 2001 publication.

[11] Also see Ren Wang and Yuan Wang, "Survey of Damage Caused by *Alternanthera philoxeroides* and its Biological Control in South China," *Journal of Weed Science*, Vol. 2 (1988), 38–40.

[12] Xiumei Li, "Research Progress on Comprehensive Control of a Serious Weed – *Ambrosia*," *Weed Science*, Vol. 1 (1997), 7–10.

[13] Jianqing Ding, Ren Wang, and Zhongnan Fan, "Distribution and Infestation of Water Hyacinth and the Control Strategy in China," *Journal of Weed Science*, Vol. 9, no. 2 (1995), 49–51.

6.3.2 Insect Species

- Fall Webworm (American White Moth) (*Hyphantria cunea*). This species is native to North America; it entered China in the late 1970s and spread to Liaoning, Shandong, and Shanxi provinces, where it afflicted more than 100 plant species. The species completely defoliates trees in orchards, parks, and along roadsides, leading to high rates of mortality of both urban and rural trees.[14] In 2005, the species was spotted in Beijing, Tianjin, and Hebei Province. By 2007, it had spread to 270,000 ha and observers feared it might threaten Beijing's vision of a "Green Olympics" in 2008.[15]
- Banana Moth (*Opogona sacchari*). This moth has a global range, and was first noticed in China in the mid-1990s. The caterpillars feed on more than 50 species of plants, including such crops as bananas, sugarcane, corn, sweet potatoes, and horticultural plants. It is an especially serious threat in southern provinces, where it is believed likely to threaten natural ecosystems.[16]
- Vegetable Leaf Miner (*Liriomyza sativae*). This insect pest also has a broad global distribution (found in 40 countries), but was not reported in China until the early 1990s. Since then, it has spread to 21 provinces and covers more than 1 million hectares. The leaf miner attacks the cotyledons of a number of plant species, and causes large crop losses.[17]
- American Rice Water Weevil (*Lissorhoptrus oryzophilus*). This insect pest originated in North America, and then traveled from Japan, through the Koreas, and was reported in northern China in the late 1980s. By the next decade, it had spread to eight provinces and damaged 300,000 ha of rice fields.[18]

6.3.3 Fish Species

- Mosquitofish (*Gambusia affinis*). The mosquitofish originated in North America; it was introduced to China in the 1920s from Taiwan in order to reduce the mosquito population. The species spread to many other provinces in China, and is suspected to have preyed on native fish populations by eating fish fry.

---

[14] Yunchao Li, Huansheng Zhang, Yisheng Han, and Denghua Xu, "Preliminary Investigation on American White Moth in Qinghuangdao," *Journal of Forestry Science and Technology*, Vol. 4 (1991), 29–30.

[15] Xiaohua Sun, "Biological Measures Used to Clear Forests," *China Daily*, September 22–23, 2007, 3.

[16] Guifang Cheng and Jikun Yang, "Preliminary Report on New Quarantine Pest Insect—*Opogona sacchari Bojer*—Found in Beijing," *Plant Quarantine*, Vol. 11 (1997), 95–101.

[17] Fuxiang Wang, "Occurrence and Comprehensive Prevention and Cure of Vegetable Leaf Miner," *Journal of Agricultural Science and Technology*, Vol. 2 (1997), 34.

[18] Hongjun Wei, "Occurrence and Continuous Control of American Rice Water Weevil in China," *Plant Quarantine*, Vol. 11 (1997), 60–62.

- The "Four Major Domestic Carps" (*Mylopharyngodon piceus*, *Ctenopharyngodon idellus*, *Hypophthalmichthys molitrix*, and *Aristichthys nobilis*). Although these species of carp are native to China, in the 1960s they were introduced into a different ecosystem – upland lakes and ponds in Yunnan province – in order to improve fish production. They quickly spread to middle and upper layers of lakes and reservoirs, exhausted aquatic grasses, and crowded out indigenous fish species.[19]

6.3.4 Other Species

- Amazonian Snail (*Ampullaria gigas*). Originating in the Amazonian Basin, this snail entered Guangdong Province in the early 1980s. An overseas Chinese from Brazil raised it for local consumption, but a market did not develop and the species was released, spreading quickly throughout the province (and entering both Fujian and Yunnan provinces too). The Amazonian snail reproduces rapidly, feeds voraciously and has damaged rice, vegetable and water crops.[20]
- Black Spot Fungus (*Ceratocystis fimbriata*). This fungus is native to Japan, and entered China in the 1930s. Since then, it has spread to 26 provinces and become a major disease of sweet potatoes. The fungus decomposes sweet potato seeds, kills young plants and also stimulates sweet potatoes to produce toxic biochemicals damaging both human and livestock health.[21]

Professor Xie's 2007 compendium on invasive species, the most comprehensive study to date, lists the most dangerous 16 invasives. Her list includes the four plant species discussed above – alligator weed, ragweed, Crofton weed, and water hyacinth – and also mile-a-minute weed, poison Darnel, smooth cord-grass, Siam weed, and Johnson grass. She lists both the American white moth and banana moth, mentioned above, of insect species, and also the loblolly pine mealy bug and red turpentine beetle. Finally, she lists the giant African snail and Amazonian snail (described above) and the American bullfrog.[22]

Estimates vary on the economic cost incurred by the state to remedy crop losses and other damages of invasive species. One study, by Xu Haigen of the Nanjing

---

[19] Junxing Yang, "The Alien and Indigenous Fishes of Yunnan: A Study on Impact Ways, Degrees and Relevant Issues," Biodiversity Working Group/CCICED, *Conserving China's Biodiversity (II)*. Beijing: China Environmental Science Press, 2001, 161–62.

[20] Hanxiong Cai and Rizhong Chen, "New Pest – Amazonian Snail," *Guangdong Agricultural Sciences*, Vol. 5 (1990), 36–38. Also see Ziyun Li, "Invasive Snail, Other Species Threaten China's Eco-Security," World Watch Institute, September 12, 2006.

[21] Qingguang Lu, "The Importance of Classical Biological Control to Biodiversity Protection," *Chinese Biodiversity*, Vol. 5 (1997), 224–30.

[22] Yan Xie, ed., *Bioinvasion and Ecological Security in China* (in Chinese). Shijiajuang, Hebei: Hebei Science and Technology Publishing, 2007, 301–18.

Institute of Environmental Sciences, estimates that invasive species have caused $2.4 billion in damages to eight major Chinese industries alone.[23] Annual damages to China's forests are estimated to total $7.6 billion.[24] A report in *China Daily* puts the overall cost at $14.5 billion annually.[25] A general report in China's *Strategy for Plant* Conservation mentions "annual economic losses by related sectors of the national economy" as $17.1 billion.[26]

In November 2009, the first international congress on biological invasions was held in Fuzhou, China. The congress adopted the "Fuzhou Declaration on Biological Invasions," which urged policymakers globally to concentrate on interactions of climate change and bioinvasions as well as the threats that invasives present to biodiversity, food security, trade, economic development and human health. Conferees mentioned that the global cost of invasive species damage was about $1.4 trillion.[27]

Estimates of costs incurred to mitigate specific invasive species are somewhat more exact. For example, large portions of the budgets of southern provinces have been allocated to remove water hyacinth. Funds to remove water hyacinth in Wenzhou, Zhejiang and Putian, Fujian in 1999 were US$133 million and US$67 million respectively. Total costs for removal across China, to prevent further losses in agricultural irrigation, cereal transportation, and aquatic production, would be far higher.[28] The vegetable leaf miner has parasitized and caused heavy damage to vegetables and melons such as cucumbers, musk melon, watermelon, summer squash, tower gourd, tomatoes, hot peppers, egg plants, cow peas, kidney beans, peas and lentils. Costs to mitigate its effects have risen to $530 million.[29]

Finally, the director of the Office of Biosafety Management in SFA's Nature and Ecology Conservation office, Wang Jie, states that invasive species bring economic losses of $29.3 billion annually. Moreover, they have invaded 46% of China's nature reserves.[30]

Clearly, costs of mitigating the effects of invasive species are quite high, but in the absence of a complete catalog of invasive species, and systematic analysis of their environmental as well as food system costs and benefits, any estimate of damage is speculative.

---

[23] Cited in Dennis Normile, "Invasive Species: Expanding Trade with China Creates Ecological Backlash," *Science*, Vol. 306, no. 5698 (2004), 968.

[24] Xinhua, "Scientists Address Alien Invasion," *China Daily*, December 14, 2007, 3.

[25] Chuan Qin, "Invasive Species Costly to Ecology," *China Daily*, May 21, 2004, 2.

[26] Editorial Committee, *China's Strategy for Plant Conservation*. Guangzhou: Guangdong Press Group, 2008, 27.

[27] *People's Daily Online*, "Invasive Species Cause $1.4 Trillion of Losses a Year: Scientists," http://english.people.com.cn/90001/90776/90883/6806481.html (retrieved November 27, 2009).

[28] Jianqing Ding and Yan Xie, "The Mechanism of Biological Invasion and the Management Strategy," in Biodiversity Working Group/CCICED, *Conserving China's Biodiversity (II)*. Beijing: China Environmental Science Press, 2001, 133.

[29] Wang, 1997, 35.

[30] See: http://www.china.org.cn/environment/features_analyses/2009-06-03/content_17880215.htm (retrieved November 27, 2009).

6.4 Immediate Environmental Stressors as Causes of the Unintentional Transmission of Invasive Species

Xu et al. report that one-half of alien invasive plants were intentionally introduced as pasture, feed stock, ornamental plants, textile plants, medicinal plants, vegetables, or as lawn plants. About one-fourth of alien invasive animals were introduced for purposes of cultivation, recreation, or biological control.[31] However, about half of China's invasive species were introduced unintentionally. The processes of transmission were related in many cases to socio-economic changes we described in Chapter 3. We consider these changes as well as domestic transportation improvements, which facilitated invasive species transmission. Most of this section, however, treats globalization (increased economic interconnections between China and the broader world), for it is the primary factor explaining the sharp increase of invasive species in the modern period.

There is a relatively long lag time from the initial establishment of an invasive species to its expansion. The dormant stage of some invasive species may be several decades, depending on the period of time required for seed production for plants, and the sexual reproduction and climate cycles.[32] This means that estimates of current transmissions amounts are likely to understate the problem.

6.4.1 Socio-Economic Change

Increase in population and greater movement and dispersion of that population are large factors in the unintentional transmission of invasive species. We have noted the extent to which population growth and urbanization have reduced arable land in China. Cities themselves often are entry points for invasive species. For example, development of cities reduces vegetative and other ground cover that might reduce the opportunities for invasion.

Economic development has similar effects. The construction of dams, irrigation projects, factories, industrial plants, and even shopping centers disturbs the natural ecosystem, and may lower the resistance of native species and even cause their extinction. Invasive species appear more likely to invade habitats altered by humans than intact ecosystems.[33] Specific economic development activities, involving the intentional adoption of invasive species, may go awry as the species grow wild and escape

---

[31] See Haigen Xu, Sheng Qiang, Zhengmin Han, Jianying Guo, Zongguo Huang, Hongying Sun, Shunping He, Hui Ding, Hairong Wu, and Fanghao Wan, *Biodiversity Science*, Vol. 12, no. 6 (2004), 626–38.

[32] See Ding and Xie, 2001, 138.

[33] Jeffrey A. McNeely, "Human Dimensions of Invasive Alien Species: How Global Perspectives are Relevant to China," in Biodiversity Working Group/CCICED, *Conserving China's Biodiversity*. Beijing: China Environmental Sciences Press, 2001, 170.

development regions. For example, in the 1950s and 1960s, the Chinese government adopted a national policy of "planting three aquatic plants" – the water hyacinth, alligator weed, and water lettuce. Efforts to promote their production resulted in rapid expansion in the wild, threatening ecosystems and agricultural production.[34]

6.4.2 *Improved Domestic Transportation Systems*

China is one of the world's most biologically rich countries, because of the size of its territory, range of climate zones, and diversity of habitats and environmental conditions. China's mountain ranges, river systems, deserts: all had the effect of isolating different bioregions, and protecting them from invasive species that might threaten indigenous species. With China's recent rapid economic development, this situation has changed.

In 2008 few bioregions in China are completely isolated. Highways, railroad lines, canals and other waterways crisscross the land. These improved linkages facilitate movement of invasive species, which easily follow transportation corridors. For example, it is reported that the brown (also called the Norwegian) rat (*Rattus norvegicus*) entered Xinjiang Province by railway from the interior of China.[35]

Recently, scientists have discovered 55 invasive species in the drainage area of the Three Gorges Dam, including the water hyacinth. Its propensity to clog pipes threatens the dam's generation of hydroelectric power. The Qinghai–Tibet railway encourages the spread of invasives to remote regions.[36]

Population growth, urbanization, economic development, and improved domestic transportation systems also have facilitated the transmission of intentionally introduced alien species. A number of such species have escaped their area of adoption in China and become invasive as well.

6.4.3 *Increased International Trade and Tourism*

The huge increase in China's trade with the world has greatly multiplied the opportunities for invasive species to enter China (and for Chinese species to

---

[34] Ding et al., 1995, 51.

[35] Danming Zhang, Fuchun Zhang, and Jianming Wang, "Invasive Migration and Selection of Habitat of Brown Rat in Arid Regions of Inland China," in Wuping Xia and Jie Zhang, *Evolution of Beasts under the Impact of Activities of Humans*. Beijing: Science and Technology Press of China, 1993, 37–38.

[36] Weixiao Chen, "Invasive Foreign Species Threaten China's Biodiversity," Science and Development Network at http://www.scidev.net/en/news/invasive-foreign-species-threaten-china-s-biodiver.html (retrieved November 27, 2009).

invade other nations). Trade products arrive in China on ships, trains, and planes. The containers of these products and packaging materials are ready hosts for invasive species such as insects (the banana moth and fall webworm), pinewood nematode,[37] and plant pests (ragweed, Crofton weed, and amaranth). Wastewater and ballast of ships are discharged in China's ports; ballast discharges introduce diseases, bacteria and viruses, which harm marine and freshwater ecosystems and may seriously damage fisheries. Ballast itself is a rich nutrient host for a variety of microbial species.

Xie et al. report that from 1986 to 1990, about 200 species of weed seeds from 30 families and 100 genera were intercepted from 350 ships at the port of Shanghai.[38] Li Yanghan et al. comment that in the late 1990s, weed seeds of 547 species and five varieties were intercepted in 12 ports of China, including Dalian, Qingdao, Shanghai, Nanjing, and Guangzhou among others. These invasive species originated in 30 different countries; they contaminated imported cotton, wool, forage, imported food and other economic plant seeds.[39]

Some of these interceptions have become controversial in China's relations with other powers. For example, in 2007 the General Administration of Quality Supervision, Inspection and Quarantine (AQSIQ) seized 21 containers with 460 t of soybeans imported from the United States because they contained live Khapra beetles. This invasive species is one of the world's most serious pests for stored grain products including wheat, rice, corn, soybeans, and also beans and nuts. From 30% to 75% of a crop can be damaged by the beetles. In this case, the shipment unloaded at Huangpu Port in Guangdong Province was returned to the United States.[40] Foreign observers, however, questioned whether the intention was to intercept a handful of invasive species in a large shipment or to retaliate against the United States for its investigation of the safety of Chinese food products and toys entering the USA (see Chapter 8).

Jenkins and Mooney make the observation that the overlapping biogeography of the USA and China – now the world's most important trade partnership – facilitates the exchange of invasive species:

---

[37] The pinewood nematode is believed to be native to the United States. It feeds on fungi within the wood and on living plant cells surrounding the resin canals of pines. The pine sawyer beetle carries the nematode in the tracheae of its respiratory system, and thus spreads them from tree to tree. Huge populations of nematodes develop in the trees, causing them to wilt and die. It appears that the pinewood nematode arrived in China in packing materials of US or Japanese containers in the early 1980s, spreading to coastal cities from Beijing to Guangzhou. See: Bruce Fraedrich, "Pinewood Nematode." Charlotte, NC: Bartlett Tree Research Laboratory Technical Report, 1999; also see abstracts, International Symposium on Pine Wilt Disease Caused by the Pine Wood Nematode. Beijing, 1995; and "Pine Wilt Disease – Pathogen Biology" at http://www.apsnet.org/education/lessonsplantpath/PineWilt/pathbio.htm (retrieved February 29, 2008).

[38] Xie et al., 2001, 104.

[39] Yanghan Li, *China Weeds*. Beijing: Agriculture Press, 1998, 1617.

[40] Zhe Zhu, "460 Tons of Tainted Soybeans Returned," *China Daily*, September 29–30, 2007, 1, 2.

> (B)oth countries contain a high diversity of native species, a large and growing reservoir of both established and incipient invaders from around the world available for further exchange, and very wide ranges of similar eco- geographic regions and life zones with many vulnerable habitats. Further, both countries still lack comprehensive, proactive, regulatory frameworks for preventing the introduction and spread of invasives.[41]

The US–China case illustrates the need for caution in assigning blame to individual countries as originating areas for dangerous invasives. For example, the compendium edited by Xie Yan lists chestnut blight (*cryphonectria parasitica*) as observed first in New York in 1904.[42] The fungus enters tree wounds, growing under the bark, and eventually kills the cambium. The cankers stunt chestnut tree growth, and there is little resistance to the disease in the U.S. Recent research suggests that the source of the pathogen in the U.S. was Japanese chestnut trees imported first in 1876. Chestnut trees in both Japan and China have developed resistance to the disease.[43]

As travel to China, and by Chinese abroad, has increased, tourists and other travelers have become pathways for the introduction of an increased number of invasive species that become hitch hikers on clothing, luggage, and vehicles. For example, fruit flies have been found on fruit carried by tourists entering China, and hoary plantain (*Plantago virginica*), a rapidly spreading weed, attaches itself to tourists' luggage.[44] Cross-border trade between China and Vietnam, Thailand, Myanmar, and Nepal has grown greatly, and rising vehicular traffic increases opportunities for species invasion.

A final point is that a number of alien species have entered China as a result of entirely natural processes. They were blown across China's borders by the winds, or carried to China's shores by ocean and river currents.

6.5 Climate Change and Invasive Species

As noted in Chapter 4, climate change has accelerated the impact of immediate environmental stressors, and we commented on the impact of climate change on traditional plant pests and plant diseases in Chapter 5. Warmer temperatures, particularly in North China, have increased spread to this region of some invasive plant pests. Changes in precipitation also can be correlated with the movement of invasive species. Drought conditions, for example, have increased the incidence of some invasive nematode species.[45] Extreme weather events too exacerbate the spread of

---

[41] Peter T. Jenkins and Harold A. Mooney, "The United States, China, and Invasive Species: Present Status and Future Prospects," *Biological Invasions*, Vol. 8 (2006), 1591.

[42] Xie, 2007, 321, 511.

[43] Sandra Anagnostakis, "Revitalization of the Majestic Chestnut: Chestnut Blight Disease," APS News Release, December 2000 at http://www.apsnet.org/online/feature/chestnut/ (retrieved February 29, 2008).

[44] Ding and Xie, 2001, 137.

[45] Personal interview with entomologist, Plant Protection Institute, Beijing, May 29, 2007.

invasives. For example, typhoons Krosa and Wipha spread water hyacinth throughout Shanghai's waterways in 2007. A local water environmental official said: "When the typhoons arrived, water hyacinths scattered in the branches of Huangpu River were collected by the strong wind and covered the river from shore to shore."[46]

Research in China concerning the impact of climate change on biodiversity as well as on food production capabilities is at an early stage, and there are few research articles on this subject.[47] Although there are some 2,000 climate monitoring stations in China, to the present there has been little species monitoring that would identify changes in spread of invasive species as a consequence of alteration in weather and climate.[48]

Nevertheless, scholars studying individual invasive species are now extending their research to treat climate change effects. We discussed this subject with an entomologist who has specialized in study of the bark beetle, clear wing moth, carpenter moth, and the Asian Longhorn Beetle. The latter has now infested poplar, willow, elm, and maple trees in North America and Europe and reached epidemic proportions globally. It is carried from country to country in wood packing materials. The scientist recounted differences in insect types related to climate change:

> There are two types of economically-damaging insects: 1) the leaf feeders or defoliators, and 2) the boring insects. Climate change impacts are greater with respect to the defoliators (for example, the vegetable leaf miner) than the borers. Those insects inside wood are less influenced by climate change because their nests inside the wood or plant tissue are not exposed. In the case of the Asian Longhorn Beetle and the Emerald Ash Borer, we note fewer climate change effects because they avoid healthy forests. They are attracted to vulnerable, stressed trees.[49]

6.6 Responses to Spread of Invasive Species in China

In the early twenty-first century, invasive species have become a global concern. Although foreign species have been noticed in China (and indeed intentionally adopted) for a very long period, only in the last decades has the government focused on their mitigation. We start this section by discussing some global recommendations on the treatment of invasive species. We then turn to domestic responses in China – through creation of law and regulation and through developing mitigation strategies. We conclude by pointing out the future directions in which study and action on invasive species seems to be progressing.

---

[46] Hongyi Wang, "Hyacinth Jams City's Waterways," *China Daily*, October 13–14, 2007, 3.

[47] Personal interviews with climate scientists in Beijing, August 28, 2006; March 14, 2007, and May 25, 2007.

[48] Personal interview with invasive species specialist, Chinese Academy of Sciences, Beijing, May 21, 2007.

[49] Interview with entomologist, Beijing, March 15, 2007.

6.6.1 Global Recommendations

A number of international conventions make mention of invasive species, and there are a series of guidelines addressing methods to reduce their spread. The most explicit reference is in the Convention of Biological Diversity (CBD). Article 8h of the treaty, to which China is a signatory, calls on parties to "prevent the introduction of, control or eradicate those alien species which threaten ecosystems, habitats, or species."

International non-governmental organizations such as the World Conservation Union (formerly known as the International Union for the Conservation of Nature or IUCN) also have worked to increase measures to prevent or combat invasive species.[50] In the late 1990s, WCU spearheaded creation of the Global Invasive Species Program (GISP). In 2005 four international NGOs met to inaugurate the GISP voluntary association. They include in addition to WCU: CAB International (CABI), The Nature Conservancy (TNC), and the South African Biodiversity Institute.[51]

GISP advocates four different management approaches to the resolution of invasive species problems: (1) Use the precautionary principle[52] in a process of expert consultation to evaluate all alien species proposed for introduction; (2) strengthen scientific research to establish sound bases for predicting which species are likely to become invasive and which species are likely to be beneficial; (3) strengthen control of the pathways through which unintentional introduction of invasive species occurs, for example through wood packing materials, ballast water, and other transmission routes associated with increased global trade; and (4) when prevention fails or is impractical, strengthen measures to eradicate or control invasive species.[53]

GISP has established a partnership network to raise awareness of invasive species issues, create linkages among governments, environmental groups, and the private sector, and to support projects to minimize the spread and impact of invasive species.

6.6.2 Law and Regulations on Invasive Species

China has general laws on plant and animal quarantine and fairly specific regulations, but no general or specific legislation that would prohibit damage to native agricultural species or local ecosystems from invasive species. The Quarantine Law on

---

[50] See C. Shine, N. Williams, and F. Burhenne-Guilmin, *Legal and Institutional Frameworks on Alien Invasive Species: A Contribution to the Global Invasive Species Programme Global Strategy Document.* Bonn, Germany: IUCN Environmental Law Programme, 2000.

[51] See: http://www.gisp.org/about/members.asp (retrieved October 21, 2007).

[52] The precautionary principle is Principle 15 of Agenda 21, adopted at the United Nations Conference on the Environment and Development in 1992. It is defined as: "In order to protect the environment, the precautionary approach shall be widely applied by States according to their capabilities. Where there are threats of serious or irreversible damage, lack of full scientific certainty shall not be used as a reason for postponing measures to prevent environmental degradation."

[53] McNeely, 2001, 175.

Import and Export of Animals and Plants has stipulations on the prevention and control of alien and invasive species. The Wildlife Protection Law of 1989 also has regulations, but these cover only the intentional importation of alien species. The most recent legislation is the List of Entry and Exit Plant Quarantine of Potentially Dangerous Epidemics, Pests and Weeds.

For example, to prevent diseases such as mad cow and foot-and-mouth diseases from entering China, the central government regulates direct and indirect importation of a range of beef products and ruminant foodstuffs from nations in which these diseases are present.[54] Further, Article 14 of the Compendium of National Ecological Conservation states: "All alien species have to undertake risk assessments. The import quarantine work has to be strengthened to prevent any alien invasive species from entering into China."[55] However, this ambitious passage is not embodied in specific action plans to curb entry. Considering the status of current legislation and regulation, Ding and Xie recommend that large-scale use of non-indigenous species be prohibited until small-scale trials of several generations show no signs of threats to native species and local ecosystems.[56] This would embed the precautionary principle in Chinese law, resembling the way in which it has been proposed that genetically-modified organisms (discussed in Chapter 7) be treated.

The implementation of laws and regulations on issues of food security is discussed throughout this volume. In the area of invasive species prevention and control, implementation recently has been strengthened. In 2004, a National Coordination and Cooperation Working Group on Prevention of Alien Organisms brought together the Ministry of Agriculture (which leads the group), the General Administration of Quality Supervision, Inspection, and Quarantine (AQSIQ), the Ministry of Environmental Protection (MEP), the State Forestry Administration (SFA), the Ministry of Science and Technology (MOST), the General Administration of Customs, and the State Oceanic Administration. SFA established a management office for Prevention of Alien Forest Detrimental Organisms, and MOA established the Research Center for Prevention and Control of Alien Invasive Organisms.[57] In August 2007 the Ministry of Agriculture issued guidelines to strengthen coastal and border inspections to prevent the spread of invasive plants and insects. The guideline specified an increase in number of inspection stations from somewhat more than 2,000–3,000, in order to reduce threats to corn and fruit crops, remarking:

> Invasive insects and plants have come across the border and are rampant in some regions. They are endangering our agricultural product safety and posing a great threat to the global competitiveness of our farming sector.[58]

---

[54] See Biodiversity Clearing-House Mechanism of China, http://english.biodiv.gov.cn/rdwt/200603/t20060323_30675.htm (retrieved October 21, 2007).

[55] Ibid.

[56] Ding and Xie, 2001, 111.

[57] Editorial committee, 2008, 27.

[58] Jiao Wu, "Ministry Ups Ante on Pest Control," *China Daily*, August 20, 2007, 3.

The guideline mentioned the risk of increased spread through provinces of agricultural pests as a consequence of growing farm trade and specifically discussed the Colorado potato beetle, first detected in China in 1993. Since then it has infested 35 counties in Xinjiang, affecting more than 10,000 ha of potato crops. Typical of the devolution of government functions since the onset of the economic reform period, the guideline ordered provincial and local authorities to establish inspection stations, and told them to make improvements in their technology and staff to increase their effectiveness.[59]

At the central government level, the greatest activity regarding invasive species has been in research and planning. Several agencies (AQSIQ, SFA, MOA, and MEP) have conducted risk analyses. MOA formulated an Initiative of Emergency Response to Major Agricultural Detrimental Organisms and Alien Organism Invasion Contingency. This ministry collected information and statistics on more than 300 species, which became the basis for China's Alien Species Database. MOA also compiled China's Major Agricultural and Forest Alien Invasive Species List. Finally, the ministry launched the "Animal and Plant Conservation System," which is one of seven major reaction systems, to invest in advanced warning and control of major agriculturally detrimental organisms.[60]

6.6.3 Mitigation Measures

Four methods have been used to prevent or control economically harmful invasive species: physical, biological, chemical, and integrated control.

6.6.3.1 Physical Control

China has a long history of removing damaging plants through manual labor. For example, in the 1990s, governments of Yunnan, Zhejiang, and Fujian provinces and the municipality of Shanghai used labor power to remove water hyacinth from local waters. In fact, in Kunming (Yunnan's capital) more than 100,000 people were employed to do this work.[61]

The government of Shenzhen used labor to eliminate the South American climber. Governments of Shaanxi and Liaoning provinces employed workers to cut off the larva web curtain of moths and to crop tree crowns in order to control fall webworm infestation.[62] Similarly, in mid-2008, more than 10,000 workers in thousands of fishing boats worked to clean a huge algal bloom carpeting the Olympic

[59] Ibid.

[60] Editorial Committee, 2008, 28.

[61] Ding and Xie, 2001, 128.

[62] Ibid., 129.

sailing venue in Qingdao, Shandong Province. Authorities attributed the bloom, which covered 13,000 km^2, variously to climate change, ocean pollution, and natural causes.[63] Developing special equipment to reduce the costs of such control efforts has not been successful, to date. Moreover, manual labor has not eliminated the insect and plant pests from these areas.

6.6.3.2 Biological Control

Many invasive species spread rapidly because they encounter no natural enemies in their new area of residence, and thus the introduction of biological control agents promises the opportunity to restore ecological balance.[64] A corresponding difficulty, however, is that biological control agents may require several years to take effect. Also, they may introduce new risks by threatening economically and ecologically beneficial organisms in the environment, becoming dangerous invasive species themselves.

In the last 2 decades, Chinese scientists have made several attempts to introduce biological control agents. In 1988, they placed a parasite from Japan in Guangdong to control the pine scale pest, which successfully reduced infestation.[65] Scientists have introduced seven inspect species in southern China to control alligator weed, common ragweed, ragweed, water hyacinth and Crofton weed; this control effort was most successful in managing Crofton weed.[66]

6.6.3.3 Chemical Control

Chemical pesticides and insecticides have been used more frequently to control dangerous invasive pest species. Their advantage is that they are relatively easy to apply and take effect quickly. However, most agricultural fields in China suffer from an overdose of chemicals, with extremely deleterious effects on the ecosystem as well as on human health. Moreover, few insecticides and pesticides are specific to just the invasive species targeted: they spontaneously eliminate native species too. In addition, most herbicides and pesticides eliminate only the upper parts of plants and not the roots, meaning they need to be used repeatedly.[67] Finally, chemical control methods are relatively expensive to use.

---

[63] http://www.guardian.co.uk/enviornment/2008/jun/30/pollution.olympicgames2008 (retrieved July 24, 2008).

[64] Jianqing Ding and Weidong Fu, "Biological Control: Conservation of Biodiversity by Using Biodiversity," *Biodiversity*, Vol. 4, no. 4 (1996), 222–27.

[65] Ding and Xie, 2001, 129.

[66] Ibid.

[67] Ding and Xie, 2001, 142.

Chemical insecticides have been used effectively to control the American vegetable leaf miner and rice water weevil. Chemical sprays have been used to control the American white moth. Too, herbicides such as Roundup have had some effect in controlling water hyacinth, alligator weed, South American climber, and Crofton weed.[68]

6.6.3.4 Integrated Pest Management

The final method of controlling invasive species is also the most recent. This method fuses manual, biological, and chemical control methods in order to emphasize the superiorities of each and minimize the deficiencies. It promises improvements in efficiency, sustainability of the control methodology, greater safety, and reduced cost.[69]

To the present, integrated control measures have been used with effect in the treatment of water hyacinth, Crofton weed, and ragweed.[70] Perhaps because the method is new, requires extensive coordination of agencies, and necessitates special training of personnel, it has not been applied widely in the treatment of invasive species problems.

6.6.4 Future Directions

Specialists in invasive species of China have identified four areas in which continuing efforts need to be expanded. First, as mentioned above, laws and regulations do not provide sufficient authority to counter the growing threat of invasive species to agricultural crops and to local ecosystems. Law alone is insufficient, however, if the implementation system is incomplete or ineffective.

Second, because the recent explosion in growth of invasive species is a consequence of globalization, cooperation among countries in sharing information about invasive species and developing effective prevention and control strategies is important. In the quarter century of rapidly increased trade between the United States and China, the countries have exchanged about the same number of invasive species. Mutual interests in controlling invasive species to protect local ecosystems have led to collaboration,[71] a valuable development.

Third, the Global Invasive Species Program and its supporters in China emphasize the critical need for increased public awareness and education. Again, the recency

---

[68] Ibid., 129.

[69] Ibid., 143–44.

[70] Ibid., 129.

[71] China Environment Forum, "An Odyssey of Ecological Invasion: Invasive Species Challenges and Collaboration Globally and Between the U.S. & China," October 19, 2005.

of the invasive species issue in China means that there is lack of knowledge on the part of the general public about both present and future risks of invasive species.[72] Information is even poorly spread among China's farming population, which is at the front line of defense against invasive species.

Finally, as we have noted above, the invasive species problem presents large challenges to China's scientists: to establish data bases on invasives, to compile case studies on effective and ineffective means of prevention and control, and to disseminate information to the broader public. Recent conferences and publications on invasive species indicate that this effort is well under way.[73]

6.7 Conclusions

Hundreds of plant species have been introduced into China in the last several millennia, and they comprise about one-quarter of all the cultivated plants in the nation.[74] These include, among the important staples, maize, wheat, and potatoes. Knowledge that China has benefited greatly from the introduction of alien species is relatively widespread. Too, for many years, scientists and government officials working in the areas of agriculture, animal husbandry, and forestry have focused on the potential harm of a small number of alien species. However, as Ding and Xie note, "(I)t was only a few years ago that the concept of invasive species was introduced into China and the potential threats to China's natural heritage are not yet widely recognized." They continue:

> Except for the attention given those species that have created great economic loss or have been listed as quarantine species by the Bureau of Animal and Plant Quarantine, little research has been conducted on invasive alien species and their impacts on natural ecosystems.[75]

In this chapter we have described the nature of invasive species, clarifying the difference between intentional and unintentional spread and economic as opposed to non-economic effects.

Our primary concern has been with those invasive species that have adverse impacts on food production. We gave examples of plant, insect, fish, and a few other alien species, illustrating their region of origin, transmission to China, and

---

[72] See Xie et al., 2001, 108.

[73] These future directions correspond closely to the action plans on alien invasive species outlined in *China's Strategy for Plant Conservation*: "Action 10.1 Establishing and improving related laws and regulations and emphasizing risk assessment of alien organisms . . . Action 10.1 Enhancing scientific research on alien invasive organisms . . . . Action 10.3 Enhancing inter-departmental information exchange and actively initiating international cooperation . . . Action 10.4 Enhancing technical training and public education of alien invasive organisms." See Editorial Committee, 2008, 28–29.

[74] Xiaofu Xie, *Plant Introduction* (in Chinese). Beijing: Science Press, 1994.

[75] Ding and Xie, 2001, 103.

effects on native species and local ecosystems. Knowledge of the damage costs to Chinese agriculture and animal husbandry are quite inexact, but provisional estimates range in the billions of dollars. (The costs in the USA are estimated to be $140 billion annually.)

About half of China's alien species were intentionally introduced; some of these did not meet the expectations of those who imported them and escaped into the wild. The other half of China's alien species arrived without an invitation, as an indirect consequence of human activity. Some invasives infested regions and areas that became vulnerable because of population concentration and economic development. Others took advantage of China's improved transportation infrastructure, traveling along new road, rail, and river corridors. The largest number of invasive species (and also the most recent) arrived consequent to the large growth of China's trade with other nations, since the onset of economic reform. Both human and natural factors explain a potential change in numbers of invasive species, associated with climate change.

The invasive species issue is global, and it presents a grave threat to the ecological and economic well-being of the planet. China participates in international conventions, such as the CBD, which focus attention on the identification and resolution of alien invasive species problems; China is also responsive to international NGOs, such as the World Conservation Union, which offer nation-states guidelines regarding invasive alien species. In response to these identified problems, China has developed law and regulations, and the issue has entered the implementation system. Four methods of controlling problem species have been employed. Unlike economically developed nations, China's labor costs are relatively low, and it can take advantage of a large labor force to counteract immediate and serious invasive species problems. Scientists, officials, and farmers also have used biological and chemical control methods, with uneven results to the present. Most promising (and to the present least used) is integrated pest management as a means of controlling invasive species damages.

It is too early in the process of control efforts to make judgments about their efficacy, but the directions charted by scientists and officials seem quite reasonable: strengthening regulations and implementation, enhancing international collaboration, increasing public awareness and knowledge of this issue, and improving scientific research and application. In the next chapter we turn to another new problem influencing food security, the development and production of genetically-modified organisms.

Chapter 7
Biotechnological Responses to Food Security Needs

Abstract This chapter begins by explaining the development of biotechnology, including motivation for China's aggressive pursuit of it, organization and funding of agricultural biotechnology research, role of the private sector, and importation of GMOs (the case of soybeans). It features the "success story" of Bt cotton, narrating its development, initial impacts, and recent problems in production. The chapter discusses other crop and plant species approved for commercialization. It then introduces the special case of genetically-modified rice, including development and testing of Bt rice, and its promise as well as resistance to its commercialization. The next section considers biosafety concerns about GMOs in China. The recent context is global controversy, because of European Union resistance to GMOs and the Cartageña Protocol on Biosafety. In China, the elite is divided on whether to approve commercialization of Bt rice, while the general public is not well informed on the issue. The final substantive section discusses the biosafety regime in China: its origin, structure, and how biosafety currently is being implemented.

Keywords Biotechnology • Hybrid rice • 863 Program • 973 Program • Monsanto • GM soybeans • Bt cotton • Bt rice • Pre-commercial clearance • Bt maize • European Union • Cartageña Protocol on Biosafety • Biosafety regime • Precautionary principle.

7.1 Introduction

Although China's arable land today is sufficient to produce food for the population, both pressures on that land and the growing population itself mean that ways to increase agricultural productivity are essential. In Chapter 3, we indicated some of the methods employed to increase agricultural output, such as improvements in farming infrastructure and greater use of irrigation. In this chapter, we turn to the current "scientific revolution" in agriculture, the combination of biological and technological techniques and methods, focusing on improvement of seed varieties.

J.H. McBeath and J. McBeath, *Environmental Change and Food Security in China*, Advances in Global Change Research 35, DOI 10.1007/978-1-4020-9180-3_7,

For hundreds of years, farmers as well as agriculturalists have sought improvements in the disease- and drought-resistance of crop species through breeding programs. They have carefully selected seeds from crops that have produced the most desirable characteristics; they have produced hybrid species through grafting superior elements of one variety onto another. What sets the new approaches apart is their invasive nature and technological requirements. They penetrate within the seed possessing the desirable property (such as insect resistance), and extract the relevant gene, which is then transplanted into the crop species. This requires the equipment of a modern research laboratory, and only those with requisite knowledge – such as agricultural scientists and geneticists – are able to do the work.

The products of this new biotechnological approach are called genetically-modified organisms or GMOs. They also are referred to as transgenic (*Zhuanji*) species. In this chapter we first treat the development of biotechnology in China, and focus on the most successful new species, Bt[1] cotton. We consider other species that have undergone testing and commercialization. A second area of focus is the special case of Bt rice, which also is the most controversial because of the prime importance of rice in China's food system. The chapter concludes with analysis of biosafety concerns regarding GMOs in China.

7.2 The Development of Biotechnology in China

Among the world's economically developing nations, China has focused greatest effort and spent the most money on biotechnological development. We explore China's motivations, the system in which the approach has formed, the limited role of private organizations, and China's trade in biotechnology products.

7.2.1 Motivations

Several factors explain why China has so aggressively pursued agricultural biotechnology. First in order of importance doubtless is food security. Some specialists predict that China's food production must increase significantly to keep pace with population growth in future decades.[2] Enhancements to agricultural

[1] Bt is the abbreviation of *Bacillus thuringiensis*.

[2] See, for example, Qifa Zhang, "China: Agricultural Biotechnology Opportunities to Meet the Challenges of Food Production," in G. J. Persley and M. M. Lantin, eds., *Agricultural Biotechnology and the Poor: Proceedings of an International Conference*. Washington, DC: Consultative Group on International Agricultural Research, 2000, 45–50. Zhang uses the figure 60%, which seems quite high. Population projections of the UN show China reaching its population peak in 2025–30, with an increase of 15–20%.

practices, some mechanization (such as of irrigation systems), and improved fertilizers have not had a sufficiently large impact on production rates.[3] The Ministry of Agriculture (MOA) has invested heavily in "super rice," and one hybrid developed in the early 1970s by Yuan Longping has been successful in increasing rice productivity.[4] However, productivity of other hybrids has not yet met expectations for greatly improving grain yields.[5] A related factor is demand for improvement in quality of food products and in particular the quality of cereal crops. The release of new high yield cultivars and hybrids has not been accompanied by much improvement in the eating and cooking quality of the products. In fact, Zhang comments "(M)ost of the widely used cultivars and hybrids have poor cooking and eating qualities, and thus are disfavored by producers and consumers."[6]

Second, concerns about environmental degradation motivate the growth of agricultural biotechnology. In Chapter 3 we discussed the extent to which intensive applications of chemical fertilizers and insecticides have eroded soils and increased soil desertification, salinity, and acidity. As we shall see, development of disease- and pest-resistant GMOs has the promise to reduce significantly the application of chemicals on China's farmlands, and thereby limit deterioration of the environment.

A third factor is China's drive to modernize, especially after Deng Xiaoping launched China's opening up to the world and marketizing reforms. As discussed in Chapter 2, Premier Zhou Enlai announced the "Four Modernizations" shortly before his death in 1976, including both agricultural and scientific/technological advances. In the early 1980s, reform leadership emphasized scientific development of agriculture, and biotechnology development fits perfectly within this area of concentration. Also, agricultural biotechnology was then at the leading edge of scientific research, and for China to be heavily investing in it lent a cachet of legitimacy to the state.

---

[3] A Ministry of Agriculture report in 1996 indicated that production rates of major grain crops had declined in the previous decade. See: Ministry of Agriculture, *Report of Agriculture Development in China* (in Chinese). Beijing: China Agricultural Press, 1996.

[4] Yuan is a scientist in the Chinese Academy of Sciences (CAS). Supporters of his work claim that the hybrid he developed "boosted the country's rice output over the ensuing 25 years by a combined 400 million tons." See Ying Wang, "Secrets of Hybrid Crops in Focus," *China Daily*, October 15, 2007, 3. In 2008, one of Yuan's colleagues, Qing Xianguo, said the innovation in hybrid rice technology would increase production 20 percent annually. See Jiao Wu, "Hybrid Rice Set to Boost Yield by 20%," *China Daily*, July 24, 2008, 3.

[5] The Ministry of Agriculture has conducted research and development on highly productive rice species for a decade. Most of the three dozen varieties of super rice are strains for just a single-season harvest. They thrive only in high-yield farmlands where natural conditions are optimal, and not under normal conditions. Further, inadequate auxiliary planting and management expertise limit productive capacity. See Huanxin Zhao, "Higher Yields Sought from the Super Rice," *China Daily*, October 7–8, 2006, 7.

[6] Zhang, 2000, 45.

A final factor, of less direct significance, is the extent to which biotechnology developments might improve the competitiveness of China's many small farms.[7] What stands out in the China case is that the state has played virtually every role in biotechnological development.[8]

7.2.2 *Organization and Funding of Agricultural Biotechnology Research*

Most of the Chinese research in agricultural biotechnology has been done in the public sector. The state established National Key Laboratories, which were designed to carry out research programs in the areas of agricultural biotechnology, plant breeding, and crop genetics. Such laboratories were built in North, Central and South China. Too, the Ministry of Agriculture, the Ministry of Education, and the Chinese Academy of Sciences supported laboratories at many universities throughout the nation.[9]

In 1986, the State Council also established dedicated foundations to support basic and applied research, including the National Natural Science Foundation of China and the Chinese Foundation of Agricultural Scientific Research and Education. Providing focus to research initiatives and programs for research and development of biotechnology were the National Program on High Technology Development (called the 863 Program) and the National Program on the Development of Basic Research (called the 973 Program). In both these programs, agricultural biotechnology was a major component.[10] A dedicated funding source was the Special Foundation for Transgenic Plants, managed by the Ministry of Science and Technology (MOST) from 1999 to 2005.

A number of provinces and local governments also established biotechnology programs. Also, China took advantage of international funding sources to support research of its scientists, including subvention of programs by the Rockefeller Foundation, McKnight Foundation, and the European Union–China collaboration programs among others. Altogether, China has nearly 200 major biotech laboratories funded by governments at different levels, with more than 40,000 technical and research personnel.

The importance of a country's effort typically is revealed in the amount of resources allocated to it. Jia and Peng indicate that the central government funded

---

[7] This factor is mentioned in Valerie Karplus's 2003 summary of China's crop biotechnology program. See "Let a Thousand GM Crops Bloom," *International Herald Tribune*, October 8, 2003.

[8] James Keeley, "Balancing Technological Innovations and Environmental Regulation: An Analysis of Chinese Agricultural Biotechnology Governance," *Environmental Politics*, Vol. 15, no. 2 (April 2006), 293–309.

[9] Zhang, 2000, 45–6.

[10] Zhang, 2000, 46. The 863 program is named for the date Deng Xiaoping introduced it, March 1986; similarly, the 973 program is named for March, 1997, when Ziang Zemin inaugurated it.

the 863 Program to the tune of 2.86 billion yuan (approximately US $381 million), shared equally between agricultural biotechnology and medical biotechnology, in the 15-year period from 1986 to 2000.[11] The Chinese government spent $112 million (USD) on crop biotech research in 1999, outspending all other developing countries (including India and Brazil) combined.[12] This commitment of the central government increased in the twenty-first century. In 2000 the leadership quadrupled funding for the period 2000–2005.

By 2003, China was spending about $350 million in plant biotechnology and $570 million in total agricultural biotechnology (plant, animal and microorganism) in terms of purchasing power parity.[13] Thus, the government has been spending nearly US $1 billion per year, which is close to the amount that the American government has spent annually on plant biotechnology research (but overall US spending is higher, because of significant private sector investment).[14]

In 2007–2008, central government planners were preparing to expand and enlarge applied research on GMOs. A specialist in this area commented:

> Approval and management of GMOs will have the same status as the recently announced plan to develop new airplanes and to land a Chinese on the moon. The multi-billion dollar projects in GMOs would include rice, corn, cotton and also wheat—those with the greatest likelihood of making a difference in food productivity. Then, genetic modification of sheep, cows, and other livestock.[15]

In 2008 the State Council approved a plan focused on cultivating high yield and pest-resistant GM crops. China would spend at least $1 billion a year on biotechnology in the next 2 decades.[16] While observers claimed that expenses on agricultural biotechnology might even double every 3 years, they are not likely to keep pace with other biotech applications, for example in medicine.

7.2.3 Role of the Private Sector in Agricultural Biotechnology

To the present, Chinese private firms have played a marginal role in research and development of agricultural biotechnology. Most of the private firms are spin-offs from public sector research laboratories, and they lack the capital to support the

[11] Shirong Jia and Yufa Peng, "GMO Biosafety Research in China," *Environment and Biosafety Research*, Vol. 1 (2002), 5–8.

[12] Jikun Huang, Carl Pray, Scott Rozelle and Qingfang Wang, "Plant Biotechnology in China," *Science*, Vol. 295, no. 5555 (January 25, 2002), 674–77.

[13] Scott Rozelle, Jikun Huang and Keijiro Otsuka, "The Engines of a Viable Agriculture: Advances in Biotechnology, Market Accessibility and Land Rentals in Rural China," *The China Journal*, no. 53 (January 2005), 93.

[14] Julian Alston, Phil Pardey and J. Roseboom, "Financing Agricultural Research: International Investment Patterns and Policy Perspectives," *World Development*, Vol. 26, no. 6 (1998), 1057–72.

[15] Personal interview with scientist and specialist in GMO development, Beijing, March 13, 2007.

[16] Jiao Wu, "Green Light for Plan to Boost GM Crops," *China* Daily, July 11, 2008, 2; also see Shanshan Wang, "More Cash for Biotech Research," *China Daily*, April 5–6, 2008, 3.

extensive laboratory and field testing now required to meet regulatory authority approval for commercialization.

Some private firms, however, have succeeded by partnering in joint ventures with multinational corporations, such as Monsanto and Syngenta. It is global corporations working in China that have made the largest private sector contribution to agricultural biotechnology development. Monsanto, a global agribusiness firm, had invested in China before the foundation of the People's Republic in 1949. In the mid-1980s, the firm returned to China, and its primary business endeavor was sale of agricultural chemicals such as herbicides and fungicides and sales of seeds.

In 1997 Monsanto introduced its Bt cotton to China (discussed below), and this business has expanded. Still, as a multinational without the ability to conduct tests of GM products in Chinese fields, Monsanto's direct reach into the agricultural biotechnology market is limited. Monsanto officials state that it has a small share of the Bt cotton being grown in China today,[17] but knowledgeable Chinese scientists say most of the Bt cotton has Monsanto's gene.[18] This corporation remains the chief foreign multinational in the biotechnology business.

Syngenta is the world's largest agricultural company, and it has been active in China for nearly two decades. In China it sells a range of agricultural chemicals, but its fastest growing business has been in selling seed varieties. The headquarters of the firm is in Basel, Switzerland, and because a sizable part of its business is within the European Union, where attitudes are adverse to GM development and regulations are prohibitive, Syngenta has not developed transgenic plants for the China market.[19]

The important point is that, to the present, central government funding has been critical to the development of agricultural biotechnology; private sector firms with the exception of Monsanto have contributed relatively little.

7.2.4 China's Importation of GMOs: The Case of Soybeans

Soybeans are an important food crop in China. They are used to make soy sauce, bean curd, and are part of many processed foods; and soya is used as an essential ingredient for livestock feed. In the 1980s and 1990s, China had imported some soybeans, but importation increased exponentially in the early twenty-first century, for two reasons. First, in anticipation of its entry into the WTO (which would require China to abolish its quotas on importation of foreign goods), China liberalized its soybean sector in the late 1990s, and prices for soybeans dropped sharply. Wholesalers increasingly found prices of imported soya from the USA, Argentina, Canada, and Brazil to be lower than domestic prices. Second, large milling companies

[17] Personal interview with Monsanto corporate officer, Beijing, May 18, 2005.

[18] Personal communication from a food policy specialist, January 24, 2008.

[19] Personal interview with Syngenta corporate officer, Beijing, May 25, 2005.

on China's East Coast, many of which had international corporate investment, increasingly relied on foreign imports.[20]

By the early twenty-first century, about 18–20 million metric tons of soya were being imported annually, and some 70% of this was transgenic. In fact, in 2003, China was the world's largest soybean buyer, and imported nearly 21 million tons, most of which was transgenic.[21] For example, Monsanto exports RoundupReady soybeans to China and has a large share of the market. The amount of imports and their market success have stimulated the central government to develop testing and production areas in the Northeast (in Heilongjiang Province).

In 2002–2004, Greenpeace, which is opposed to GMOs internationally because of concerns that they will contaminate native species (and China is a global center for biodiversity of soya) and may adversely affect human health, raised concerns about the safety of soybean and soya imports to China.[22] Exporters of soya to China, such as Monsanto, believe these allegations to be erroneous. One corporate official remarked:

> The imported soybeans are not used as seed. They are used as grain products, for crushing; they are not allowed into the field. The oil crushing companies buy them. They unload the beans into warehouses, and they don't go into the fields. Greenpeace's charge of a threat from GMO soybeans is false.[23]

Government concerns about the biosafety of GM soybeans (as well as pressures from domestic competitors for the soybean market) have led to a number of scientific studies. For example, Wang and Liu found a gene flow from transgenic soybeans to wild soybeans, but the frequency was low because wild soybeans are self-pollinating.[24] To the present, then, the risk emanating from imported GM soybeans has been deemed to be minimal.

7.3 The "Success Story" of Bt Cotton

As mentioned in Chapter 4, China is one of the world's largest producers of cotton, and it is a very important cash crop. Cotton is more affected by insect pests, specifically the bollworm, than any other crop grown in China. The cotton bollworm incidence dramatically increased in the early 1990s, and this pest has a devastating

[20] Personal interview with food scientist, Chinese Academy of Agricultural Sciences, Beijing, May 21, 2007.

[21] Tianfu Han and Wensheng Hou, "Research Progress and Application Potential of Transgenic Soybeans in China," in Dayuan Xue, ed., *Risks and Regulation of Genetically Modified Organisms* (in Chinese). Beijing: China Environmental Science Press, 2005, 114.

[22] Personal interview with NGO representative, Beijing, May 22, 2007.

[23] Personal interview with multinational corporation executive, Beijing, May 15, 2005.

[24] Chaoyong Wang and Yan Liu, "Monitoring of Gene Flow from Transgenic Glyphosate-resistant Soybean to Wild Soybean," in Dayuan Xue, *Environmental Impacts and Safety Regulation of Genetically Modified Organisms* (in Chinese). Beijing: China Environmental Science Press, 2006, 157–61. See also Youfu Xia and Li Li, "On Comprehensive Impacts of GM Soy Importing on Biosafety in China," in Xue, 2006, 68–79.

impact on cotton yields. (Reportedly, it causes $5 billion in damages globally every year.) It was in this context that China adopted transgenic cotton, which is the most successful GM crop in China today (and the only GMO under large-scale commercialization).

7.3.1 Development of Bt Cotton

The Bt cotton grown in China today has varied sources: several dozen strains of transgenic cotton (incorporating a gene resistant to bollworm infestation) have been developed by scientists in the Chinese Academy of Agricultural Sciences (CAAS), and nearly a dozen strains were developed by Monsanto. Approval of the commercialization of Bt cotton in the late 1990s was relatively quick, for several reasons. Transgenic cotton already had been approved for growing in the United States. Also, transgenic cotton did not appear to present a food safety risk (with the exception of cotton oil and cakes, as noted below), nor did it seem likely to cause agronomic problems.

Pray et al. note that the approval process took just 2 years, which in retrospect seems quite short. Monsanto was required to conduct environmental impact trials for 2 years as well as mouse feeding trials (to ascertain the safety of cotton oil and seed cake). The Biotechnology Research Institute of CAAS had to demonstrate the effectiveness of its technology with several years of field trial data. Scientific panels called by government agencies opined that the mixed cropping system used by cotton farmers in North China made it unnecessary to establish separate refugees for non-Bt cotton.[25]

Bt cotton was approved for commercialization in 1997, which was before controversy developed internationally concerning the safety of agricultural biotechnology. After this point, and for trade reasons as discussed below, China tightened its regulatory requirements (also discussed in Chapter 8). The Ministry of Health's Nutrition and Food Safety Institute conducted food safety testing in 2000 of cotton seed oil and cake. The Institute of Plant Protection of the Ministry of Agriculture also received funding to conduct research on environmental effects of GM crops.[26]

By the early twenty-first century, Bt cotton varieties had been planted in 12 provinces. In 2001, about 2 million hectares had been planted in Bt cotton, half using the CAAS varieties and the other half the Monsanto varieties.[27] In 2009, more than 15 million hectares of Bt cotton grow in China, which is about 70% of the total area of cotton production. Monsanto's share of production has declined,

---

[25] Carl E. Pray, Bharat Ramaswami, Jikun Huang, Ruifa Hu, Prajakta Bengali, and Huazhu Zhang, "Costs and Enforcement of Biosafety Regulations in India and China," *International Journal of Technology and Globalisation*, Vol. 2, nos. 1/2 (2006), 148.

[26] Ibid.

[27] Jia and Peng, 2002, 5.

and today company officials report it has a relatively small share, about 10% of total production.[28]

In fact, Bt cotton varieties with the Monsanto gene account for a very large share of the total varieties grown by Chinese farmers. This introduces the sensitive issue of intellectual property rights (IPR), a major source of friction in trade relations between the United States and China. Many domestic seed companies in China generate Bt cotton seeds through back-crossing Monsanto varieties with local varieties.[29] Thus, Monsanto has played a very important role in Bt cotton production.

7.3.2 Initial Impacts of Bt Cotton

Although a few critics of Bt cotton alleged that its use did not increase the income of farmers (because Bt cotton seeds are from 100% to 250% more expensive than non-Bt seeds) or enhance productivity, most analysts have found evidence of positive impacts: in reducing the use of pesticides, reducing farmers' costs, and increasing cotton yields. Huang et al. found that farmers adopting Bt cotton sprayed two-thirds less than those who did not adopt the transgenic variety. Also, farmers reduced their expenditures on pesticides by 82%. And, because they no longer needed to spray frequently, they cut labor costs. Moreover, yields of Bt cotton were from 7% to 15% higher.[30]

The reduction in pesticide use is a very important consequence of agricultural biotechnology, because of the highly toxic nature of insecticides used in rural China. Insecticides are used more prevalently than either herbicides or fungicides, and China is one of the largest pesticide users in the world. One heavily used insecticide is HCH (called 666 in Chinese) and marketed as Lindane. It is an organochlorine, which does not wash out of the environment (it has a half life of 8–10 years). This insecticide has chronic toxicity, and enters the human nervous system, causing hormonal changes, leading to poisoning and death of farmers. Huang et al. illustrate the poor farming practices that lead to human damage:

> (A) vast majority of farmers have not changed the way that they handle and apply pesticides in recent years. Most pesticides are mixed by hand, applied without any protective clothing or breathing apparatus, and residues are discarded in irrigation ditches and other commonly used water sources.[31]

---

[28] Personal interview with Monsanto corporate official, Beijing, May 25, 2007.

[29] Personal communication, Chinese agricultural scientist, January 24, 2008.

[30] Jikun Huang, Ruifa Hu, Hans van Meijl and Frank van Tongeren, "Biotechnology Boosts to Crop Productivity in China: Trade and Welfare Implications," *Journal of Development Economics*, Vol. 75 (2004), 27–54. See also, Yuejing Su, Ruifa Hu, Jikun Huang, and Cunhui Fan, "The Determinants of Farmer Bt-Cotton Technology Adoption Behavior in China," *Cotton Science* (in Chinese), Vol. 16, no. 5 (2004), 259–64.

[31] Jikun Huang, Ruifa Hu, Carl Pray, Fangbin Qiao, Scott Rozelle, "Biotechnology as an Alternative to Chemical Pesticides: A Case Study of Bt Cotton in China," *Agricultural Economics*, Vol. 29 (2003), 55–67.

Another frequently used insecticide is an organophosphate, with similar effects.

In 2007 the Ministry of Agriculture adopted regulations prohibiting the use of these pesticides, but many farmers remain unaware of them. Moreover, because these insecticides are perceived to be effective, there is a market for them, which business firms supply even though sales are now illegal.[32] The development of Bt cotton, by reducing the need to use insecticides, has obvious benign effects on human health as well as the environment.

One other area of impact is worth noting, and that is the welfare effect in using Bt cotton. Agricultural economists such as Huang Jikun and his colleagues argue that introduction of Bt cotton reduces costs of cotton production and therefore cotton prices.[33] Anderson and Yao extend this analysis by considering the impacts on China's domestic prices as well as export sales of textiles and clothing. They contend that the cheaper domestic cotton supply would lead to an expansion of exports of finished fabrics, but they would sell at somewhat lower prices. Nevertheless, they project gains for China of $300 million through using Bt cotton.[34]

The initial impacts of developing Bt cotton, then, seemed to be largely positive. The most recent scientific study found Bt cotton efficacious in controlling bollworm infestation. Wu Kongming and colleagues from the Chinese Academy of Agricultural Sciences analyzed data from 1997 to 2007 in six northern provinces. They found dramatic reductions in bollworm infestation, especially during the period 2002–2006. They concluded that Bt cotton was responsible for the long-term suppression.[35] However, they did not test for hybrid varieties or secondary pests.

7.3.3 Potential Problems in Bt Cotton Production

As Bt cotton extends throughout China, its initial blush of popularity has faded somewhat. NGOs such as Greenpeace claim that Bt cotton's performance is overrated, and that yields are less good than supporters contend. To the present, this charge has not been verified by field studies.

A more incisive criticism of the application of Bt cotton concerns its disease-resistance nature. Some critics allege that over time, the pest resistance properties of Bt cotton will be diluted as pests adapt to the transgenetic variety and increase their tolerance to it.[36] This became an issue with use of some tested but not approved Bt hybrid varieties. When these varieties were tested against the approved Monsanto

[32] Personal interview with NGO representative, Beijing, May 22, 2007.

[33] Huang et al., 2003, 1079.

[34] Kym Anderson and Shunli Yao, "China, GMOs and World Trade in Agricultural and Textile Products," *Pacific Economic Review*, Vol. 8, no. 2 (2003), 167–68.

[35] Shujuan Lin, "Scientists Report Pest-Resistant Crop," *China Daily,* September 19, 2008, 3.

[36] Personal interview with NGO representative, Beijing, May 22, 2007. See also, Paul Thiers, "Book Review: Biotech Boosterism and Chinese Realities," *China Development Brief*, June 1, 2004, 2.

and CAAS Bt varieties, they were found effective in killing bollworms early in the season, but the effectiveness of the varieties waned as the season progressed.[37]

Farmers' use of "illegitimate" or "illegal" Bt varieties makes it difficult to test the efficacy of the Bt varieties. A number of provincial and local government cotton breeding programs (and some private farms) have backcrossed the approved CAAS and Monsanto Bt genes into the best of their local varieties. Surveys conducted from 2001 to 2003 indicate that in North China, approximately 20–30% of the Bt cotton fields were planted with the "illegal" varieties.[38]

In fact, the most serious problem encountered in Bt cotton production has been secondary pests and in particular the *mirid* (an insect of the heteropteran family). What transpired was that after the adoption of Bt cotton, the plant environment changed. Bt cotton was effective in controlling bollworms, whose population declined. The mirid population, unnoticed under conditions of broad-scale cotton bollworm infestation, increased.[39]

Unfortunately, mirids cannot be controlled with agricultural biotechnology, as there is no gene to control their infestation of cotton plants. The mirid population likely will increase under conditions of climate change. At higher temperatures and humid conditions, there are more mirids, and cotton growing regions have experienced higher temperatures from May to August (with strong rainfall) in recent years. In the absence of a biotechnological solution, many farmers have turned back to pesticides to mitigate the latest iteration of insect infestation. This limits the potential of Bt cotton.

7.4 Approval of Other Crop and Plant Species

In the 1990s there was an explosion of experimental research and testing on agricultural transgenic species. Chinese scientists by 2001 were studying nearly 150 different organisms and examining more than 200 candidate genes to transform these organisms.[40] However, the certification process changed as GM products came under increased international scrutiny, and as international NGOs and the European Union developed strong policies against their adoption. This international criticism led to the development of a Biosafety Committee in China and tight procedures governing approval for testing and commercialization (discussed below).

---

[37] K. M. Wu, "Impacts of Bt Cotton on Status of Insect Pests and Resistance Risk of Cotton Bollworms in China." Conference on Resistance Management for Bt-Crops in China: Economic and Biological Considerations. Raleigh, NC: North Carolina State University, April 2002.

[38] Pray et al., 2006, 153. Also, personal interview with agricultural policy analyst, CAS, Beijing, May 29, 2007.

[39] Personal interview with entymologist, Chinese Academy of Agricultural Sciences, Beijing, March 27, 2007.

[40] Jia and Peng, 2002, 5.

By 2009, the Chinese government had approved applications for about 60 transgenes, including those influencing insect-resistance, shelf-life, virus resistance, herbicide tolerance, and quality improvement. No cereal or oil crop has been approved for commercial use as of this date.

The small number of crops approved for commercialization or approaching that stage includes tomatoes, sweet and chili peppers, petunias, nitrogen-fixing bacteria, vaccines for animal use, poplar, and papaya. (A tobacco transgenic crop was approved for commercialization in 1992. This was the first such approval in the world, but it was withdrawn when tobacco exporters in the USA and other nations objected.) A shelf-life altered tomato variety has been planted in a small area in Hubei Province.[41] Color-altered petunias and virus-resistant green peppers have been commercialized in small areas of China. Green peppers could be commercialized on a large scale, but as one biotechnology research professor opined, "The yield and quality are not as good as the natural products."[42]

A variety of poplar (*yangshu*) is now being introduced in large sections of the North and Northwest (particularly in Shaanxi and Inner Mongolia), because it has properties appropriate for growth in dry regions to stop desertification. It was approved for commercialization in 2005. Although this variety does not counteract all tree diseases, it has had good results for resistance to caterpillars. Similarly, in 2007 a variety of papaya was being commercialized with disease-resistance properties.[43]

In general, though, the momentum behind development of agricultural biotechnology slowed considerably in the early twenty-first century, because of biosafety and other concerns regarding Bt rice.

7.5 The Special Case of Genetically-Modified Rice

Rice remains China's primary food crop, and for this reason, development of transgenic rice varieties, but especially their commercialization, has been fraught with controversy. We review the development and testing of GM rice varieties, the promise of Bt rice, and the sources of resistance to its commercialization.

7.5.1 Development and Testing of Bt Rice

Since the late 1990s, a large number of transgenic rice varieties and hybrids has been developed in government research laboratories. The technical details of genetic manipulation go beyond the scope of this volume, but the varieties include

---

[41] Ibid., 5–6

[42] Personal interview with biotechnology research institute professor, Beijing, May 30, 2007.

[43] Personal interview with plant pathology research professor, Beijing, May 18, 2007.

(1) several transgenic Bt rice varieties and hybrids that are resistant to rice stemborer and leaf roller, as well as leaf hoppers; (2) transgenic rice with resistance to bacterial blight and rice blast; (3) transgenic rice with herbicide tolerance; and (4) varieties with drought and salinity tolerance.[44]

These varieties underwent field trials and environmental tests, following the new, tightened biosafety regulations of China. Four rice hybrids have advanced to the pre-production phase. Three of these varieties contain genes, either Bt or other, providing resistance to stem borers. The fourth hybrid has genes providing resistance to bacterial blight, a prevalent rice disease in central China.[45]

These varieties of GM rice have met the stringent standards required for commercialization of biotechnological products in China, but they have not been approved, for reasons mentioned below.

7.5.2 The Promise of Bt Rice

Huang et al. conducted studies on the likely impact of transgenic rice varieties on yield and agricultural inputs, were they approved for commercialization. They concluded that adoption of GM rice would significantly increase yields of rice. It would reduce the use of pesticides, and thereby also reduce labor involved in rice cultivation. Overall, the higher price charged for transgenic seeds would be more than recovered through these cost savings. Finally, the commercialization of GM rice would enhance welfare in China by $4.2 billion in 2010, which is four times higher than the welfare benefits of Bt cotton (a much smaller agricultural sector).[46] Increased yields would likely, in the estimation of scientists, reduce the cost of rice.

The yield increases possible with GM rice are welcome news to observers of the decline in rice output of China from 1999 to 2003 (a decline of 16% as measured by USDA and FAO).[47] Supply problems triggered a 27% increase of rice prices in early 2004, showing the dependence of price on productivity. Some rice growing land has been converted to production of cash crops, such as vegetables and fruit trees. Rice productivity itself has, in the estimate of plant scientists, increased no more than 1–2% at the most annually (by using non-transgenic hybrids).[48] The slow rate of growth is notwithstanding much development of new crops (some 15,000 since the 1950s), high yield hybrid rice, and a new "super-hybrid wheat."[49]

[44] Pray et al., 2006, 140.

[45] Ibid., 140–41.

[46] Huang et al., 2003, 1080–81.

[47] Heping Jia, "China Ramps Up Efforts to Commercialize GM Rice," *Nature Biotechnology*, Vol. 22, no. 6 (June 2004), 642.

[48] Personal interview with plant scientist, CAAS, Beijing, May 19, 2007.

[49] Jin Fu, "Crop Caution," *China Daily*, May 19, 2008, 12. These figures do not account for variation in yields because of insect pests and plant diseases.

Increased concern with climate change has stimulated research into rice varieties adaptable to suboptimal growing conditions. A scientist in Shenzhen has developed a hybrid that mimics cord grass flourishing in salt marshes. His preliminary field trials with the hybrid rice showed success in growing in land areas encrusted with crystalline salt.[50] A larger group of researchers has focused on developing GMs with strong drought-resistance properties. A team of scientists from Wuhan has identified a rice gene that enhances the ability of abscisic acid to close the stomata (leaf openings) to reduce evaporation. Such an innovation might lead to greater seed production under drought conditions.[51]

Because of the promise of GM rice, farmers have grown "illegal" varieties in their fields, hybrids which have not yet been approved for commercialization. There is some question as to how the rice entered markets, but in 2005 the global NGO Greenpeace reported it had discovered GM rice in markets of Hubei Province. It sent samples to a German biotechnology firm, revealing positive results for Bt rice with a genetically-engineered pesticide. Greenpeace also found two rice samples in wholesale yards of Guangzhou.[52]

Greenpeace alleged that the GM rice came from scientists at Huazhong Agriculture University in Wuhan, who had field tested the rice. They also charged the New Technology Company of that university with having advertised rice seed samples. Both sources were alleged to have distributed 29 t of rice seeds.[53]

An official with the State Environmental Protection Administration questioned whether the Bt rice could really be considered "illegal," because the rice at issue was approved for the pre-production phase:

> The policy is unclear as to what the pre-production phase is … The question is how big of an area is planted with Bt crops … No other country has a pre-production stage. So it is not technically illegal to be planting Bt rice now. The government scientists request that farmers plant it, but it is not clear who will buy it. The government doesn't give money to farmers. So it is in between a law testing and a commercialization. The farmers plant and harvest the Bt rice. Greenpeace ventured upon a market and saw sales. Now, there are regulations that the government has to buy back seeds and crops from the farmers. Now it is better controlled.[54]

Whatever the process of transmission from laboratories to farms, it is clear that the results of field testing of Bt rice have been good, and farmers in China are interested in taking advantage of the GM seeds.

---

[50] Hong Chen, "New Rice Hybrid Grows on Salty Lands," *China Daily*, October 14, 2008, 4.

[51] Jing Li, "Scientists Seek Stronger Crops," *China Daily*, February 19, 2009, 4.

[52] *China Development Brief*, "Greenpeace Targets Retail Giant in Fight against GM Rice." August 10, 2005. An NGO representative told the authors that Greenpeace representatives informed the local agriculture department of the Ministry of Agriculture, which discovered the Bt rice in stores. Local government officials proceeded to fine the sellers and destroyed some rice fields in which Bt rice was growing. Personal interview, Beijing, March 18, 2007. See also, David Barboza, "China's Problem with 'Anti-Pest' Rice," *New York Times*, April 16, 2005, A14.

[53] Ibid.

[54] Personal interview with SEPA official, Beijing, May 27, 2007.

7.5.3 Resistance to Commercialization of Bt Rice

Four factors retard the commercialization of genetically-modified rice varieties: concerns about impacts on human health, biodiversity issues, market acceptance, and whether there is demonstrable need for GM rice. Of greatest importance is uncertainty regarding the human health and environmental effects of GM rice. Greenpeace employs the *precautionary principle*. Its belief is that if products are not completely safe for human consumption and use in environments, then they should not be in the market and people should not be able to consume them. A Greenpeace briefing paper points out that genetically-engineered rice "could have unexpected and hidden effects that could have an ecological impact or be unsafe for human consumption."[55] Specifically, Greenpeace believes that genetic disruptions could lead to production of unexpected toxins and allergenic proteins. Other NGOs, for example the Organic Consumers Association, have made similar observations questioning the safety for humans and environments of GM rice and other crops.[56]

Scientists who have worked on the development of GM rice varieties in general are supportive of commercialization. Said one biotechnology research scientist: "We have done 5–6 years of experiments on it, and they show no problem (affecting human health)."[57] The debate on human and environmental effects of GM rice has even entered the pages of *Science*, with arguments both strongly supportive and in violent opposition to this GM crop's commercialization.[58] Because of this controversy, one of our respondents, a SEPA official, commented: "Even if the risk is 1 percent, why take it?"[59] It is these health and environmental concerns that have motivated the European Union to restrict importation of GM crops, discussed below.

The second factor retarding commercialization is China's status as an origin point for biodiversity in rice species, and the fear that GM varieties may contaminate endemic and wild species. The Greenpeace briefing paper notes that although rice is largely self-pollinating, pollen is affected by wind speed and direction and can travel up to 100 m.[60] If Bt rice is cultivated in the same region where wild rice grows, hybrids likely will develop and the genetically-induced trait will "swamp" native species. Either this will improve the competitiveness of the wild species, possibly leading to the endangerment of rice crops as weeds; or hybrids may overwhelm natural wild varieties, leading to their extinction.[61]

---

[55] Greenpeace, "Rice at Risk," Briefing paper, April 2005, 2.

[56] The Institute of Science in Society, "Genetically Engineered Rice Experiments in China." Retrieved October 10, 2006 from http://www.organicconsumers.org/ge/ricechina120204.cfm

[57] Personal interview with biotechnology research professor, Beijing, May 30, 2007.

[58] See Jikun Huang, Ruifa Hu, Scott Rozelle, and Carl Pray, "Insect-Resistant GM Rice in Farmers' Fields: Assessing Productivity and Health Effects in China," *Science*, Vol. 308 (April 29, 2005), 688–90, and responses in "Debate Over a GM Rice Trial in China," *Science*, Vol. 310 (October 14, 2005), 231–33.

[59] Personal interview with SEPA official, Beijing, May 27, 2007.

[60] Greenpeace, 2005, 2–3.

[61] Ibid, 3.

Greenpeace's objections are based on its global campaign and the professional literature, and not on its own research in China. Chinese scientists recently have paid attention to the effects of GM crops on biodiversity loss. In the case of Bt rice, they have considered transgene escape and environmental consequences such as pollen flow, fitness performance and crop-to-crop as well as crop-to-wild gene flow.[62] As might be expected, evidence of contamination varies, dependent on location of GM rice-planted fields with respect to native species.

The third factor retarding commercialization is the question of market acceptance of GM rice products. In Europe, there has been relatively strong consumer resistance to GM products. In general terms (and this issue is discussed further below), Chinese consumers are less well-informed about GM products and Bt rice in particular. Although they are wary about such products, and a minority (ranging from 20% to 33%) would not purchase them if available, large-scale resistance to them has not yet developed. A recent study of consumer acceptance of GM rice (conducted in 2002) indicated that at least 60% of respondents would purchase biotech foods without price discounts; however, about 20% of consumers would only accept natural, non-biotech foods.[63] Yet, we should be cautious in using this statistic, for Greenpeace has claimed that 50% of Chinese consumers prefer non-GM food; only 26% would be willing to try one of the GM varieties.[64] And some manufacturers and retailers are labeling non-GM foods as "GM-free" to give consumers the impression that genetically modified goods are unsafe.[65]

The final factor retarding commercialization is the question of need for GM rice varieties, and their efficacy in countering disease problems. The Greenpeace briefing paper argues that GM rice resistant to bacterial blight disease is unnecessary, as conventional breeding and changes in farming practices can eliminate this plant risk.[66] Yet bacterial blight is just one rice disease for which transgenes have been developed. A larger argument against commercialization is that the current status of rice production in China is adequate. This argument has three parts. First, opponents of commercialization contend that there is no problem of grain security in China now; there has been sufficient production in recent years. Second, notwithstanding price spikes in 1999–2003, the price of rice has been stable since then (with the exception of inflation in most food prices in 2008). This indicates that a sufficient quantity of rice is reaching the market. Third, critics of GM rice maintain

---

[62] Baorong Lu and Jun Rong, "On Exotic Gene Escape of Transgenic Rice and Environmental Safety," in Dayuan Xue, *Environmental Impacts and Safety Regulation of Genetically Modified Organisms* (in Chinese). Beijing: China Environmental Science Press, 101–09; and Minsheng You, Sheng Lin, and Dingcheng Huang, "Risk Assessment of Transgenic Rice," ibid., 122–38.

[63] See William Lin, Agapi Somwaru, Francis Tuan, Jikun Huang, and Junfei Bai, "Consumers' Willingness to Pay for Biotech Foods in China: A Contingent Valuation Approach," *AgBioForum*, Vol. 9, no. 3 (2006), 166–79.

[64] See Lei Xiong, "Farmers' Fear, Food Future Genetically Modified," *China Daily*, March 1, 2007, 12.

[65] Weitao Li, "Farmers Cotton on to Biotech Industry," *China Daily*, September 11, 2007, 15.

[66] Greenpeace, 2005, 4–5.

that the insect problem has not been serious in recent years, and farmers can counter the insects with available pesticides. Unlike the cotton bollworm that devastated more than one-third of China's cotton crop in the early 1990s, insect pests do less damage to rice. In turn, farmers use just one-fourth the amount of pesticides to control insects on rice, as compared to the amount used on corn.[67]

No country in the world produces gene-altered rice for commercial sale, and this is an additional factor accounting for the controversial nature of the issue in China. In November 2006, the State Committee for the Safety of Agricultural Transgenic Living Things declined to authorize the commercial production of GM rice, for the fourth time since 2004.[68] In 2007–2009, scientists still conducted research on transgenic rice, and a MOA official said "We are at the last stage of safety evaluation."[69] Given dissent of the leadership on commercialization (discussed below) as well as continuing global controversy on the issue of GM products, it does not seem likely that China will proceed with full-scale commercialization of Bt rice soon.

A positive decision on GM rice doubtless will trigger the commercialization of other food crops, including GM corn, soybeans, barley, and wheat. Bt corn has been widely planted in some countries, such as the United States, in the last decade. In China, field and environmental testing of Bt corn has demonstrated high resistance to plant pests and economic benefits to growers.[70] Scientists have developed virus-resistant GM wheat, which has been undergoing field and environmental release trials.[71] They are conducting experiments on genetically-modified soybeans, barley, and potatoes as well.

7.6 Biosafety Concerns About GMOs

Although we have treated the resistance to commercialization of GM food products, we have not emphasized three additional factors that condition ultimate acceptance: global resistance, lack of elite consensus on GMOs, and limited public information. We consider each here.

7.6.1 Global Resistance to GMOs and the Cartageña Protocol

China tightened its biosafety regulations in response to developing animus in Europe against GMOs in the late 1990s. For example, European Union legislation

[67] This discussion is based on interviews the authors conducted in China in 2006–09.

[68] See Li Deng, "China Cautions about GM Rice" (in Chinese), *21st Century Economic News*, July 10, 2008.

[69] Jiao Wu, "Transgenic Rice Seeds Still Await Go-ahead," *China Daily*, January 26–27, 2008, 1.

[70] See Hui Liu, Shuxiong Bai, Zhenying Wang, and Kanglai He, "Effects of Transgenic Corn Resistant to Insect Pests on Diversity of Arthropods," in Dayuan Xue, 2006, 191–202.

[71] Rozelle et al., 2005, 94.

has called for a phasing out of antibiotic resistance marker genes in GM crops. Most of the GM rice varieties tested in China use such genes, which are susceptible to gene transfer to pathogenic bacteria. Too, EU rules were imposed for labeling, shipment and tracing of GM food.[72]

As China joined the WTO in late 2001, leaders paid much greater attention to biosafety requirements of other nations, while developing China's own new, tougher regulations. In 2001 the State Council promulgated the "Regulation on the Safety Administration of Agricultural GMOs." The new regulation, implemented by the Ministry of Agriculture, included management regulations, labeling requirements for marketing GMOs, as well as import and export regulations for GMOs and GMO products.[73]

In 2002 the central government announced a ban on new foreign investment in agricultural biotechnology ventures. While government officials explained the policy change by saying it was necessary to prevent foreign companies from threatening locally developed GMOs, the ban appeared to be more of a trade than a biosafety issue. The ban reflected apprehension of leaders that other countries might impose restrictions on China's agricultural exports if farmers harvested large quantities of transgenic crops, along with traditional varieties.[74] The ban on foreign investment in GMOs had the effect of segregating GMO crops from traditional varieties.

In the reform period, China's trade has expanded more rapidly than the overall growth in GDP. Trade in agricultural products is part of this trend, but the composition of agricultural trade has changed. Grain exports, note Huang and Rozelle, have fallen from about 20% of agricultural exports in the 1990s to about half of that in the early twenty-first century. Correspondingly, trade in horticultural, animal and aquatic products has increased, as has trade in processed foods.[75]

Most of China's grain historically and at the present is dedicated to the domestic market, and that is the destination for China's developing GMOs. However, WTO membership is quite important to China, and it carefully follows trade policy developments of its major trading partners, such as the European Union and the United States. When the policies of these two units are contradictory – as they are in the case of GMOs – China needs to exercise caution.

China's ratification of international conventions also influences its GMO policy. In 1995 China joined the Convention on Biodiversity, which pledged China to write a biodiversity action plan and develop measures to reduce threats to biodiversity. An ancillary protocol to this treaty – the Cartageña Protocol – which China ratified in 2005, regulates the transboundary movement of GMOs, especially those potentially posing risks to biological diversity. China is the world's largest importer

[72] Lei Xiong, 2007, 12.

[73] Pray et al., 2006, 144–45.

[74] Yimin Ding and Jeffrey Mervis, "China Takes a Bumpy Road From the Lab to the Field," *Science*, Vol. 298, no 5602 (December 2002), 2317–19.

[75] Jikun Huang and Scott Rozelle, "Trade Reform: The WTO and China's Food Economy in the Twenty-First Century," *Pacific Economic Review*, Vol. 8, no. 2 (2003), 146.

of soybeans, and in most recent years has also imported substantial quantities of maize. This puts it under treaty requirements of developing a detailed documentation regime. Similar requirements would be necessary were China to commercialize and export Bt rice and maize.[76] Clearly, China's signing this convention moved it closer to observance of the precautionary principle. Wan Bentai, director of the Nature Conservation Department of SEPA commented: "We will take effective measures to control and supervise the research, development, use and movement of GMOs and follow international practice in the field."[77]

Effectively, this is a call for caution in China's approach to agricultural biotechnology.

7.6.2 *Lack of Elite Consensus*

While the development from the mid-1980s of an agricultural biotechnology emphasis seemed to express elite consensus, that agreement has frayed within recent years. This lack of consensus can be attributed primarily to the increasing focus on risk posed to human life and environmental safety by GMOs. Political leaders are knowledgeable about biosafety, and they now have 10 years of testing experience with major crops. But no government leader wants to take the lead on an issue of such potential risk.

Ministries of the state also are divided on GMO development. The most supportive agency is the Ministry of Science & Technology (MOST). It supplies most of the funding for GM laboratory work and testing, and uniformly has urged commercialization of GM crops. The Ministry of Agriculture, which is China's lead agency on issues of agricultural development, has supported GMO development. Indeed, much of this work is done under CAAS auspices or in laboratories supervised by MOA.

However, scientists and scholars working for MOA are as likely to oppose GMO development as support it. Some are opposed on scientific principles, but a larger number have conflicts of interest. They may not be involved in GMO research and resent the diversion of funding into it. Alternatively, they may be engaged in GMO testing and studies, but want a larger share of the research funding pie. Also, they may resent the formation of new businesses, spun off of their research laboratory discoveries, over which they have no control.[78]

---

[76] Jikun Huang, Deliang Zhang, Jun Yang, Scott Rozelle, and Nicholas Kalaitzandonakes, "Will the Biosafety Protocol Hinder or Protect the Developing World: Learning from China's Experience," *Food Policy*, Vol. 33 (2008), 1–12. For a broader view of the impact of globalization on agricultural biotechnology, see Peter Newell, "Globalization and the Governance of Biodiversity," *Global Environmental Politics*, Vol. 3, no. 2 (May 2003), 56–71.

[77] Quoted in Jing Li, "World Biosafety Standard Adopted," *China Daily*, May 20, 2005, 5.

[78] Personal interview with agricultural policy analyst, Beijing, May 27, 2007.

Finally, the National Development and Reform Commission (NDRC) is fully ambivalent on GMO development. It supports the research, but opposes commercialization. In the words of one close observer of this agency's behavior: "It is afraid of the uncertainty, about the effects on children, about future risks."[79]

In short, the state's initial strong support for agricultural biotechnology has fractured, largely because of biosafety concerns.

7.6.3 Limited Public Information on GMOs

We discussed market factors above, and noted differing interpretations as to how many consumers readily would accept GM products. This raises the question: How informed are Chinese consumers about GM products? Huang et al. designed and administered a consumer survey in 11 cities of five provinces in eastern China in 2002 and 2003, in order to test market acceptance of GMOs. They found that some two-thirds of respondents had heard about GM foods, which the authors explain by the relatively long history of hybrid breeding technology in China. But when respondents were asked if they had ever eaten GM products, only 18% of respondents said they had. In fact, in 2002 China imported half the soybeans consumed domestically, and most of these soybeans were genetically-modified products. Further, about half of the respondents said that GM fruits were sold in the market, when they were not.[80] This information about the general public's actual knowledge of GM products needs to be kept in mind when viewing statistics, such as those from the Huang et al. study, to the effect that the acceptance rate of GM products in 2002 was 88%, among the highest in the world.

Chinese elites are familiar with GMO issues, but members of the general public are not. Newspapers, TV and radio stations do carry news of GM developments, but mostly of breakthroughs. A second consumer survey of 1,000 respondents, directed just to residents of Beijing, attempted to ascertain the familiarity of the public with regulatory changes and biosafety requirements. Some 45% of the respondents to this survey in 2004 did not know that the state had implemented a labeling system for GMOs 2 years previously. Of those who were familiar with the labeling system, another 43% could not identify the GMO labels.[81]

China remains an economically developing country, and just a small percentage of its residents – under 4% – are college-educated. Although the regime has liberalized

---

[79] Ibid.

[80] Jikun Huang, Huanguang Qiu, Junfei Bai and Carl Pray, "Awareness, Acceptance of and Willingness to Buy Genetically Modified Foods in Urban China," *Appetite*, Vol. 46 (2006), 147–48. See also: William Lin, Agapi Somwaru, Francis Tuan, Jikun Huang, and Junfei Bai, "Consumer Attitudes Toward Biotech Foods in China," *Journal of International Food & Agribusiness Marketing*, Vol. 18, no. 1/2 (2006), 177–200.

[81] Yuqing Wang, "Labeling System for Genetically Modified Organism (GMOs) in China and its Recognition Status in Beijing Market," in Dayuan Xue, 2006, 244–46.

its controls over the media during the reform period, newspapers and TV stations are still censored (or self-censored). Given a substantial degree of public ignorance concerning GMO issues, the central government can mobilize opinion in support of GMO products, if the decision is made to commercialize them. Yet the Internet is a threat to state censorship, and China has the largest number of Internet users in the world. One of our respondents put the general public's lack of knowledge about GMOs in the context of type of medium relied on (Internet, personal communications, newspapers):

> If information about GMOs comes from the Internet, or from friends, then favorable perceptions of GMOs decline precipitously. Some 90 percent of the Internet blogs on Bt varieties are negative. The greater knowledge of the technology, the greater acceptance of it. Those who don't know are more likely to oppose it. Also, government media are more positive than non-government media.[82]

This suggests that as people become more informed, they become more receptive to GMOs, an acceptance that is reinforced by government media such as newspapers and TV broadcasts. If uninformed, they are subject to media impacts, and then the source of medium – government-owned newspapers and TV or Internet blogs – makes a large difference in opinions.

7.7 China's Biosafety Regime

China's first regime in the area of food security concerns the safety of genetically-modified organisms or GMOs (also known as living-modified organisms or LMOs). We describe the origins of this regime, its structure, and implementation (including related issues).[83]

7.7.1 Origin of the Biosafety Regime

As noted above, China's investment in agricultural biotechnology began in 1985, and the 1990s was a period of extremely rapid growth in scientific laboratory development as well as field-testing. Indeed, by 1996, some 47 genetically-modified plants had been transformed by use of genetic engineering technology. This figure included seven grain crops, five economic crops, ten vegetables, 11 fruit trees, three forest trees, two fodder crops, and five medicine and ornamental plants, and work

[82] Personal interview with journalism professor, Beijing, May 18, 2007.

[83] For a general introduction, see Tianjie Ma, "Wielding the Double-Edged Sword: The Chinese Experience with Agricultural Genetically Modified Organisms," Woodrow Wilson center, China Environmental Forum, 2008.

on 103 genes (excluding marker genes).[84] This rapid development was almost entirely the product of active government engagement in sponsorship of scientific research in government laboratories of biotechnology.

Initially, work on GM products proceeded under relatively little government regulation. The Ministry of Science and Technology (MOST) adopted the first regulation only in 1993, entitled "Safety Administration Regulation on Genetic Engineering." It stated the general principles for research work, and broadly spelled out safety categories, risk evaluation, application and approval processes, safety control principles, and legal responsibilities.[85] MOST asked the relevant ministries to formulate and issue regulations, and within 3 years, the Ministry of Agriculture (MOA) approved the "Implementation Regulation on Agricultural Biological Genetic Engineering." This too was loosely drawn and included neither labeling requirements nor restrictions on importation or export of GM products. Under MOA's regulation, a National Agricultural GMO Biosafety Committee was formed in 1997 to provide the ministry with expert advice on biosafety regulation.[86]

This was the official beginning in construction of the Chinese biosafety regime, and its evolution over the next 4 years was strongly influenced by two developments: the onset of resistance to GM products, especially in Europe, and the developing requirements of the global biosafety protocol. Within the European Union, NGOs, such as Greenpeace, and many scientists raised objections to the safety of GMOs. In response, the EU began the development of detailed regulations on GM imports.[87] Although China's exports of products including LMOs to Europe were then (and are today) tiny, Europe is a large market for other Chinese products including processed foods, and European resistance to GMOs was a factor in the tightening of China's system.

The second factor was also international. In 1995, China joined the Convention on Biodiversity (CBD), and biosafety is an important part of this agreement. International discussions on implementation of the CBD, conducted by the United Nations Environmental Program, focused on potential threats of GMOs to biodiversity, to the environment in general and to human health. By the fifth Conference of the Parties to CBD in 2000, meeting at Cartageña, Colombia, an international protocol on biosafety had been developed.[88] China signed the protocol and in 2005 became the 120th country to ratify the Cartageña Protocol on Biosafety.

---

[84] Changyong Wang, Zhidi Yu, and Dehui Wang, "China: Risk Assessment and Risk Management," IUCN Regional Biodiversity Programme, Asia. Retrieved May 30, 2006 from: http://www.rbp-iucn.lk/biosafety/CouStatus_China.htm

[85] Carl E. Pray, Bharat Ramaswami, Jikun Huang, Ruifa Hu, Prajakta Bengali, and Huazhu Zhang, "Costs and Enforcement of Biosafety Regulations in India and China," *Journal of Technology and Globalisation*, Vol. 2, nos. 1/2 (2006), 144.

[86] Ibid.

[87] See Heike Baumuller, *Domestic Import Regulations for Genetically Modified Organisms and their Compatibility with WTO Rules: Some Key Issues*. Winnipeg, ONT: International Institute for Sustainable Development, 2003.

[88] Wang et al., 2006, 2.

An additional factor was China's entrance into the World Trade Organization (WTO) in late 2001. While WTO membership does not require strict country regulations on GMOs, it does require uniformity and non-discrimination in the application of national standards. As Newell notes, China may work with other developing countries to oppose restrictive patenting provisions while it works with food exporting nations to promote liberalization of world markets for agricultural products.[89]

7.7.2 *Structure of the Regime*

China's biosafety regime has three national components, provincial elements, and new, tightened regulations developed in 2001. The first component is the Allied Ministerial Meeting, which the State Council established as the primary coordinating body. This body includes representation from the MOA, the NDRC, MOST, the Ministry of Health, the Ministry of Commerce, AQSIQ, and SEPA. Its responsibility is to coordinate major issues related to the safety of agricultural GMOs. It examines and approves applications for the commercialization of GMOs, ascertains which GMOs need to be labeled, and it also writes import and export policies for agricultural GMOs.[90] An Office of Agricultural Genetic Engineering Biosafety Administration within MOA handles routine operations, which gives MOA the leading role in the biosafety regime.

The second component is the Biosafety Committee. Composed of nearly 60 agricultural scientists, the committee meets biennially to process all biosafety assessment applications pertaining to "experimental research, field trials, environmental release, pre-production trials, and commercialization of agricultural GMOs."[91] Based on its assessments, the committee makes recommendations to the administrative office in MOA. Although this office is responsible for final approval, the committee is the primary agent in biosafety management.

The third component is the MOH, which is responsible for food safety management of biotechnology products. It sponsors an Appraisal Committee, which includes experts in food health, nutrition, and toxicology; they are responsible for the review and assessment of GM foods.[92]

By 2005, all provinces of China had established their own provincial biosafety management offices, which operated under the departments of agriculture of the provinces. The function of these offices is to gather statistics on GMOs tested in the province and to monitor the performance of research and commercialization.

---

[89] Peter Newell, *Domesticating Global Policy on GMOs: Comparing India and China*. Brighton, UK: IDS Working Paper 206, Institute for Development Studies, 2003.

[90] Pray et al., 2006, 145.

[91] Ibid.

[92] Ibid.

They screen all applications for GM research, field trials and commercialization, and may disapprove those they find wanting. Only applications approved by the provincial management offices may be submitted to the Biosafety Committee, giving provinces a strong role to play.[93]

In 2001, the State Council adopted a new regulation on biosafety called the "Regulation on the Safety Administration of Agricultural GMOs." It replaced the 1993 regulation of MOST, and influenced the MOA to tighten its regulations in the areas of biosafety management, trade, and labeling of GM products. Pray et al. note that:

> The changes included an extra pre-production trial stage prior to commercial approval, new processing regulations for GM products, labeling requirements for marketing, new export and import regulations for GMOs and GMO products, and local and provincial-level GMO monitoring guidelines.[94]

Oosterveer suggests that China's new biosafety regime approaches the precautionary position taken by the European Union.[95] Indeed, the early regulation of 1993, which represented a product-based approach, has been transformed into a regulatory system that focuses on process.

7.7.3 *Biosafety Implementation*

As discussed above, the process for approval of Bt cotton in China was rapid, requiring just 2 years of testing and environmental impact trials (although more tests were required *after* the approval of Bt cotton in 1997). However, the approval process for commercialization of other crops has been quite slow, and as we pointed out, genetically-modified seeds for major food crops such as rice, maize, and wheat are in the pipeline and have remained there for years.

The study by Pray et al. compares the costs of compliance with biosafety regulations of China to India, and finds that compliance costs are higher in India for private firms, while costs are lower for government research institutes. Moreover, the Chinese government absorbed many of the safety testing costs for GMOs.[96] Enforcement of biosafety regulations and in particular restricting the use of GM seeds not approved by the Biosafety Committee, initially was problematical in China. However, this difficulty was overcome through replacement of unapproved Bt cotton seeds with approved ones produced inexpensively through government laboratories. Moreover, the initial conflict between provincial agricultural departments and MOA over the approval process was resolved by requiring the provinces to establish their own biosafety committees and coordinate their approvals with the

[93] Ibid. Personal interview with multinational seed company representative, Beijing, May 23, 2005.

[94] Pray et al., 2006, 145.

[95] Peter Oosterveer, *Global Governance of Food Production and Consumption: Issues and Challenges*. Cheltenham, UK: Edward Elgar Publishing, 2007, 137.

[96] Pray et al., 2006, 150.

MOA before regional variety trials were conducted.[97] This problem in implementation concerned Bt cotton, which raises fewer threats to human safety than that of food crops. As mentioned, the introduction of field testing of Bt rice, and the lack of control over the products generated in field tests, led to sales of unapproved Bt products in some marketplaces.

Implementation of the biosafety regime in China has also encountered the two problems that are perhaps generic in the development of transgenic products. First, multinational agricultural biotechnology firms such as Monsanto object to the length of time and expense required to move their GMOs from laboratory and field testing to commercialization. In China, Monsanto complained that it had waited 7 years for approval of its Bt maize variety, without any recouping of its investment and none likely in the immediate future. In 2009, Monsanto announced that genetically engineered seeds will become the cornerstone of its corporate strategy, despite protests by Greenpeace. It is now working with the China National Seed Group to market insect-resistant corn and also plans to sell soybean and oil seeds to farmers.[98] Yet, with the exception of Monsanto, multinationals in China play a much less important role in development and commercialization of GM crops than they do in most other countries. It is government scientists who do most of the research and testing.

Second, NGOs object to government regulatory policies that do not completely express the precautionary principle, allowing commercialization only after extended tests indicating the absence of any risk. This is the position of Greenpeace-China, even after the significant tightening of regulations following 2001. It also objects to the composition of the Biosafety Committee, alleging that two-thirds of the members are biotechnology scientists, and that as a result the approval process is biased.[99]

What is clear from this review is that China has adopted a cautious approach to the regulation of GMOs. International forces influenced China in the development of its biosafety regime, but increasingly, concerns about the potentially adverse impacts to human health have slowed the approval process considerably.

7.8 Conclusions

China entered the biotechnology era later than most nations, beginning its program only in the mid-1980s. Food security was the primary reason for China's growing investment in genetically-modified organisms, but there were other motivators as well: lessening damage of pesticides to the environment, participating in leading-edge technology (and thereby enhancing the legitimacy of the state), and improving competitiveness of China's millions of small farms.

---

[97] Pray et al., 2006, 152–54.

[98] Nancy Larson, "Monsanto Sows Seeds of Controversy," *China Daily*, July 6, 2009, 4.

[99] Personal interview with member of the Biosafety Committee, Beijing, May 28, 2007.

As a socialist state, China's investment in GM research was almost entirely public. Under the 863 Program, the state established regional biotechnology research centers and through MOST funded research institutes at state universities. The investment grew to outpace that of any other developing nation and by the 1990s equaled the public investment in agricultural biotechnology of the United States. Private sector firms played a very small part in this enterprise, with the exception of a few multinational agribusinesses, such as Monsanto.

In 1997 China approved the commercialization of Bt cotton, and it has been the major success of the agbiotech effort. In 2009, nearly 70% of China's cotton is grown from Bt cultivars. Bt cotton has increased cotton yields while significantly reducing pesticide spraying (and associated labor efforts). Because China has been the world's largest user of chemical pesticides, with all the accompanying costs to human health and environmental degradation, a two-thirds reduction in pesticide use in cotton-growing areas must be regarded as a significant accomplishment. However, continued use of Bt cotton has revealed a rise in incidence of secondary pests, making it necessary to curb enthusiastic praise for this GMO.

In addition to Bt cotton, the central government also approved another half-dozen GMOs: transgenic varieties of tomatoes, bell peppers, petunias, and insect resistant poplars and papayas. None of these approvals seems likely to have more than a marginal impact on food productivity.

Instead, we focused attention on genetically-modified rice varieties, because rice remains China's major food crop. A relatively long period of field and environmental testing brought four varieties (three with insect-resistance properties) to the penultimate stage in the state's new biosafety regime – pre-commercial clearance. Bt rice has remained in this stage of limbo since 2002, because of growing resistance globally and locally to GM products. Areas of resistance include concerns about the potential impact of transgenic products on human health, on the environment, and on biodiversity, particularly of wild rice species. These concerns have dampened enthusiasm globally for agricultural biotechnology, and influenced the behavior of Europeans and the EU particularly. While market acceptance for GMO products seems high in China, the current need for Bt commercialization has been questioned.

In the pipeline behind Bt rice stand transgenic varieties of maize, wheat, barley, soybeans and even potatoes. Indeed, Chinese scientists (and scientists working for a few multinational agribusiness firms, such as Monsanto) have found ways to technically improve on seeds for all of China's primary cereal crops and staples. It is perhaps a testament to China's current situation of food security that the state has not found it necessary to commercialize any of them.

It is impossible to say when, if ever, these staples will be commercialized. Clearly, the global environment for the production of GM crops is more inhospitable in the early twenty-first century than previously. While global opinion did not influence China during the Maoist era, as China opened to the world during the current reform period, it became subject to actions and policies of other states, especially the economically developed ones.

Too, China's leaders and government agencies are now divided as to whether GM crops such as rice, maize, and wheat should be commercialized. Because the public is not well-informed on GM issues (or, for that matter, state regulations), leaders have the latitude to wait until consensus is reached. Meanwhile, the magic bullet representing agricultural technology is encased in the state revolver, with the safety on.

China now has a biosafety regime, developed in response to international concerns regarding GMOs represented in China itself by scientists and an NGO, Greenpeace. Some international observers believe that China's biosafety regime, which certainly has slowed the commercialization of GM varieties (and particularly of food staples) approaches the European precautionary principle.

In the next chapter, we turn to the broad legal and institutional framework to address food security issues. In that chapter we treat the food safety regime, which has been hastily constructed to handle China's latest food threat – tainted products.

Chapter 8
The Legal and Institutional Framework to Address Food Security Needs

Abstract This chapter begins by discussing the authorities under which China's food and food security system operate: the 1982 constitution, the framework of laws, and both regulations and policies. It introduces the most important central ministries, classifying them into agencies emphasizing food production, food consumption, system-wide control functions, and related agencies. Then, devolution of authority and enforcement powers to sub-national governments is examined, showing weakened government authority at the local level while the central government attempts to improve integration and coordination throughout the system. The role of non-governmental organizations (NGOs) in food security is treated, with a focus on Greenpeace's campaigns in China. The chapter then analyzes the important network of laws and institutions dealing with food security. The most recent is China's new food safety regime. Discussion centers on how China handled international criticism of its "tainted products," through revising the structure of the food safety system, revision of laws and regulations, and adoption of corrective measures.

Keywords 1982 Constitution • Ministry of Agriculture (MOA) • Ministry of Environmental Protection (MEP) • State Food and Drug Administration (SFDA) • Devolution • Non-governmental organizations (NGOs) • Greenpeace • Tainted products • Food safety regime

8.1 Introduction

Social scientists use the concept "regime" to describe and analyze the network of laws, regulations, and policies concerning an important issue area – for example, climate warming – and the institutions (even including non-governmental organizations [NGOs]) that focus resources toward the resolution of problems in this area. While food security is most certainly a major issue area in China today, with an

J.H. McBeath and J. McBeath, *Environmental Change and Food Security in China*, Advances in Global Change Research 35, DOI 10.1007/978-1-4020-9180-3_8,

associated set of problems to which we make reference throughout this volume, there is not yet a "food security regime." Instead, food security is treated as a generic concept, and the particular questions pertaining to it may be discussed systematically.

We could describe the elements of a "grain security regime," because the Chinese government historically and today has interpreted the essential grains – particularly rice and then wheat and maize – as the most important constituents of China's food system. Yet grain security is not in common use either as an overarching regime. Thus, in most of this chapter, we discuss the laws, regulations, and institutions (including NGOs) that generally have been most influential in resolving problems of China's food supply.

In Chapter 7, we analyzed the biosafety regime established to reconcile China's developing agricultural biotechnology program with the international conventions in which China participates as well as the demands of some of China's major trading partners. We conclude this chapter by discussing a second regime. It is a regime-in-formation – the food safety regime – specifically designed to address the quality of food products, both those exported abroad and those consumed domestically.

8.2 Constitutional and Legal Provisions

The context in which rules are made in China differs sharply from that in liberal democracies. China remains an authoritarian state system, and the direction of policy is determined by the Chinese Communist Party whose leading members head the most important offices of the state. At the 17th Congress of the CCP in October 2007, President Hu Jintao proposed no more than cosmetic political reforms, while maintaining that the CCP must remain "the core that directs the overall situation and coordinates the efforts of all quarters."[1] Nevertheless, China has the full panoply of rules addressing food production, the food system, and the several dimensions of food security. We consider the 1982 constitution, the framework of laws, and both regulations and policy statements.

8.2.1 The 1982 Constitution

China has had four constitutions since the establishment of the People's Republic in 1949. The most recent and authoritative is the constitution adopted at the Fifth Session of the Fifth National People's Congress in December 1982. The 1982 Constitution established a pattern of law emanating from the National People's

[1] Joseph Kahn, *New York Times*, October 16, 2007, A7.

Congress; it also recognized the State Council as "the highest organ of state administration (Article 85)."

The constitution has a large number of economic provisions. Resources are state owned: "All mineral resources, waters, forests, mountains, grassland, unreclaimed land, beaches and other natural resources are owned by the state.... (Article 9)" The 1982 Constitution assigned ownership of rural areas, where most of China's cereal crops are grown, to collectives or the state, and allowed the people "to farm plots of cropland and hilly land allotted for their private use, engage in household side-line production and raise privately owned livestock" (Article 10). This section was amended in 1988 to incorporate "the responsibility system, the main form of which is the household contract that links remuneration to output."

Amendments in 1999 further clarified the household responsibility system by specifying the "dual operation system" in which centralized operation would be combined with "decentralized operation on the basis of operation by households under a contract." Further, the degree of privatization which practically had occurred was legitimized through adding this phrase:

> The non-public sectors of the economy such as the individual and private sectors of the economy, operating within the limits prescribed by law, constitute an important component of the socialist market economy.[2]

The party's constitution also influences socioeconomic development, and at the 17th Party Congress, the central committee said the constitution would be amended to incorporate Hu Jintao's "scientific development." This vague concept is used to promote more balanced, equitable, and sustainable development, and its incorporation adds Hu's thoughts to those of Mao, Deng Xiaoping, and Jiang Zemin.[3]

8.2.2 *Framework of Laws*

A number of laws have addressed different dimensions of food security in China. Laws offer the strongest degree of authority beyond the constitution. They are enacted by the National People's Congress (NPC), China's supreme legislative body, on the recommendation of the State Council. Although Chinese law is quite broad and often more in the nature of a policy statement than a statute, the law is increasingly transparent and enforceable.[4] Laws pertaining to food security are numerous – probably

---

[2] *Constitution of the People's Republic of China*. Beijing: China Legal Publishing House, 2001, 121.

[3] Joseph Kahn, "New China Hierarchy May Limit President's Power," *New York Times*, October 13, 2007, A1.

[4] In early 2008, the State Council issued a white paper highlighting China's efforts in promoting a rule of law under a "socialist legal system with Chinese characteristics." The paper noted that the NPC had enacted 229 laws currently in effect, while the State Council had enacted 600 administrative regulations. See Xinhua, "White Paper Highlights Rule of Law," *China Daily*, February 29, 2008, 1.

more than three dozen. They can be classified broadly into three types: production of food, consumption of food, and control of the food system and its products. Admittedly, this classification does not produce neat, mutually exclusive, categories, but it is serviceable. We also use it in introducing central government ministries administering food security laws, regulations, and policies.

Examples of laws concerning the production of food include the Agricultural Law, the Seed Law, the Land Management Law (mentioned in Chapter 3), the law on Water and Soil Conservation, and the Water Law. Because forests and grasslands also produce food products, legislation relevant to these areas also should be included, for example the Forest Law of 1985 (amended in 1998) and the Grassland Law of 1985.

Examples of laws concerning food consumption emphasize the safety of food products, both for the domestic public and those in nations to which Chinese products are exported. They include generic legislation such as the Law on Protection of Consumer Rights and Interests, the Standardization Law, and the Product Quality Law. More specific laws pertaining to food safety include the Law on the Quality and Safety of Agricultural Products, the Food Hygiene Law, and the Law on Animal Disease Prevention.

The last area includes much of China's generic environmental protection legislation, which is related to food products. These include the Environmental Protection law (adopted in 1979 and amended in 1989) and the new Environmental Impact Assessment (EIA) law that took effect in 2003. They also include legislation focused on pollution of land, air, and water, all of which potentially affect agricultural production, to wit: the Water Pollution Prevention and Control Law of 1984 (revised in 1996; amended in 2008), the Solid Waste Pollution Prevention Law of 1995, and the Air Pollution Prevention Law of 1987. The Flood Control Law affects general conditions of agriculture. Finally, laws specifically controlling food production as well as consumption include the Law on Import and Export Commodity Inspection, the Law on Animal and Plant Entry and Exit Quarantine, and the Frontier Health and Quarantine Law.

The sufficiency of legislation in the area of food security does not appear to be at issue, especially given recent enactments in response to food safety crises. Two observations flow from a review of legislation. First, there is no comprehensive legislation on food security in China, covering production, consumption, and control functions. In this respect, China is not unusual as no country has such legislation in 2009. Second, there is considerable overlap among the laws, meaning that redundancy is systemic.

8.2.3 Regulations and Policies

Regulations (*fagui*) are a less authoritative basis for action than laws. They do not undergo a public review process, as occurs increasingly for proposed legislation through the National People's Congress. In most cases and until very recently, regulations have not been available publicly, which raises the question of what the limits to action are. Yet regulations do reflect negotiations among affected bureaucratic interests. Provisional regulations may, on occasion, be issued by individual ministries; but in most cases they are reviewed and issued formally by the State Council.

In the general area of food production, there are numerous regulations covering such subjects as: (1) use of chemicals, pollutants, and pesticides on farms; (2) food production, manufacturing, and processing standards; (3) the licensing of food and drug manufacturers and retailers; and (4) the planting of trees in flood control areas.

Concerning food consumption, the regulations are even more numerous. Some 13 specific administrative regulations and ten departmental rules govern different aspects of the food safety system. In the area of standards, a recent report indicates that China has promulgated more than 1,800 national standards on food safety, and more than 2,900 standards for the food industry, including 600 which are compulsory.[5]

Then, with respect to both environmental and food products control, we find an equally large number of regulations. For example, with respect to GM products, regulations govern genetic engineering and the implementation of safety assessments for agricultural GMOs, their labeling, and import safety.

Regulations specify the general terms in which major legislation is written. They also may be an ad hoc response to crisis situations, as occurred in 2007 when government ministries needed to respond instantly to criticism of the safety of China's food products. Most ministries and agencies do not consult with relevant publics before developing and issuing regulations; also, they lack transparency.

Too, ministries and agencies will issue action plans, and these virtually have the effect of regulations and laws. The action plans typically are embedded in Five-Year Plans, based on which budget allocations are made. For example the Eleventh Five-Year Plan, covering the period 2006–2010, emphasizes agriculture and food production. In addition, ministries and their leaders make policy statements, and in terms of their signaling power, they have an effect equal to that of regulations. A recent example is Premier Wen Jiabao's address to the National People's Congress and statements regarding the minimum area required for grain production (discussed in Chapter 3).

Policy addresses of chief leaders indicate policy emphases and shifts. For example, Hu Jintao, China's president as well as general secretary of the party's central committee, used the keynote speech at the party's 17th National Congress in October 2007 to pronounce the "Scientific Outlook on Development." He included reference to greater efforts in reducing discharges of major pollutants:[6] "[We will] promote a conservation culture by forming an energy- and resource-efficient and environment-friendly structure of industries, pattern of growth and consumption."[7]

[5]Mingxin Bi, "White Paper: China Builds Food Safety Law Regime, Technological Guarantee System," *Xinhua News Agency*, August 17, 2007.

[6]Tian Le, "Hu: Well-off Society is Top Goal," *China Daily*, October 16, 2007, 1.

[7]Ibid., 6.

At this party congress, delegates also added a basic line to the constitution, "turning China into a prosperous, strong, democratic, culturally advanced and harmonious modern socialist country," while mentioning the word "religion" in an amendment.[8]

8.3 Central Ministries

China's long history of centralized government control insures that ministries of the state will play major roles in food security issues. The situation in China, like that in most countries with respect to food production, distribution, and consumption, is that many departments and agencies are involved, and their areas of responsibility overlap. At least 12–15 ministries, commissions, and agencies have some responsibility in this large area. We present brief summaries of ministries/agencies following the classification system outlined above – agencies with missions emphasizing food production, food consumption, and system-wide control. We add to this scheme agencies with secondary responsibilities in food security.

8.3.1 Agencies Emphasizing Food Production

Four agencies have direct responsibilities over the production of food: the Ministry of Agriculture (MOA), the State Forestry Administration (SFA), the State Oceanic Administration (SOA), and the State Grain Administration. In addition, both the Ministry of Land and Natural Resources (MLR) and Ministry of Water Resources (MWR) have duties to preserve lands and waters, without which food could not be produced. We give brief synopses of each.

1. Ministry of Agriculture. The MOA is a large and relatively powerful agency with a mission to improve agricultural productivity in China. It is in charge of both crop and animal production and health, and its orbit includes freshwater fisheries as well. Explicit economic functions of the ministry include rural development, township and village enterprises and development planning; too, it has responsibilities with respect to the alleviation of rural poverty. Its technology portfolio encompasses agricultural mechanization; science, technology and education; and research into agricultural biotechnology. The MOA has a very large number of service units. These include applied research such as the Natural Resource and Environment Research Center and the Chinese Academy of Fisheries

[8] Tian Le, "Scientific Outlook on Development Enshrined," *China Daily*, October 22, 2007, 1.

Sciences.[9] In their review of MOA's programs, Waldron et al. identify the ministry's formal affiliation with:

> Five technical inspection and regulatory units, five social organizations, four education and training units, six information, publishing and propaganda units, eight science, research, technology and extension units, and sixteen exchange, development and consulting units.[10]

2. State Forestry Administration. The SFA is an administration, meaning that it does not have cabinet rank, notwithstanding its direct location under the State Council. Although the primary mission of SFA is to safeguard and enhance China's forests, it also has food production responsibilities. A decade ago, SFA developed the *Muban liangyou* (wood staples and oil) policy, to promote forestry products humans could consume. In addition to oils, this included tree nuts (e.g., chestnuts) and edible bamboo.[11] SFA has a large service unit called the Chinese Academy of Forestry conducting specialized research on silviculture and sustainable forest management. It has been involved in the implementation of the *Tuigeng huanlin* policy discussed in Chapter 3.[12]
3. State Oceanic Administration. SOA is a specialized agency, nearly autonomous, but for administrative purposes in the Ministry of Land and Natural Resources. Its jurisdiction extends offshore China's coasts to the 200-mile limit, and thus it supervises oceans fishing and mariculture.
4. State Grain Administration. The SGA is an administrative unit operating under the National Development and Reform Commission (discussed below). It controls national grain distribution and manages China's central grain reserves. This entails managing grain supply and demand, planning procurement and storage, and supervising the implementation of laws and regulations on national grain distribution.[13]
5. Ministry of Land and Natural Resources. MLR is relatively new as a national ministry, emerging from the major government reorganization in 1998. Earlier, there were separate ministries of mines and land survey/administration. In the view of some observers, MLR was modeled on the US Department of the Interior. The ministry has a supervisory responsibility for mining activities in China, and it collects fees for lands leased for mines (which makes it a relatively rich agency). In the context of food production, it establishes and implements metrics for the measurement of China's arable lands, and it has some regulatory authority

---

[9] Personal observations of several MOA research and science centers, Beijing, March 18, 19 and May 24, 28, 2007.

[10] Scott Waldron, Colin Brown and John Longworth, "State Sector Reform and Agriculture in China," *China Quarterly*, 186 (June 2006), 287.

[11] Personal interview with professor of forest ecology, Beijing, March 15, 2007.

[12] Personal interview with former staff member, Chinese Academy of Forestry, Beijing, May 21, 2007.

[13] See SGA's website: http://www.chinagrain.gov.cn/d1eindex.html; also see Tony Saich, *Governance and Politics of China*, 2nd ed. New York: Palgrave, 2004, 249–51.

concerning land pollution.[14] This ministry too has a number of service units, for example the Environmental Economics Research Unit, which conducts studies on land sufficiency questions.

6. Ministry of Water Resources. MWR is China's water management agency, with broad regulatory authority (including establishing permit systems) over water use and pricing. The agency also constructs water works (such as rural irrigation systems and reservoirs; major dams typically are constructed by independent commissions, drawing personnel from MWR and other agencies). MWR also has responsibilities with respect to the discharge of wastewater to potable water areas, as well as monitoring reservoirs and dams of hydropower stations.[15] MWR has a number of service units, including those conducting applied research, for example the China Institute of Water Resources and Hydropower.

8.3.2 *Agencies Emphasizing Food Consumption*

Two agencies have direct authority in the monitoring and supervision of the safety of foods consumed: the State Food and Drug Administration and the Ministry of Health. A third agency has generic responsibility – the State Administration for Industry and Commerce. A research and policy body, the National Institute of Nutrition and Food Safety, has responsibilities in this area as well.

1. The State Food and Drug Administration (SFDA). Founded in 2003, the SFDA is modeled on the US Food and Drug Administration; it coordinates and oversees other health, food, and drug agencies. At its creation, the older State Drug Administration was merged within it, and its centrality is indicated by its operation directly under the State Council. The mission of the agency is "supervision on the safety management of food, health food and cosmetics and [it] is the competent authority of drug regulation."[16] The SFDA drafts laws, regulations and policies pertaining to health standards and food safety and has responsibilities in communicable disease prevention and treatment, food health, and education.

---

[14] Personal interview with official, Ministry of Land and Natural Resources, Beijing, May 27, 2007. Similar comments could be made about enforcement of environmental laws at the local level. See Jiao Wu, "Environment Watchdog Calls for Sharper Teeth," *China Daily*, May 10, 2007, 3. Even integrated national campaigns face implementation difficulties at the local level. For example, in 2006 seven government departments sought to ban sewage outlets at all source water protection areas by the end of the year. By 2007, some provinces had not completed the first phase (defining source water protection areas); others had not begun the process. Deputy Director of SEPA's supervision department lamented: "It shows that an embarrassing situation does exist where government orders on environmental protection are difficult to be implemented at local levels." See Dingding Xin, "Setback for Water Safety Campaign," *China Daily*, June 15, 2007, 3.

[15] Personal interview with official, Ministry of Water Resources, Beijing, May 30, 2007. See also: http://www.mwr.gov.cn/english1/departments.asp (retrieved September 15, 2007).

[16] See "About SFDA, " State Food and Drug Administration website: http://www.sfda.gov.cn

In early 2008, the SFDA was absorbed into the Ministry of Health, and no longer has chief responsibility for food safety, drug standards, or development of a national medicine system. It gained supervisory responsibility for sanitation of the catering industry.[17]

2. The Ministry of Health (MOH). The MOH is an old ministry, established at the founding of the People's Republic, and has the broadest authority in drafting health laws, regulations, and policies, and health law enforcement. It enforces food safety and hygiene inspections, and investigates food quality violations. Within this agency is the Institute of Food Safety Control and Inspection, which identifies unsafe foods and food production and handling practices.[18] As mentioned, in the early 2008 government restructuring, MOH became China's food and drug safety czar. The SFDA now reports to the Ministry of Health and not to the State Council.[19]
3. The State Administration for Industry and Commerce (SAIC). The SAIC also is under the direct control of the State Council. It has general authority to regulate market activity. Its Consumer Protection Bureau drafts and enforces standards and regulations for products in the marketplace. It investigates the quality of marketed goods and distribution of fake or substandard goods. It also issues business licenses.[20]
4. National Institute of Nutrition and Food Safety. This institute is a research agency and without regulatory authority. It is affiliated with the Center for Disease Control and Prevention, and conducts research studies on nutrition, food hygiene, prevention of food-borne diseases, and improvement of nutrition and fitness.

8.3.3 Agencies Emphasizing System-Wide Control Functions

Two agencies have comprehensive authority in the area of regulation, the Ministry of Environmental Protection and the Administration of Quality Supervision, Inspection, and Quarantine. An additional two agencies –the Ministry of Science and Technology and the National Development and Reform Commission – have general authority in planning and control functions with significant impacts on the food system and food security.

1. General Administration of Quality Supervision, Inspection, and Quarantine (AQSIQ). AQSIQ is a ministry of the state; it oversees the export and importation of food products and administers quarantines at the national and local levels.

---

[17] Juan Shen, "MOH Will Oversee Food, Drug Safety," *China Daily*, March 12, 2008, 1.

[18] See Ministry of Health website: http://english.gov.cn/2005-10/09/content_75326.htm

[19] Shan, "MOH Will Oversee…," 2008.

[20] See State Administration of Industry and Commerce website: http://gsyj.saic.gov.cn/wcn/WCMData/pub/saic/english/default.htm

Departments of the agency most directly involved with food safety issues are the Department of Supervision on Animal and Plant Quarantine, the Bureau of Import and Export Food Safety, and the Department of Supervision of Food Production.[21]

2. Ministry of Environmental Protection (MEP). China first developed an environmental agency, the National Environmental Protection Administration, in 1988. The agency's name changed to the State Environmental Protection Administration in 1998, when it was elevated to sub-ministerial status. It has an expansive mandate, with responsibility for each of China's major environmental problems, including air, water, and land pollution, acid rain and climate change.[22] Too, it has specialized departments of nature and ecology conservation, biosecurity, and biosafety. SEPA has a number of affiliated research and policy institutes, such as the Nanjing Institute of Environmental Science. In March 2008, SEPA not only became a full-fledged Ministry of Environmental Protection, but also was labeled one of five "super ministries." This significantly elevated its status and appeared designed to show the serious attention the central authorities placed on pollutional control and sustainability issues.[23]
3. Ministry of Science and Technology (MOST). This ministry has regulatory authority over the quality of products in the marketplace, and can punish sellers who have violated product quality standards. It also regulates product packaging and enforces standards on product ingredients.[24] The agency sponsors (and funds) technological innovations to improve food production, as we noted in Chapter 7 on development of GM foods in China.
4. National Development and Reform Commission (NDRC). This ministry is the descendant of the State Planning Commission. It is a powerful agency, given its role in guiding the overall reform program of the state. It also has missions and

---

[21] See "Standards Organizations and Related Bodies" at: http://www.standardsportal.org/prc_en/ov_standards_orgs.aspx

[22] See Xiaoying Ma and Leonard Ortolano, *Environmental Regulation in China: Institutions, Enforcement, and Compliance*. Lanham, MD: Rowman & Littlefield, 2000, 78–81.

[23] Jiao Wu, "Question is Not Big or Small, but Efficiency," *China Daily*, March 12, 2008, 8. State Council member Hua Jianmin said MEP would "boost environmental and ecological protection efforts and accelerate the building of a resource-saving and environment-friendly society." MEP was expected to gain consolidated authority over pollution prevention (now distributed among several agencies), full responsibility for water pollution, responsibility for biodiversity management and assessment of ecological protection zones. See: Jiao Wu and Jing Fu, "Ministry Will Give More Weight to Green issues," *China Daily,* March 13, 2008, 5; Jing Fu, "Uphill Task," *China Daily*, March 24–30, 2008, 4; and Xiaohua Sun, "Environment Chief Vows to Add Muscle," *China Daily*, March 25, 2008, 3. The State council allowed it to establish two supervision and monitoring departments for pollution control, and to recruit 50 additional employees. See Jing Li, "Environment Ministry Adds 2 Departments," *China Daily*, July 11, 2008, 2.

[24] See Ministry of Science and Technology: Product Quality Law, 2000 at http://www.most.gov.cn/eng/policies/regulations/200501/t20050105_18422.htm

offices respecting environment and natural resources and rural development, which link it to food security concerns.[25] Also, it houses the National Energy Bureau, which has been the main force in integrating national energy management.[26]

8.3.4 Related Agencies

A number of other ministries and agencies treat food security issues as part of more general missions. These include the Ministry of Foreign Affairs and Ministry of Commerce. In addition, much of the research in food production and agricultural technology is done under the auspices of the Chinese Academy of Sciences and Chinese Academy of Agricultural Sciences. We treat them briefly and also introduce the customs agency.

1. Ministry of Foreign Affairs (MOFA). This ministry is responsible for international conventions in which China participates as well as participation in the United Nations and affiliated organizations such as the Food and Agricultural Organization (FAO). Invariably, it heads delegations at international conferences treating food security issues.
2. Ministry of Commerce (MOC). This ministry administers regulations governing the food trade, foreign investments, and food distribution. It was created in 2003 from the formerly powerful Ministry of Foreign Trade and Economic Cooperation,[27] and thus oversees both domestic and international market activities.[28]
3. The Chinese Academy of Sciences (CAS). As noted above, most of the ministries have service units involved in areas such as research, standards, testing and inspection, as well as enterprise units. Additionally, an important institution is CAS for the conduct of pure as well as applied natural science research. Several of the CAS institutes have played important roles in formulating policy recommendations, for example the Center for Chinese Agricultural Policy.[29] Too, the CAS has several key laboratories engaged in research on climate change among other topics.
4. The Chinese Academy of Agricultural Sciences (CAAS). A second national set of institutes does work specifically in agricultural sciences. CAAS has units such as the Institute of Plant Protection that focuses on plant and pest diseases.

---

[25] See National Development and Reform Commission website: http://www.sdpc.gov.cn

[26] Jing Fu, "Energy Bureau Gets Nod to Increase Size," *China Daily*, June 28–29, 2008, 3. The National Energy Bureau seems likely to be the basis for formation of a cabinet-level energy ministry in the near future.

[27] Saich, 2004, 135.

[28] See the ministry's website at: http://english.mofcom.gov.cn/

[29] Personal interview with research professor, CAS, Center for Chinese Agricultural Policy, Beijing, May 29, 2007.

It also has broader units such as the Institute of Environment and Sustainable Development, which considers climate change impacts on agriculture in China.[30] CAAS played an important role in increasing China's grain output through its development of hybrid rice.[31]

5. China Customs. The General Administration of Customs also is a ministerial-level agency reporting directly to the State Council. It controls inward and outward means of transportation, goods and articles. Also, it collects customs duties and taxes, and enforces laws against smuggling.[32] It works closely with AQSIQ in prohibiting export or import of contaminated food products and those violating China's phytosanitary standards.

This does not complete the list of agencies involved with China's food security and safety conditions. Virtually every ministry, commission, and central agency of the state has a unit with duties respecting food production, distribution, or consumption. Because food security does not yet have a regime, no ministry has a clear lead-agency role (although MOA is involved in most issues), and this lack of administrative integration is the case cross-nationally. Below we discuss some mechanisms that induce coordination.

8.4 Devolution of Functions to Sub-national Governments

Although China is the world's third largest nation-state in area, with vast differences across quite diverse biogeographical regions, it remains a unitary state. In the reform era, to accommodate differences among regions (as well as lessen central government fiscal responsibility), the state has devolved (transferred) administrative power from the center to the provinces (or autonomous regions) and to counties, townships and municipalities. This has introduced additional complexity in the administration of national laws and regulations.

8.4.1 Grants of Authority

Several of the central ministries and administrative agencies introduced above are hierarchically organized, with a national office in Beijing, but required to have departments in each of the provinces (or autonomous regions) and bureaus at the local levels too.

[30] Personal interview with staff, Institute of Environment & Sustainable Development, Beijing, May 25, 2007.

[31] According to CAAS's estimate, the hybrid rice species developed in the early 1970s increased China's rice output by 400 million metric tons in the next 25 years. See Jing Fu, "Agriculture Looks to Scientific Future," *China Daily*, November 12, 2007, 2.

[32] See the Customs' website: http://english.customs.gov.cn/default.aspx

This applies clearly to all of the agencies emphasizing food production except SOA, and also to the Ministry of Health, and to MEP. In fact, in these agencies there are more employees at the provincial and local levels than at the national.

The sub-national departments and bureaus operate in a problematical administrative context, for they serve two masters: for example the MOA, MWR, or MEP in Beijing, and the provincial governor (or local mayor). Because administrative control tends to follow the source of funding, and the central government allocates less to most food security functions (with the possible exception of agriculture) than provinces and municipalities, there is no clear line of authority from the center to the site where problems of food production and consumption must be resolved. (In Chapter 9, we discuss this from the perspective of bureaucratic fragmentation.)

Moreover, national laws and even many regulations are vague and leave much discretion to local administrators, who correspondingly will be influenced by their personal relationships (*Guanxi*) as well as administrative rules. Given the relatively weak bureaucratic position of local administrators of mainline agencies, there is little incentive for officials to rigorously enforce the rules, and they adopt pragmatic orientations. A respondent from the Ministry of Land and Natural Resources described the vertical transmission of authority in these terms:

> Each province, county, township has a Land and Natural Resources unit. Although their officials receive salaries from the local government, the national ministry still can insure some uniformity. It establishes the regulations and makes definitions for such things as "land pollution." It also inspects provincial or local offices for compliance. The problem is that the farther down the bureaucratic chain one goes, the lower the salaries, benefits, and perquisites. For this reason, the general perception is that local governments are the most corrupt and least likely to follow the rules.[33]

Provincial and local offices of national ministries are relatively well-supplied with personnel. In the larger provinces, for example, employees of agriculture departments and bureaus number in the thousands. In the major central government administrative reorganization of 1998, the number of employees in MOA was cut by about half.[34] Cuts also were made to provincial and local agriculture departments, but many of the personnel were absorbed into service and enterprise units. The MEP has 300 central employees. Across all levels there are 2,900 environmental inspection agencies, and a total of 53,000 employees.[35]

Devolution often means impediments to the central direction of policy. In an hugely diverse nation such as China, there are good arguments for the devolution of administrative power. One administrator expressed positive views toward devolution of power on implementation of food grains policy:

> Although the personnel and basic instructions on work come from the provinces (or local areas) in most cases, still they must observe national laws. There are many differences

[33] Personal interview with official, Ministry of Land and Natural Resources, Beijing, March 10, 2007.

[34] Waldron et al., 2006, 284; see also John Burns, "'Downsizing' the Chinese State: Government Retrenchment in the 1990s," *China Quarterly*, 175 (2003), 623–56.

[35] Jing Li, "Environment Ministry Adds 2 Departments," *China Daily*, July 11, 2008, 2.

> across the country; the areas are very diverse. Local officials know best how to administer the law. There are some regions, such as in the West, where they lack financial resources and face development pressures. In these cases, leadership makes a difference.[36]

Negative views, however, are more common in discussing failures in implementation of national policy. A seasoned observer of relationships between local governments (*xiangzhen*) and farmers made these comments about the regulatory environment:

> Except for a few state- or collectively-managed large farms, the local governments now usually do very little to affect the farmers' practices (e.g., what they grow, their use of fertilizers and pesticides, etc.), because most of the crops are consumed by the farmers themselves. Those crops sold in the markets usually are free from inspection by environmental and health officials. Scientific farming skills are not taught by the government, because local officials are poorly paid and poorly motivated. In most cases, farmers are not organized into groups capable of improving their farming, management, and marketing skills. This greatly hinders the improvement of China's food productivity and food safety.[37]

Of course, devolution is problematical in any hierarchical system relying on command-and-control implementation methods. Recently, central ministries such as MEP have worked to develop market-based instruments, such as green credit, insurance, and securities policies. Yet, these do not promise immediate relief, as MEP Vice Minister Pan Yueh remarked:

> Every environmental-economic policy faces difficulties. But they all have something in common – the established distribution of interests will be rearranged. It will spark conflict between locals, government, and industry. They will try their utmost to protect vested interests. And that's the core problem.[38]

Clearly, providing incentives for local levels, and for the private sector, to comply with policy is essential, but it is unclear whether the current mix of incentives is likely to be effective.

8.4.2 Coordination Methods

Given the large number of agencies involved in China's food security policies, and the devolution of administrative functions, both horizontal and vertical coordination are important. Four methods have been employed to improve integration in government policy and practice.

First, the State Council is able to bring different ministries and agencies together, as we shall see in the discussion of the food safety regimes below. Sustainable development planning, pursued under State Council leadership, also assists policy integration. The Chinese government seemed to pay less attention to sustainability

[36] Personal interview with official, SEPA, Beijing, January 5, 2005.

[37] Personal communication from social science policy analyst, Beijing, January 31, 2008.

[38] Xiaohua Sun, "SEPA Shifts Policy Focus to Incentives," *China Daily*, February 27, 2008, 13.

policies than economically developed states because of their apparent cost. However, in 2004, sustainable development entered economic planning as SEPA and the National Bureau of Statistics worked jointly to create a "green GDP" measure, which would subtract resource depletion and other environmental externalities from the GDP in order to illustrate the relationships between the environment and the economy.[39] While popular with national environmental officials, the pilot encountered obstacles in several provinces and municipalities.

Release of the first Green GDP Report for 2004 (issued in September 2006) indicated problems in its development. The report estimated that economic losses caused by environmental pollution amounted to $66 billion or about 3.05% of China's GDP for that year. Critical observers pointed to the difficulty of forming accurate figures for economic losses resulting from environmental pollution.[40] The National Bureau of Statistics complained of flaws in both theory and method and argued that inaccurate data could adversely affect local governments participating in the project.[41]

Planning is ongoing to better measure environmental costs of rapid economic development at provincial and local levels. The Chinese Academy for Environmental Planning is cooperating with Yale and Columbia universities (which developed the Environmental Performance Index) to design an evaluation system for environmental performance at the provincial level. Project coordinator Cao Zhiguo says: "Unlike the Green GDP calculation, which focused mainly on economic statistics, the indicators adopted by the evaluation system will cover a wider range, for instance ecology and health."[42] The China Council for International Cooperation on Environment and Development, an advisory panel including foreign and Chinese experts, has called local governments' poor environmental awareness a "major risk." Its latest recommendation is that the goals of a low carbon economy should be considered when the Twelfth Five-Year Plan (2011–2015) is drafted.[43]

A concept related to Green GDP, called the "circular economy," focused on low energy consumption, low pollution and high efficiency.[44] In 2008, lawmakers prepared to enact a circular economy promotion law. Environmental advocate and law professor Wang Canfa described it positively:

> The new law requires all the links from production to consumption to follow the rules of the circular economy. Everyone, from the government to society, from manufacturers to consumers, is legally bound to saving resources and protecting the environment. However, energy intensive industries will be influenced the most.[45]

---

[39] Chuan Qin, "Highlighting Green Factor in Economy," *China Daily*, July 20, 2004, 7.

[40] Xiaohua Sun and Dingding Xin, "Local Officials Not Keen on Green Arithmetic," *China Daily*, December 11, 2006, 1.

[41] Xiaohua Sun, "Green GDP Shows the Red Signal," *China Daily*, March 23, 2007, 1.

[42] Jing Li, "Broader Environmental Checks on Table," *China Daily*, October 7, 2008, 4.

[43] Jing Fu and Jing Li, "Green Warning," *China Daily*, December 8, 2008, 5.

[44] Chuan Qin, "Circular Economy Enhances Development," *China Daily*, September 29, 2004, 1.

[45] Xiaohua Sun, "'Green' Law Likely from Next Year," *China Daily*, August 26, 2008, 14.

These are more applied iterations of the idea of sustainable development, including grain security, and they are designed to integrate the planning of powerful agencies and institutions.

Second, although the revenue flow to the central government has declined following marketizing reforms and devolution, it still has the largest budget in China. It can leverage its fiscal power to improve the management of food security policies nationwide. We shall see in Chapter 9 the extent to which fiscal policy has been used to alleviate poverty in rural China.

Third, some integration of policy is attained through the use of national campaigns. Unlike the Maoist era, when mobilization campaigns were essential components of public administration, the regime has been quite selective in their use in the reform era.[46] A compulsory national tree-planting campaign, which started in 1981, has been a main factor in the success of China's afforestation program,[47] and this program has significantly reduced erosion.

Fourth, central initiatives to increase government transparency and combat corruption may facilitate coordination. In April 2007, Premier Wen Jiabao announced State Council Regulations on Open Government Information, to take effect by mid-2008. The regulations require officials to expeditiously release information involving "immediate interests of individuals and groups," and empower citizens to request "special information" to be released within 30 days. Priority areas included government spending, procurements, and investigations into environmental pollution, workplace safety, and public health. The regulations extend to local governments. County and village authorities are required to make public data on land acquisitions, land use, and fiscal accounting. Clearly, the 2004 SARS epidemic, as well as protests in rural areas over illegal land seizures and expropriations as well as environmental pollution, stimulated this policy change, as did food safety concerns.[48]

The National People's Congress also passed two laws to insure dissemination of accurate and timely information about public emergencies and outbreaks of animal diseases. The first, the Emergency Response Law, prohibited news media from publishing false reports on public emergencies, but required responsible governments to provide coordinated, accurate and up-to-date information. If they did not, officials could be criminally charged or disciplined.[49] The second law amended the Law on Animal Disease Prevention by requiring prompt reports to

[46] Kenneth Lieberthal, *Governing China: From Revolution through Reform*. New York: W. W. Norton, 1995, 199.

[47] Zhida Zhang, ed., *Ten National Ecological Forest Projects* (in Chinese). Beijing: China Forestry Press.

[48] Huanxin Zhao, "Statute to Make Government Open, Clean," *China Daily*, April 25, 2007, 1; also see editorial, "Exposure of Pollution," *China Daily*, April 30, 2007, 4.

[49] The first use of this law occurred after the May 12th earthquake in Sichuan Province. The Chinese Earthquake Administration released news on the quake 18 min after it struck, including instructions for all-out efforts to rescue the injured and regular bulletins on the death toll and relief activities. See editorial, "Transparency Works," *China Daily*, May 15, 2008, 6.

the public on outbreaks of animal diseases.[50] Then, in April 2008, the NPC Standing Committee announced:

> In principle, all draft laws submitted to the National People's Congress Standing Committee for review will be released in full text to solicit public opinion. The draft laws will be published on our official website (www.npc.gov.cn). If they're closely related to the interests of the people, the drafts will also be published in major newspapers.[51]

Following this announcement, the NPC publicized the draft food safety law.

These coordination measures notwithstanding, the central government still suffers from an "implementation deficit" in the administration of food security policies and laws. One example is the food manufacturers' continued use of red dye to color their products, after the state banned this in 1996. Sudan I red dye causes adverse health effects and has been linked to cancer. In 2005, reports of its continued use appeared in several parts of China: in noodles and vegetables in southern provinces; in chili sauce manufactured by Heinz in Beijing; in medicines in Shanghai; and by Kentucky Fried Chicken in its 1,200 restaurants. Officials gave two reasons explaining why the 1996 ban had not been enforced: (1) too many agencies oversaw food production, creating confusion and inefficiencies; and (2) food control agencies lacked the trained personnel and equipment needed to detect the dye earlier.[52]

We discuss other food safety cases below and their role in the development of the food safety regime. At this point it suffices to say that there are serious gaps in the implementation process. Of course, China is not alone in revealing implementation problems, which apply universally to developing nations and to many economically developed countries as well.

8.5 Non-governmental Organizations (NGOs) and Food Security

An important difference between liberal democracies and authoritarian governments is the presence in the former of robust civil societies. Typically, a host of civic organizations – called non-governmental organizations or NGOs – focus the public's attention on government performance. Although China's politics and society are significantly more liberalized now than during the Maoist era, civic organizations still lack the autonomy and freedom they possess in countries such as Germany, Australia, and the United States. We consider the environment in which NGOs work in China, and then give examples of those that have directed inquiry and publicity toward food security issues.

---

[50] Zhe Zhu, "Laws Passed to Ensure Openness," *China Daily*, August 31, 2007, 2.

[51] Zhe Zhu, "NPC to Make All Draft Laws Public," *China Daily*, April 21, 2008, 2.

[52] Hu Yan, "Red Dye a 'Food for Thought' for Chinese," *Renmin Ribao (People's Daily)*, March 31, 2005.

8.5.1 The NGO Environment in China

The Communist Party of China is a survivalist institution that sees any grouping as potentially in opposition to the regime, but also as a reflection of perceived challenges to state authority.[53] The Tiananmen incident of 1989, which reflected the organized efforts of students – joined by labor and other mass elements – to liberalize the regime, signified the need on the part of the state to closely regulate social associations. In response, in October 1989, the state issued new regulations for social associations which required that they be sponsored by an agency of the state and meet requirements pertaining to their membership and financing.

In 1998, new regulations[54] from the Ministry of Civil Affairs reinforced the requirement that NGOs be sponsored by a state agency, and excluded the opportunity for the social association to register as a business affiliate (as early forming NGOs in the mid-1990s, such as Global Village of Beijing, did). Membership in the group was limited to 50, and it had to demonstrate fiscal responsibility. A group was required to seek separate registration in each place it operated, which effectively prohibited the growth of national associations not directly part of a national bureaucratic agency. Also, only one social association could register in the same area for a specific activity, such as environmental protection. Schwartz notes: "As a result, NGOs tread carefully, avoiding strong criticism of government environmental protection failures."[55]

Altogether, the 1998 regulations have had a dampening effect on the development of social associations in China, especially those with no friends in the bureaucracy. The crackdown on the *Falungong* in 1999 had an even more chilling effect. The requirement that a new association find a sponsor in the state bureaucracy is a daunting challenge, because few officials are eager to take the risk of becoming responsible for events and activities performed by an NGO. Moreover, the refusal of one agency to serve as a sponsor typically influences other potential sponsors. Finally, few voluntary associations have the financial means to meet the state requirements. In late 2005, the Ministry of Civil Affairs announced that foreign NGOs would need to follow the same registration requirements mandated for domestic groups, including affiliation with a government office. Further, in response to the "color revolutions" in countries of the former Soviet Union, NGOs

---

[53] This discussion follows the treatment of NGOs in Gerald A. McBeath and Tse-Kang Leng, *Governance of Biodiversity Conservation in China and Taiwan*. Cheltenham, UK: Edward Elgar, 2006, 171–72, and 181.

[54] Jonathan Schwartz, "Environmental NGOs in China: Roles and Limits," *Pacific Affairs*, Vol. 77, no. 11 (Spring 2000), 38.

[55] Ibid. Also see Tony Saich, "Negotiating the State: The Development of Social Organizations in China," *China Quarterly*, 161 (March 2000), 129; and Jude Howell, "New Directions in Civil Society: Organizing around Marginalized Interests," in Jude Howell, ed., *Governance in China*. Lanham, MD: Rowman & Littlefield, 2004.

(and especially those with international headquarters and local organizations receiving funding from abroad) have been under increased scrutiny from the state.[56]

Thus, China's NGOs are at an early stage of evolution when compared to other Asian states. The oldest, dating just to the 1980s, are government-organized NGOs (often called GONGOs), and they exist in order to extend the reach of the central and in some cases provincial and local governments. Most NGOs have been established for little more than a decade. The largest single groupings of NGOs are those focusing on China's many environmental issues. Beijing has three dozen or so environmental NGOs,[57] and there are many grassroots environmental NGOs. These have developed in provincial cities confronting rapid environmental change, for example in areas impacted by large dam construction.

In 2005, the All-China Environmental Federation (ACEF)[58] conducted a survey of environmental protection NGOs. The survey identified 2,768 NGOs with a total membership of 224,000. About 90% of the organizations were GONGOs or grassroots groups organized by student volunteers; less than one quarter had registered with the Ministry of Civil Affairs. About 30% had ample resources, some $370 million collectively. The richer organizations also were likely to have paid, full-time staff. The two-thirds lacking fixed funding sources had part-time staff, most of whom were unpaid volunteers. Although 80% of the NGO members were less than 30 years old, they were well educated: 90% had college or graduate degrees.[59]

Nevertheless, recent cases show some increased and effective public pressure, aided by NGOs. For instance, in early June 2007 thousands of residents in Xiamen demonstrated against a government-approved chemical project to be constructed near a residential area.[60] Demonstrators at SEPA offices in Beijing protested a planned waste incineration power project in the capital city. Both protests focused on failure of local governments to listen to citizen complaints. Subsequently, the relevant departments suspended the projects pending further environmental assessments. Environmental NGOs called these incidents victories for the "people's voice," and established a website (www.greenlaw.org.cn) to disseminate information

---

[56] See Nick Young, "NGOs Will Have to 'Negotiate the State' for Some Time Yet," *China Development Brief*, Vol. IX, no. 5 (June 2005), 3; and "Under Scrutiny," *China Development Brief*, Vol. 9, no. 7 (September 2005), 1.

[57] Increasingly, Beijing environmental NGOs have joined together on broad issues. For example, on World Water Day, 21 NGOs urged the public to boycott products produced by companies causing pollution. Led by Ma Jun, director of the Institute of Public and Environmental Affairs (and author of *China's Water Crisis*), the group black-listed companies releasing waste water failing to meet environmental standards. Although most were small to medium-size firms, several multinationals – such as Panasonic Battery, Pepsi Cola, American Standard – drew attention. See Xiahua Sun, "NGOs Call on Consumers to Buy Green," *China Daily*, March 22, 2007, 3. See also Xiaohua Sun, "Multinationals Blacklisted for Pollution," *China Daily*, August 21, 2007, 1.

[58] The ACEF itself is a GONGO, one of six operating directly under SEPA.

[59] Fanchao Li, "NGOs in Difficulty, Survey Shows," *China Daily*, April 24, 2006, 2.

[60] Dangruo Sung, "Hsiamen PX Project: Move to Another Site," *People's Daily*, December 19, 2007, 1.

on environmental laws and regulations.[61] Spread of new information technologies, such as the Internet (see Chapter 9), mobile phones and text messaging, as well as the central government's proclaimed commitment to transparency, appeared to weaken somewhat local government resistance to change.[62]

A very small number of NGOs have focused directly on food security issues. We present the most involved NGO, Greenpeace, and then consider a few other NGOs with programs related to food safety and poverty alleviation issues.

8.5.2 *Greenpeace's Food Security Campaigns in China*

Greenpeace is an international NGO which has had a Hong Kong office since the 1980s. In 2002 it established a program office in Beijing, but it is not yet registered as a civic organization with the state. In Chapter 3 we discussed the Greenpeace campaign against multinational corporations threatening "paradise forests" and particularly against the activities of APP in Yunnan Province. Four other Greenpeace campaigns bear more directly on food security issues: GMOs in China, illegal insecticides, climate change, and water sufficiency.[63]

Greenpeace's campaign against agricultural biotechnology is global. In China, it has had two foci: from 2002 to 2004, the NGO concentrated on the importation of GM soybeans, and thereafter the focus shifted to Bt rice, with a minor emphasis on Bt cotton. Greenpeace has three objectives in the campaign against the importation of soybeans. First, it pointed out the threats to native production of soybeans from large-scale imports (from the United States, Argentina, and Brazil, most of which are genetically-modified). The Greenpeace contention was that China was a center of biodiversity in soya, and the importation of GM soya needed to be severely curtailed. Second, Greenpeace argued that the government needed a comprehensive labeling system. Under the 2003 biosafety law, processed foods including soybean oil such as cakes were not required to be labeled, which did not advance Greenpeace's goal of providing informed choice for consumers. Third, Greenpeace sought to improve the security of soya by working with a research institute in the Northeast and with local governments there.

As noted in Chapter 7, Greenpeace objected to the widespread cultivation of Bt cotton in China, stating that the yield of Bt cotton was less than non-GM varieties, and that pest resistance to it was growing. Moreover, it objected to the inability to control all potential effects of the use of this GM crop. The campaign against field-testing of Bt rice is the latest initiative. It was Greenpeace that notified the Ministry

---

[61] Zhe Zhu and Xiaohua Sun, "SEPA Calls for More Public Involvement," *China Daily*, June 22, 2007, 3.

[62] Weifeng Liu, "Transparency Key to Public Faith," *China Daily*, July 16, 2007, 2.

[63] This section is based on personal interviews with Greenpeace representatives in Beijing, April 27, 2004, August 23, 2006, and May 22, 2007.

of Agriculture concerning market sales of Bt rice. In its public relations campaigns, it has repeatedly emphasized the risks associated with transgenic products.

The Greenpeace campaign against illegal insecticides began in 2006. As mentioned briefly in Chapters 3 and 7, the Ministry of Agriculture regulations in 2006 and 2007 prohibited the use of organochlorines and organophosphates, because of their adverse impacts on human health. However, many farmers are poorly educated about dangers of different types of insecticides. They did not know that the insecticides were illegal, and they bought them because prices were low and marketers promised effective results. The Greenpeace campaign has had two objectives: to investigate illegal production of the pesticides and notify the MOA about them. Second, it has found outlets in the media for a public information campaign on pesticides.

The Greenpeace climate warming campaign has both global and local aspects. The main goal has been to spread awareness about the effects of greenhouse gas emissions on climate, and to urge a reduction in emissions. In China, Greenpeace has focused on the impact of climate change in the present. One representative noted: "We want to create a sense of urgency. Most think it's a long-term problem, but we want to educate people to create a sense of urgency."[64]

In its campaign against fossil fuels, Greenpeace has emphasized clean energy, and publicized China's experience of wind farms in Guangdong, Shandong, Xinjiang, and Inner Mongolia. In 2007–2008, Greenpeace was developing a campaign on water sufficiency. This too was designed for the Chinese environment and not a local spin-off of an international effort. The intended focus was reduction of water pollution.

Greenpeace's campaigns definitely have raised awareness of environmental and food safety issues in China. Wexler et al. in their analysis of NGO advocacy in China attribute the success of Greenpeace to the strategic partnerships it has formed with Chinese agencies (especially with MEP), its effective courtship of the mass media, and its targeting foreign corporations and not the Chinese government. They also opine: "The recipe may be more effective for the fact that it has been largely prepared by ethnic Chinese campaigners from Greenpeace Hong Kong."[65]

8.5.3 Activities of Other NGOs

Several of China's environmental NGOs have broadened their agendas to include food security issues. For example, WWF has developed a China Freshwater Program with several pilot projects. One set of demonstration projects included the Dongting Lake in Hunan, the Jianghan Plain in Hubei, and the Poyang Lake in

[64] Ibid., May 22, 2007.

[65] Robyn Wexler, Xu Ying, and Nick Young, *NGO Advocacy in China. China Development Brief*, September 2006, 66.

Jiangxi. It involved creation of new nature reserves, restoring flows between the Yangtze and lakes in Hubei and Anhui provinces, and establishment of a wetland conservation network. A second set of projects focused on improving livelihoods of communities living on land reclaimed from lakes, for example through promotion of sustainable fishing, which reduced pressure on wetlands, thereby improving water quality. WWF has engaged in policy advocacy by promoting integrated river basin management in China, raised awareness through public education campaigns on water conservation, and built capacity in local organizations.[66] Recently, WWF launched the Yellow Sea Ecoregion Support Project. Designed for a 7-year period, the plan will support local communities' sustainable conservation activities, to improve management of critical habitats adjoining the Yellow Sea.[67]

WWF has had projects to reduce desertification through preservation of the natural vegetative cover. As mentioned in Chapter 4, it also has participated in activities to mitigate climate change effects. Its approach has been to cooperate with the National Development and Reform Commission in developing both long- and mid-term investment strategies to increase energy efficiency (solar, biofuels).[68] The US-based Environmental Defense Fund has been active in air pollution control programs in the Beijing-area, often in concert with MEP. It too shares interests in energy efficiency and climate change mitigation with WWF.

China's oldest domestic NGO, Friends of Nature, has conducted small campaigns against pollution in urban and rural areas, siting of incinerators in residential areas, and preservation of rare and endangered species. As discussed in Chapter 4, it too has advocated enhancing civil society in China through addressing climate change at the grassroots level.[69]

Although individuals have established blogs on food safety issues, few consumer advocacy organization have been formed to date. A Beijing Consumer Association has paid attention to consumer rights. It commissioned a survey in early 2008 on consumer attitudes. More than half the respondents indicated they had had problems with the quality, authenticity, and service related to daily necessities. More than 58% chose to do nothing when their rights were violated. The association accepts cases on defective products. In 2007, it took 21,000 cases, 12% fewer than the previous year.[70]

A few NGOs have directed attention to the alleviation of poverty in rural China. One example is the FuPing Development Institute, which focuses on poverty alleviation, environment and sustainable development, civil society construction, education and employment. Established in 1998 by Tan Min, former Chinese economist in the World Bank, it consists of a professional institute and household service center.

---

[66] WWF, "Water for People, Water for Life." Beijing: WWF-China Freshwater Program, 2006.

[67] Xiaohua Sun, "WWF to Help Preserve Yellow Sea Ecosystem," *China Daily*, September 8–9, 2007, 2.

[68] Personal interview with WWF representative, Beijing, May 21, 2007.

[69] Personal interview with FON program director, Beijing, May 24, 2007.

[70] Ying Wang, "Consumer Rights Still Neglected," *China Daily*, March 13, 2008, 3.

It works mostly in regions of North and West China with high rates of poverty. One program is to assist families of migrant laborers (*nungmingong*) by providing adult education and training services. A second is a micro-credit program. A third is to provide computers (and training in their use) to isolated rural villages. Its most successful program is to train household workers, many of whom are illiterate female migrants to China's cities.[71]

Limitations on NGO activities may be relaxed somewhat in the future, for the director of the bureau for NGO administration in the Ministry of Civil Affairs has suggested that it would recommend the revision of laws and policies to encourage the development of foreign and domestic NGOs. Among its recommendations was a simplification of the onerous registration procedure and establishment of a foundation to recognize and reward NGOs demonstrating "good performance."[72]

A promising development with respect to transparency and citizen action occurred in 2008 when the National People's Congress amended the Water Pollution Prevention and Control Law. A new stipulation allows victims of water pollution to seek civil compensation directly from polluters. The amendments increase opportunities for public participation and "social groups" (NGOs) in the process. If victims are numerous (more than ten), they can file class action law suits. The law does not specify the kinds of cases in which class actions are allowed, and this ambiguity may limit its utility in the Chinese court system. Moreover, the legislation pertains less directly to food security than outright tainted products cases described in this chapter. Yet this is the first serious attempt to introduce such explicit stipulations into Chinese civil laws and procedures.[73]

Finally, China's NGOs played notable roles in the relief efforts following the May 12, 2008 earthquake in Wenchuan County, Sichuan. One entrepreneur gathered a dozen college students from the Liangshumin Rural Construction Center (a Beijing NGO dedicated to assisting village development) to establish relief and rescue centers. Another eight Beijing NGOs brought food, tents, medical supplies, and other necessities to quake victims. Among them was the environmental organization Green Earth Volunteers, whose director, Wang Yongchen, is a recipient of the Condee Nest environmental prize. Several private foundations that have given grants to fund public welfare projects provided support to grassroots organizations engaged in disaster relief. Altogether, more than 120 NGOs from different parts of China participated in relief efforts.[74]

In an analysis of reconstruction efforts, Jessica Teets finds that civil society in China resembles in important respects civil societies in many democracies. However, she argues that:

---

[71] Personal interview with program officer, Beijing, May 27, 2007.

[72] Xiaofeng Guan, "NGOs Have More Room to Develop," *China Daily*, May 25, 2007, 1. Also see Zhe Zhu, "Heavy Fines Await Polluters," *China Daily*, February 27, 2008, 2.

[73] Jingyun Li and Jingjing Liu, "Quest for Clean Water in China's Newly Amended Water Pollution Control Law," Woodrow Wilson Center, China Environmental Forum, 2009; also see Zhe Zhu, "Pollution Victims Get Protection," *China Daily*, December 25, 2007, 1.

[74] Xinhua, "NGOs Play Key Roles in Quake Relief," *China Daily*, June 6, 2008, 2.

Necessary reforms to improve the trust and capacity of civil society organizations are increasing the trust of both government and society, strengthening auditing procedures, improving resource capacity, and reforming laws about social group status, role in the policy process, donations and registration.[75]

8.6 China's New Food Safety Regime

Cases of tainted products "made in China" hit the world news in 2007–2009, drawing instant attention to the safety of Chinese food products – both for international and domestic consumption. We discuss examples of food safety problems, revisions to the structure of the food safety regime, revisions to laws and regulations, and corrective measures.

8.6.1 Tainted Products

In 1995, on the recommendation of the State Council, the National People's Congress (NPC) enacted the Food Hygiene Law, but no single agency was clearly responsible for the myriad health problems of food safety. Then, also at the recommendation of the State Council, in 2003 the NPC established the State Food and Drug Administration (SFDA). It was charged with supervision of the varied food safety regulations, by bringing them under one control structure. As we note below, this did not occur.

From 2003 to 2007, tainted products continued to appear in Chinese marketplaces. We mentioned red coloring dye, which has adverse health effects, and its relatively widespread use some 9 years after it was banned from production and sales. However there is a long list of tainted products: carcinogenic turbot (a fish species), raw snails causing meningitis, poisonous mushrooms, vegetables covered in pesticide residue, adulterated pickled vegetables, lead-contaminated noodles, substandard baby food formula, soy sauce manufactured from human hair, fake drugs and liquor, and frying oil laced with chemicals, among others.

Not all of these cases from 2004 to 2007 were featured in the Chinese media, but average Chinese long have been suspicious about the quality of food purchased in rural markets, small shops as well as supermarkets. China's State Food and Drug Administration released survey results in 2007 indicating that nearly two-thirds of respondents were concerned about food safety.[76]

---

[75] Jessica C. Teets, "Post-earthquake Relief and Reconstruction Efforts: The Emergence of Civil Society in China? *China Quarterly*, no. 198 (June 2009), 345.

[76] See *China Digital Times*: http://chinadigitaltimes.net/2007/03/apologies_for_food_safety_in_china_1.php

Contaminated food products from China became an international issue in April 2007, when the American government protested that pet food manufactured in China had killed hundreds of dogs and cats in the United States. The United States' claim was that pet food ingredients including wheat gluten had been contaminated by the chemical melamine, and that the products had been manufactured by Xuzhou Anying Biologic Technology Development company of Jiangsu Province.[77] As has been customary in cases of product quality and trade concerns, China's General Administration of Quality Supervision, Inspection and Quarantine (AQSIQ) denied that China had exported contaminated pet foods. However, within 1 week the agency discovered that the source of contaminated pet food was indeed China, and it admitted responsibility.

In ensuing weeks and months, this crisis prompted the regime to pursue three somewhat contradictory paths. It launched comprehensive inspections of food quality throughout China and punished those responsible for contaminated goods. Simultaneously, China blamed the United States and other countries for exporting contaminated goods to China. Throughout, however, the state changed and revised the administrative structure for supervision of food safety, enacted new laws and regulations, and spelled out the corrective measures it would take to improve food safety.

Shortly after the pet food case surfaced, AQSIQ ordered recalls of products confirmed to endanger public health.[78] A national inspection by AQSIQ resulted in the seizure of $26 million worth of contaminated or substandard food products. The contaminants included industrial raw materials such as dyes, mineral oils, paraffin wax, formaldehyde and the carcinogenic malachite green used in production of flour, candy, pickles, biscuits, black fungus, melon seeds, bean curd and seafood. Most of these products had been manufactured by small, unlicensed food processing plants with fewer than 10 employees, and the processing plants were shut down (and those with licenses had them revoked).[79] The most dramatic action was the forced resignation of Zheng Xiaoyu, from the directorship of the State Food and Drug Administration. Subsequently, he was arrested for taking bribes from several firms in exchange for the issuance of state licenses insuring food product security. Zheng was the major casualty of the food safety embroglio, as he was executed on July 10, 2007 for corruption.

[77] Li Cao, "Firms Deny Responsibility for Poison Petfood," *China Daily*, April 4, 2007, 3.

[78] Chuanjiao Xie and Yinan Hu, "Food Recall System to Be Set Up," *China Daily*, May 29, 2007, 1.

[79] Zhe Zhu, "Industrial Raw Materials Found in Food," *China Daily*, June 27, 2007, 3. AQSIQ also revoked business licenses of companies exporting tainted wheat gluten (the companies evaded quality inspection by labeling their products as chemical ingredients, which were not subject to compulsory inspection); and it closed down the factory of a Chinese manufacturer that had mislabeled the industrial use of glycerin, subsequently used in toothpaste in central America. See Zhe Zhu, "Inferior Food Exporters Punished," *China Daily*, July 21–22, 2007, 2. See also Yang Yang and Jennifer Turner, "Food Safety in China," Woodrow Wilson Center, China Environmental Forum, 2007.

The second path followed by the regime included attempts to reassure the public that China's food was safe and that other countries had similar, or worse, problems of food contamination. In early July 2007, AQSIQ reported that its tests of 3,384 different kinds of food products showed that about 86% passed national safety tests. It pointed to records indicating that in the previous 22 years, safety of food products had steadily increased. MOA officials indicated that farm products too were safer than they had been previously, but they admitted that excessive use of food additives and pathogenic bacteria such as the coli groups were still found in samples.[80]

Given the increased degree of nationalism in China of recent years, it was not surprising that the regime lashed out at foreign critics of China's food safety, and blamed foreign producers for problems as severe or more serious than those in China. The AQSIQ Minister Li Changjiang said in mid-July 2007 that "some foreign media, especially those based in the US, have wantonly reported on so called unsafe Chinese products. They are turning white to black."[81] While conceding problems in some domestic food producing firms, he continued: "One company's problem doesn't make it a country's problem. If some food products are below standard, you can't say all the country's food is unsafe."[82] Li cited records demonstrating that more than 99% of Chinese food exports to the USA in the previous 3 years had met quality standards, and this was the same or higher than the equivalent figure for US food exports to China.[83] The regime also gave extensive publicity to apologies from the multinational toy manufacturer Mattel. In August and September 2007 the corporation ordered three high profile recalls of 21 million Chinese-made toys including Barbie doll accessories and toy cars, because of concerns regarding lead paint and tiny magnets that could be swallowed by children. Mattel executives later said that most of the toys were defective because of Mattel's design flaws, not due to problems in Chinese manufacturing.[84] The response of American journalists to these attacks was that China was attempting to deflect attention from its serious food safety problems by attempting to turn the table on the USA and other countries. Said the *Washington Post*, "The result has been an aggressive campaign to save the "Made in China" label by presenting an alternate view on consumer safety and globalization."[85]

---

[80] Zhe Zhu, "China's Food Quality Up to Standard," *China Daily*, July 4, 2007, 3.

[81] Zhe Zhu, "Foreign Media Rapped over Food Reports," *China Daily*, July 16, 2007, 1.

[82] Ibid.

[83] Ibid.

[84] Zhe Zhu, "US Toymaker Apologizes to China over Flawed Goods," *China Daily*, September 22–23, 2007, 1.

[85] Ariana Eunjung Cha, "Who's Poisoning Whom?" *The Washington Post National Weekly Edition*, July 23–29, 2007, 21. In September, the *Post* reported that officials of neighboring states – Indonesia, Philippines, and Malaysia – were bullied by Chinese officials when they complained about the safety of Chinese products. When the head of Indonesia's food and drug agency complained that dried fruit from China was spiked with industrial chemicals and children's candy was carcinogenic, she was told that Indonesia should lower its safety standards. See Ariana Eunjung Cha, "Grief over Tainted Goods," *Washington Post National Weekly Edition*, September 10–16, 2007, 22.

Within 4 months of the event, the public in China was divided as to the safety of most food sold in China. A *China Daily* survey of 1,654 people found that 47% believed food was safe, contrasted to 41% who found it unsafe and 12% who declined to comment.[86] Whether they found food to be safe or unsafe, most of the survey respondents believed more education on food safety issues was essential. They also believed it was essential to make improvements in the food safety regime,[87] a subject to which we turn.

8.6.2 Revisions to the Structure of the Food Safety Regime

The tainted products crisis stimulated a revision to the food safety regime. In its White Paper on *China's Food Quality and Safety*, published on August 17, 2007, the State Council significantly revised the institutional framework for food safety. It assigned responsibility to provincial and local governments, and realigned the central government's ministries with tasks in this area.

The State Council divided food safety supervision into four "links," managed by the Ministry of Agriculture (MOA), the Central Administration of Quality Supervision, Inspection, and Quarantine (AQSIQ), the Ministry of Industry and Commerce, and the Ministry of Health (MOH). MOA was placed in charge of the supervision of primary agricultural products; AQSIQ was to supervise quality and daily hygiene of food processing; the industry and commerce ministry was to supervise food circulation and distribution;[88] and the MOH was to oversee the catering industry, food shops, and restaurants. SFDA was left in charge of the integrated supervision and coordination of food safety as well as investigations of major incidents and the assessment of penalties. AQSIQ retained authority over imported and exported agricultural products and other foodstuffs.[89]

---

[86] Jie Liu, "Most consumers Unfazed by Food Scandals," *China Daily*, August 17, 2007, 15.

[87] A survey of more than 200 Chinese suppliers by Global Sources (a business-to-business media company) found that 62% were increasing spending on quality control, to address food safety concerns. Two-thirds of the respondents expected their exports to increase in the future. See Qi Zhang, "Quality Control Spending to Increase," *China Daily*, September 26, 2007, 3.

[88] Commerce Minister Bo Xilai announced the ministry's own method to engage in the "special war" against poor product quality: "Two chains, a system and a network." The chains referred to supervising the entire process of both industrial and food production. System referred to the product recall and accountability system, while network meant a comprehensive quality monitoring system in each part of society. Further, the minister said 12 of 20 detailed goals would require comprehensive treatment: licensing of 100% of food producers, monitoring 100 percent of agricultural product wholesale markets, and inspection of 100% of suppliers of raw materials. See Zhe Zhu, "Minister Confident of Quality Victory," *China Daily*, August 25–26, 2007, 3; see also Zhe Zhu, "Food Safety Struggle Bearing Fruit," *China Daily*, October 9, 2007, 2.

[89] State Council, *China's Food Quality and Safety*, August 17, 2007, 9 (reprinted in *China Daily*, August 18–19, 2007, 5–7).

Simultaneously, the State Council named Vice-Premier Wu Yi (a firm and internationally respected administrator) as the head of a 19-member cabinet-level task force to oversee product quality and food safety. The panel was composed of high-ranking officials from 16 central government departments including ministers of MOH, MOA, and AQSIQ. The AQSIQ minister dispatched one-third of the members of his central office of 150 to 14 provinces for a 3-month inspection of food production enterprises.[90] Wu Yi said of the campaign: "This is a special war to protect the safety and interests of the general public, as well as a war to safeguard the 'Made in China' label and the country's image."[91]

At the expiration of the 4-month campaign and task force, the proposed replacement was a national food safety risk evaluation committee. Some experts sought a restructuring of the mechanism for supervising food safety – with the integration of at least six major government departments with overlapping jurisdictions, responsibilities, and enforcement powers.[92] This did not occur, for bureaucracies in China, as elsewhere, are extremely jealous of their turf. A proposal of planners to create "mega departments," thereby reducing to a smaller number the 28 ministries and more than 40 functional organizations operating directly under the State Council,[93] gained greater support.

As discussed above, in March 2008, five super-ministries were created, one of which was the environmental protection ministry. Some efficiency – and increased effectiveness – may result from the transfer of the SFDA to the jurisdiction of the Ministry of Health, but this ministry's status was not elevated.

8.6.3 Revision to Laws and Regulations on Food Safety

In its White Paper, the State Council also announced that China's legal and regulatory regime was now complete, and that it provided a "sound foundation and good environment" for guaranteeing food safety, improving food quality and regulating food imports and exports. The national laws included the Product Quality Law, Standardization Law, Metrology Law, Law on the Protection of Consumer Rights and Interests, Law on the Quality and Safety of Agricultural Products, Food Hygiene Law,[94] Law on Import and Export Commodity Inspection, Law on Animal

---

[90] Zhe Zhu, "Food Safety High on Agenda," *China Daily*, August 18–19, 2007, 1.

[91] Wei Jiang, "Play Safe," *China Daily*, October 22–28, 2007, supplement 1.

[92] Zhe Zhu, "Committee to Evaluate Food Safety," *China Daily*, December 27, 2007, 2.

[93] Jiao Wu, "Mega Departments to Help Improve Efficiency," *China Daily*, December 27, 2007, 1.

[94] Later in 2007 the State Council proposed a substitute to this basic law. It would include a labeling system making food producers responsible for ingredients, additives, and expiration dates; a food recall system; and strict inspection of food imports and exports. See Zhe Zhu, "Committee to Evaluate Food Safety," *China Daily*, December 27, 2004, 2.

and Plant Entry and Exit Quarantine, Frontier Health and Quarantine Law, and the Law on Animal Disease Prevention.[95]

As part of this effort, the State Council approved a draft food safety law to raise standards and regulate supervision. The legislation encompassed food production, processing, consumption and regulation. Specifically, it mandated better release of information on food safety issues, larger fines for rogue firms, and punishment for errant officials. Further, it guaranteed the public's right to sue and claim compensation. Finally, it required food safety risk analysis and monitoring systems and more detailed inspection of food imports and exports.[96]

Specific administrative regulations of the above agencies were revised, as were those of related agencies, such as the State Forestry Administration's regulations on Import and Export of Endangered Wild Fauna and Flora. Heavier fines were assessed to tighten the enforcement system.[97] Moreover, detailed departmental rules were revised as well. The White Paper describes the nature of the new, seemingly comprehensive, system:

> Now, a food quality and safety standard system covering all categories, featuring a relatively rational structure and being fairly complete, has taken initial shape in China. Food safety standards cover the place of origin of agricultural products, quality of irrigation water, rules for the rational use of materials put into agriculture, rules and procedures for animal and plant quarantine, good agricultural practices (GAP), standards of maximum amount of pesticides, veterinary drugs, pollutants and spoilage organisms allowed in food, standards for food additives and their use, hygiene standards for food packaging materials, standards for special dietary food, standards for signs or labels on food packages, standards for the management and control of the safe production of food and standards for testing methods concerning food.... So far, China has promulgated over 1,800 national standards concerning food safety, and over 2,900 standards for the food industry, among which 634 national standards are compulsory.[98]

Indeed, this is a dizzying number of revisions, certainly designed to impress consumers with the safety of China's food products.

---

[95] State Council, 2007, 18.

[96] Zhe Zhu, "Draft Food Safety Law Approved," *China Daily*, November 1, 2007, 1. Preparing the new food safety law was difficult. Administrators proposed adding a product identification and tracking system, which would require that each product be labeled with a unique number or code. More than 20 firms objected that the new system would increase production costs, including multinationals such as Nestle, Mars, Coca-Cola and Pepsico. They submitted petitions opposing the proposed change to the NPC Standing Committee's legislative affairs commission and to the State Council's legislative affairs office. The China National Food Industry Association also objected to the cost of the proposed change, its necessity (as it did not deal with quality issues concerning raw materials), and its fairness (because it did not apply to small food plants). Although AQSIQ extended the deadline, it did not rescind the requirement. See Zhe Zhu, "Food Safety System Put in Draft Law," *China Daily*, April 9, 2008, 3.

[97] Zhe Zhu, "Heavier Fines to Ensure Food Safety," *China Daily*, December 28, 2007, 1.

[98] State Council, 2007, 26. See also Zhe Zhu, "More Legislation to Help Combat Shoddy Products," *China Daily*, January 5–6, 2008, 1.

8.6.4 Corrective Measures

Nearly an half million enterprises in China are engaged in the production of food products. Large enterprises, numbering about 26,000 firms, occupy nearly three-quarters of the market in output and sales. The problems in recent years as noted are with smaller enterprises having fewer than ten employees. Such firms have less than 10% of the food products market share, but number over one-third million. Food enterprises contribute about 7% of China's national industrial output value, and recent growth has exceeded 20% annually.[99]

Of equal importance is the value of food products in China's import and export markets. In 2006, China exported 24 million tons of food worth $27 billion, some 15% more than in the previous year. It exported foodstuffs to 200 countries and all of the world's regions. The top ten consumers of food products from China were: Japan, the United States, South Korea, Hong Kong, Russia, Germany, Malaysia, the Netherlands, Indonesia, and the UK.[100] Clearly, China sought to protect its export markets for food products, and for this reason the White Paper announced 11 corrective measures that the regime would implement immediately:

1. Intensifying supervision of the quality and safety of agricultural products. This measure would, among other subjects, intensify control of residue of high-toxic pesticides[101] in vegetables and carefully screen materials used in farming, production, and marketing. Too, it extended registration to fruits.[102]
2. Establish and rigorously implement market access systems for food quality and safety. This measure would revise and upgrade the production licensing system, the compulsory inspection system, and the market access labeling system.
3. Intensify state supervision through sample surveys of food quality. State agencies would increase the frequency and coverage of sample surveys, while continuing to focus on items of daily consumption, such as dairy products, meat products, tea, beverages, grain and edible oil.
4. Intensify the "rectification" of food workshops. This measure addresses small food processing firms. Some would be shut down or merged; others would be subject to more stringent supervisory measures to prevent food safety accidents.[103]

[99] Ibid., 3–4.

[100] Ibid., 7.

[101] Several months into the campaign, the Ministry of Agriculture banned use of five pesticides, including methamidophos. It revoked licenses of seven firms producing highly poisonous pesticides and confiscated nearly 500 tons of illegal pesticides. See Zhe Zhu, "Curbs Put on Pesticide Production," *China Daily*, October 30, 2007, 1.

[102] For example, fruit from unregistered orchards or packaging plants was barred from export. Fruit exporters were required to maintain records on the origin, volume and destination of fruit shipped. Also, use of agricultural chemicals and fruit disease records were scheduled for monitoring as well. See Zhe Zhu, "Fruit Registration Rule Extended," *China Daily*, September 21, 2007, 3.

[103] See also Quanlin Qiu, "Food Firms Set Strict New Safety Standards," *China Daily*, November 9, 2007, 3.

5. Promoting the "responsibility system" for regional food safety control. A three step process would be used to improve regional food safety: (a) designating personnel responsible for regional food enterprises; (b) investigating rural food production enterprises' working conditions; and (c) making local governments accountable for quality supervision and inspection in their areas.
6. Improving supervision of the "food circulation" sector. This measure refers to the food distribution system and its modernization, including increasing the number of checkpoints in the system, verifying invoices, licenses, and establishing efficient food product recall systems. For example, the State Administration for Industry and Commerce (SAIC) established a tracking system for all food markets and supermarkets above the county level. This new system required invoices and documents to be attached to goods, and its inspectors began shutting down unlicensed food vendors.[104]
7. Improved supervision of food safety in the catering business. This measure would improve food hygiene inspection and supervision in all places food is prepared (including in schools).
8. Elimination of counterfeit or fake and inferior food products from the market. State agencies would improve monitoring and inspection of the manufacture and sales of fake and faulty food (for example, food products using non-food materials and misusing food additives). A related campaign focused on development and implementation of fishery production standards, which became an issue when the United States banned four types of Chinese seafood products because it claimed they contained illegal drug residues.[105]
9. Further development of a risk-warning and emergency-response system. This measure would improve efforts to collect and analyze information on potentially tainted food products, so that timely warnings could be issued and rapid responses made.
10. Establishment and improvement of a food recall system. Regulations of AQSIQ required producers to recall unsafe toys and food products. If they failed to do so, the government department would order a recall and fine producers three times the value of the products.[106]
11. Improvement of the food safety credit system. The final measure seeks to improve the reputation of food producing enterprises through improved performance record keeping, publication of "honor rolls" and "blacklists," and enlistment of chambers of commerce and trade associations to promote self-discipline in the industry.

Altogether, the White Paper spelled out a large number of institutional and regulatory changes to China's food safety regime, with specific measures to correct problems of quality, lack of standardization and transparency, and poor enforcement.

---

[104] Chuanjiao Xie, "Food Safety Drive Stepped Up," *China Daily*, August 19, 2007, 1.

[105] Jiao Wu, "New Fishery Standards in Pipeline," *China Daily*, December 25, 2007, 4.

[106] Zhe Zhu, "Toy and Food Recalls Introduced," *China Daily*, September 1–2, 2007, 1; see also Zhe Zhu, "Regulation on Recalls in Pipeline," *China Daily*, November 23, 2007, 1.

Major officials of the state, such as Premier Wen Jiabao, pledged immediate attention to correct deficiencies.[107] The SFDA announced that China would spend $1.16 billion to build or upgrade the infrastructure of food and drug inspection,[108] while both SFDA and AQSIQ set up hot lines for complaints about unsafe food and drug products.

Although it might seem premature to comment on the efficacy of the regime's response, officials claimed success. The vice-minister of AQSIQ remarked in late 2007:

> During the first eight months of this year, China's exports grew by 23.3 percent, which shows that our exports have not been hit by these recalls …. Some Chinese manufacturers said they had received even more orders and their workers are busy trying to meet them.[109]

An inspection of 37 major cities at year's end found:

- 95.3 percent of vegetables free of pesticide residues,
- 98.4 percent of meat products without traces of clenobuterol hydrochloride (a pig feed additive damaging to human health),
- 99.8 percent of aquatic products were free of banned fish food supplements, and
- No pig products in the largest cities were contaminated with banned drugs.[110]

At the conclusion of the 4-month national campaign, Vice-Minister of Agriculture, Gao Hongbin, announced: "We have successfully completed our mission to improve quality and safety." Yet Gao acknowledged remaining challenges to agricultural production, particularly from the large number (about 220 million) of farming families,[111] which were difficult to regulate.

Clearly, the focus of leaders was on the 2008 Beijing Olympics. Another AQSIQ Vice-Minister, Pu Changcheng, expressed confidence that food for players and officials at the Olympics would be safe:

> The tasks of the rectification campaign have been fulfilled completely and all its objectives reached. Illegal practices of using non-food materials and/or recycled food to make and process food products have been eliminated. Abuse of food additives such as preservatives and colors too, has been stopped.[112]

---

[107] See, for example, Zhe Zhu, "Quality Problem High on Agenda," *China Daily*, July 28–29, 2007, 1.

[108] Juan Shan, "Gov't Pumps in $1B to Raise Drug, Food Safety," *China Daily*, August 9, 2007, 1.

[109] Chuanjiao Xie, "World 'Still Trusts' Chinese Products," *China* Daily, November 7, 2007, 2.

[110] Dingding Xin, "Food Safety Campaign Yields Positive Results," *China Daily*, December 29–30, 2007, 1.

[111] Jiao Wu, "Food Safety Mission 'Complete'," *China Daily*, January 9, 2008, 3.

[112] Zhe Zhu, "Best Quality Food During Olympics Assured," *China Daily*, January 15, 2008, 2. Also see: Zhe Zhu, "Food at Olympics is Safe: Officials," *China Daily*, February 22, 2008, 1; and Zhe Zhu, "Food Safety at Games Top Priority: Quality Chief," *China Daily*, April 19–20, 2008, 2.

These assurances did not convince teams planning to participate in the 2008 Beijing Olympics. US athletes planned to bring a good part of their food with them.[113]

Yet an important change was the establishment of a local presence for the US Food and Drug Administration (FDA) in China (and also in India). Secretary of Health and Human Services Michael Leavitt promised that up to 10 FDA staff would be stationed at three locations in China, including the US Embassy. This would further enhance the safety and quality of food and drugs exported to the USA.[114] The first testing center opened in late June 2008 in Zhuhai, Guangdong Province. The center applied food safety standards accepted by both China and the U.S. Foods passing tests are directly exported to the USA, and are exempt from further testing by the FDA.[115]

8.6.5 More Tainted Products

New concerns about China's food products suggested that the regime had more work to do to guarantee food safety. In late January 2008, the Japanese government reported that 175 people had been sickened by insecticide-tainted dumplings from China. This prompted supermarkets to pull Chinese-made meat products from their shelves.[116] In addition to continued concern about counterfeit

[113] *New York Times*, February 8, 2008.

[114] Juan Shen, "Leavitt Confirms FDA China Office Plans," *China Daily*, May 17–18, 2008, 8.

[115] Qiwen Liang, "Joint Food Safety Test Center Starts Operation," *China Daily*, June 20, 2008, 2.

[116] All the products came from the same company, Tianyang Food Processing, in Hebei Province (outside Beijing). See Martin Facklere, "Insecticide-Tainted Dumplings from China Sicken 175 in Japan," *New York Times*, February 2, 2008, A14. China's AQSIQ disputed the report and claimed that no harmful chemicals were found in samples of the frozen dumplings. China quickly dispatched an investigation team to Japan to probe the incident. See Zhe Zhu, "Officials to Visit Japan over Food Poisoning," *China Daily*, February 2–3, 2008, 1. The AQSIQ also dismissed reports that dissatisfied Chinese workers had deliberately contaminated the dumplings, and called for a Sino-Japanese joint investigation team on food safety issues. (See Zhe Zhu, "Dumplings 'Were Not Sabotaged'," *China Daily*, February 14, 2008, 1) Then, AQSIQ revealed that the producing firms were wholly-owned subsidiaries of the Nicky Foods Co. of Osaka, Japan, that they followed Japanese production procedures, and that they were supervised by Japanese staff. Perhaps, AQSIQ opined, the raw materials used had not been purchased from designated export-oriented farms. (See Zhe Zhu, "Japan Side 'to Blame' for Tainted Foodstuffs," *China Daily*, February 23–24, 2008, 1) Next, Chinese police reported that the poisoning issue was a case of sabotage, done in Japan, and not a food safety problem. See Zhe Zhu, "Dumplings Poisoned by 'Saboteurs'," *China Daily*, Febrruary 29, 2008, 1.

and unsafe drugs produced in China, this suggests that regulatory reform is incomplete.[117]

China's latest food safety crisis emerged in early September 2008, when an infant died and thousands developed urinary tract problems including gall stones from using powdered milk contaminated with melamine.[118] Multiple causes explain the crisis, with greed topping the list. State controls on milk prices cut margins of dairy farmers, operators of private milk collecting stations, and retailers. Producers and middlemen, squeezed by rising feed prices and price controls, diluted milk with water and melamine, a toxic substance used to make plastics and fertilizer. Its nitrogen content produced high protein readings, enabling the milk powder to pass inspections. The Sanlu Group, China's largest dairy company, was the primary firm involved in the crisis. It learned about sickness attributed to its powdered milk, but covered up the news, and local regulators were complicit.

The earliest report of problems with contaminated milk powder surfaced in December 2007, 10 months before the central government took action. In March 2008, several consumers complained to Sanlu,[119] to provincial authorities, and to the health ministry in Beijing – without response. In August their complaints did

[117] In late October 2007, the *New York Times* reported that pharmaceutical ingredients exported from China often were produced by chemical companies neither certified nor inspected by Chinese drug regulators. Thus, they had no incentive to restrict the export of unapproved, adulterated, or counterfeit ingredients. The account pointed out: "Yet in China, chemical manufacturers that sell drug ingredients fall into a regulatory hole. Pharmaceutical companies are regulated by the food and drug agency. Chemical companies that make products as varied as fertilizer and industrial solvents are overseen by other agencies. The problem arises when chemical companies cross over into drug ingredients. 'We have never investigated a chemical company,' said Ms. Yan, deputy director of policy and regulation at the State Food and Drug Administration. 'We don't have jurisdiction.'" See Walt Bogdanich, "Chemicals Flow Unchecked from China to Drug Market," *The New York Times*, October 31, 2007, A3. Then, in February 2008, US investigators focused on reports that more than 400 patients taking the blood-thinner heparin suffered complications and more than a dozen died. The raw material for the drug is made from pig intestines, and most was produced at a large Chinese factory called Changzhou SPL, a majority of whose shares were held by an American firm. The factory purchased its raw materials from many tiny, unclean family workshops; neither the small workshops nor the large factory had been inspected by either Chinese or US authorities, representing yet another regulatory void. (David Barboza, "Twists in Chain of Supplies for Blood Drug," *New York Times* February 28, 2008, A17; also see Walt Bogdanich, "Blood Thinner Might be Tied to More Deaths," *New York Times*, February 29, 2008, A21) The SFDA promised to address the situation by introducing higher standards for companies producing the ingredients for pharmaceuticals, and, with the assistance of the NDRC and MEP, monitor the large number (some 80,000) of chemical firms producing drug ingredients. See Juan Shen, "Tighter Control on Drug Firms," *China Daily*, April 9, 2008, 3; and Gardiner Harris, "U.S. Identifies Tainted Heparin in 11 Countries," *New York Times*, April 22, 2008, A15.

[118] Xinhua, "Milk Food recalled as Baby Dies," *China Daily*, September 12, 2008, 1. This was the same chemical found in Chinese pet food that sickened thousands of dogs and cats in the USA in 2007.

[119] In 2004, 13 infants in Anhui died of malnutrition and 171 others were hospitalized after consuming substandard milk powder falsely labeled as being produced by Sanlu. See Xinhua, "Firm Probed over 'Bad' Baby Milk," *China Daily,* September 11, 2008, 4.

reach Fonterra, a New Zealand dairy products multinational that owned a 43% interest in Sanlu. Fonterra asked Sanlu to recall the contaminated product, but only after the company had entreated New Zealand officials to make representation to Beijing authorities, were products recalled (in early September 2008).[120] An additional 21 firms were involved (of a total of 109 milk food firms in China), including three of China's largest dairy exporters, Mengniu and Yili (based in Inner Mongolia) and Shanghai-based Bright Dairy. Also involved were local regulators and particularly, officials in the government of Shijiazhuang (capital of Hebei Province) where Sanlu is based; some officials had ownership interests in Sanlu subsidiaries. Many observers suggest that central government warnings in preparation for the August 2008 Beijing Olympics created a chilly climate for whistle blowers on any issue. Agencies were advised to do everything essential to produce a "harmonious" Olympics, and journalists particularly were advised to not report anything politically sensitive.

Once the problem had been made public, the government response was effective. ASQIQ and health officials identified the 22 companies selling contaminated milk powder. Initially, they recalled and destroyed all Sanlu powdered milk, and then made a blanket recall of all dairy products made before September 14.[121] Premier Wan Jiabao visited childrens' hospitals and blamed both greed and official misconduct, remarking "The social impact is vile and the lesson profound."[122] Then, in response to complaints and tips, AQSIQ and health officials broadened investigation into other products suspected of using melamine (it had been banned 1 year previously), including frozen yogurt, White Rabbit candy, and both animal and fish feed. They arrested and tried nearly two dozen company officials, chemical factory owners, and middlemen.

The focus of post-crisis reform was on the dairy industry and melamine. Dairy products are an $18 billion industry in China, but had been poorly regulated. Regulations for supervision of the entire dairy chain, from raising cows, collection and purchase of raw milk, and production and sale of dairy products to their export and import were crafted, and regulations on animal feed were tightened.[123] The nation's milk procurement stations were brought into one supervision system.[124] Large firms, such a Mengniu and Yili, previously exempt from food product inspection because of their reputation for producing high-quality, globally competitive products, lost this privilege and their "famous brand" status as well.

---

[120] Jim Yardley, "Chinese Baby Formula Scandal Widens as 2nd Death is Announced," *New York Times*, September 16, 2008, A1.

[121] Zhe Zhu and Louise Ho, "Melamine Found in More Milk," *China Daily*, September 17, 2008 and Zhe Zhu and Bolin He, "Pre-Sept 14 Dairy Products Ordered Off Shelves," *China Daily*, October 15, 2008, 1.

[122] Xinhua, "Milk Scandal Reeled In," *China Daily*, September 22, 2008, 1. Wen later said that China would "draw a lesson" from the affair.

[123] Zhe Zhu, "Dairy Sector Regulation Beefed Up," *China Daily*, October 7, 2008, 1.

[124] Xiaofeng Guan, "Supervision of Dairies to be Unified," *China Daily*, October 1, 2008, 3.

Production controls were instituted for melamine. The regime spent $73 million to increase surveillance and monitoring of agricultural products. The draft food safety law, inspired by the earlier food safety crisis, was under discussion by the National People's Congress, and it too was toughened by setting stricter food quality standards and imposing greater responsibility on government.[125] AQSIQ sent 1,644 teams and 387 working groups to inspect the production process of dairy plants,[126] followed 6 weeks later by a 4-month nation-wide campaign, involving nine central government departments, to insure that all food products were free of non-edible substances and excessive levels of additives.[127]

The crisis killed six infants and sickened nearly 300,000. The Chinese government uses draconian punishments as means to stimulate enforcement and obedience of officials and citizens, and that was the case in this crisis too. The director of AQSIQ, Li Changjiang, was forced to resign within 2 weeks of the onset of investigation, for having failed to properly supervise the dairy market. The mayor, vice-mayor, three officials, and the party chief of Shijiazhuang, center of the scandal, were removed from their posts. The CEO of Sanlu Group, Tian Wenhua, who admitted to having known about the contaminated products since May 2008, was stripped of her party post and membership and sentenced to life in prison; her firm went bankrupt. A middleman and the head of a local dairy firm were sentenced to death, while two producers of melamine-laced "protein powder" faced spending the rest of their lives in jail; 15 others received prison terms of from 2 to 15 years. The companies were required to pay restitution to the families of deceased infants and to those children seriously injured until they turned 18 years of age, in the amount of $160 million. Meanwhile, in October 2008, China's dairy exports sank 92% compared to the previous year.[128]

In the view of some observers, action of the government was misdirected. Arthur Kroeber, owner of the Beijing consultant firm Dragonomics, wrote to his clients: "These after-the-fact administrative measures miss the point." From his perspective, the problem was in the behavior of the communist party, which remains involved in pricing control, company management and the flow of information. "The party views control of all three as necessary to its rule.... Further major scandals are thus inevitable."[129]

[125] Zhe Zhu and Xiaohuo Cui, "Food Safety Law to be Stricter, More Onus on Government," *China Daily*, October 24, 2008, 1.

[126] Xinhua, "Milk Buying System to be Overhauled," *China Daily*, September 24, 2008, 1.

[127] Zhe Zhu, "4-Month Campaign to Ensure Food Safety Launched," *China Daily*, December 9, 2008, 1.

[128] Zhe Zhu, "Dairy Goods' See Massive Fall," *China Daily*, December 2, 2008.

[129] Quoted in Jim Yardley and David Barboza, "Despite Warnings, China's Regulators Failed to Stop Tainted Milk," *New York Times*, September 27, 2008.

8.7 Conclusions

Although China, like other nation-states, does not have a fully articulated food security regime, it does have a large body of laws, regulations, and policies pertaining to the security of its food production, control, and consumption systems. We indicated the range of rules, and we also pointed out the primary and secondary institutions of government involved in food security processes and outcomes. Directed, state-centered administration of laws and regulations has been more difficult in the reform period than earlier because, since the early 1980s the state has devolved important government functions to the provinces, autonomous regions, and to local governments. This has led to problems of vertical coordination of government policy (while horizontal coordination is difficult because of the insularity of the ministries of the state). The result has been implementation deficits and frequent failure to enforce laws and regulations uniformly.

We also examined the non-governmental sector in China, which is more influential in 2009 than previously, notwithstanding stringent registration requirements and high levels of scrutiny by state security agencies. Too, China has substantially increased its participation in international organizations, such as WTO, and ratified environmental conventions, for example the Cartageña Protocol, which makes its governmental processes subject to the microscope of international analysts and organizations.

China now does have a "regime" operating in the realm of food safety. Flaws in China's food safety regime became apparent internationally in 2007, as Chinese-manufactured pet foods, laced with melamine, killed hundreds of American dogs and cats. After a brief period of denial, the regime quickly apologized (and put to death the former head of the State Food and Drug Administration); it then set about revamping the food safety system. Responsibilities were sorted out among central agencies; laws, regulations, and rules were amended; and the regime specified concrete measures to remedy the flaws. Significantly, the State Council placed a redoubtable vice-premier in charge of an action force of high officials to clean up the system, and funded the reform efforts appropriately.

Yet these transformative efforts were not sufficient to insure the safety of China's dairy products, as seen in the most recent food scandal of 2008–2009. The good reputation of the "Made in China" label is critical to China's continued economic success, and the response of the regime indicates that it has taken the reform of food safety very seriously indeed. Yet the task is gargantuan, as it involves intricate government to government and public–private sector relationships at all levels. It also involves the legitimacy of the party-state itself.

Chapter 9
Issues in Implementing Food Security in China

Abstract This chapter begins by analyzing the basis in knowledge of China's food system – the intellectual establishment, research and development expenditures in the agricultural sciences, and training to improve farming practices. A second challenge is administrative coordination, both vertically and horizontally. The third challenge concerns modernization of the infrastructure for food production, emphasizing transportation and storage. The fourth area concerns access to food and poverty. It describes the significant reduction of absolute poverty from the late 1970s to the present, examines recent poverty alleviation programs, and assesses work remaining to be done – made difficult by the lack of agreement on the number of poor in China today. The fifth challenge is public participation, and discussion centers on media reportage of food security issues and citizen protests. The pattern indicates limited growth of civil society. The chapter concludes by describing changes in agricultural trade and issues related to China's entrance into the World Trade Organization.

Keywords Agricultural universities • Research and development expenditures • Farming practices • *Tiao/kuai guanxi* • Transportation systems • Food storage • Poverty line • Poverty alleviation • Migrant labor force • Media • Civil society • Citizen protests • World Trade Organization (WTO) • Sanitary and Phytosanitary (SPS) agreements

9.1 Introduction

Producing food and distributing it to China's 1.3 billion consumers is a large and complex enterprise, but through the actions of governments and developing markets, since the late 1970s the enterprise has been successful. Throughout this volume, we have pointed out the immediate and near-term environmental stressors that have influenced food production. These problems have not been resolved, and some

J.H. McBeath and J. McBeath, *Environmental Change and Food Security in China*, Advances in Global Change Research 35, DOI 10.1007/978-1-4020-9180-3_9,

of them – for example, land and water pollution and climate change – seem more likely to worsen than ease in the future.

In this penultimate chapter, we consider challenges that China faces now and in the near-term with respect to food production and distribution. First, we examine the knowledge base concerning food security, the budget allocations for continued research and development, as well as needs regarding farming practices. Second, we return to problems of government coordination, and both the integration of diverse agencies at the central level as well as coordination from the center to provinces and local governments. Third, we consider the need to modernize China's agricultural infrastructure, both its transportation and storage systems. Fourth, we turn to the large challenge of alleviating conditions of poverty, mainly for rural households, but with some comments on urban poverty. Finally, we consider the challenges presented as a result of China's entry to the World Trade Organization, for example, China's ability to meet global production standards.

9.2 The Knowledge Base

The first challenge pertains to the sufficiency of existing knowledge about current and potential threats to food production and distribution. It includes discussion of agricultural universities, research institutions, state financial support for research and development, and knowledge gaps and their potential impacts.

9.2.1 The Structure of Knowledge Concerning the Food System

China's education of intellectuals (defined in China as those with some college education) varied greatly in correspondence with periods of ideological conflict, but never approximated the state's needs – until the onset of the economic reform era. Graduate education developed quite late, and has a history of less than 3 decades, with the first significant number of doctoral degrees in agricultural studies granted only after economic reforms began. White and Li note that from 1949 to 1980, the average rate of admission to graduate study was just 550 students a year.[1] The number of students receiving specialized training in the agricultural sciences and engineering, was quite small for a country of China's population size.

Although some early CCP leaders had post-secondary training, their personal credentials as intellectuals had little positive impact on the status of intellectuals in the party. In fact, the term "intellectual" was one of opprobrium during the Maoist

[1]Lynn T. White III with Cheng Li, "Diversification among Mainland Chinese Intellectuals," in King-yuh Chang, ed., *Mainland China After the Thirteenth Party Congress*. Boulder, CO: Westview Press, 1990, 450.

era, and Mao himself called intellectuals "the stinking ninth category." The Cultural Revolution (1966–1976) had many targets, but intellectuals were a consistent object of attack and a persecuted social group. Some intellectuals were arrested and sent to labor camps in western China or the Northeast. Millions did menial labor in the countryside. Through the Cultural Revolution, intellectuals were subject to humiliation, underwent constant criticism and thought reform, and were denied the opportunity to exercise skills of the mind.[2]

Research and scientific investigation was virtually paralyzed during the Cultural Revolution. In 1975 Deng Xiaoping observed:

> Out of the 150,000 scientific and technical cadres in the Academy of Sciences, no one dares go into the research laboratories. They are all afraid of being disparaged as "white" specialists. The young are frightened and the old are frightened.[3]

The attack on intellectuals broadened to include colleges and universities in China. By the mid-1970s, enrollments had declined to one third what they had been 10 years previously. Meisner notes that "Virtually all academic, scientific, and cultural journals had been suspended in 1966, and few were permitted to resume publication during the remainder of the Maoist era."[4] A common phrase for this era in China's development was the "Ten Lost Years," indicating that China lost a decade to political and ideological attacks on its intellectual establishment.

In Chapter 2 we discussed the "Four Modernizations" program authorized by Zhou Enlai but activated only after Mao's death in 1976. Immediate goals included the training of 800,000 professional research workers, founding of new colleges and universities and designation of 88 "key" universities filled by students who had passed competitive examinations. Also, thousands of students were sent abroad for advanced study, and research institutes were re-established.[5]

In the 3 decades of reform, China's university system and public research establishment have expanded and improved enormously. In the area of agricultural sciences alone, by the early twenty-first century China had 58 agricultural colleges and universities, 365 agricultural secondary schools, and, at the county-level, some 2,600 agricultural television and broadcasting schools. Agricultural research institutes of various descriptions numbered about 1,200.[6] Similar growth occurred in forestry education, ecology studies, and in the several dimensions of agricultural and food policy.

---

[2] See Maurice Meisner, *Mao's China and After: A History of the People's Republic*. New York: The Free Press, 1986, 387.

[3] Quoted in Roger Garside, *Coming Alive: China After Mao*. New York: Mentor, 1982, 64.

[4] Meisner, 1986, 388.

[5] Jeffrey Schultz, "Conclusion: The Four Modernizations Reconsidered," in Richard Baum, ed., *China's Four Modernizations: The New Technological Revolution*. Boulder, CO: Westview Press, 1980, 269.

[6] Qiaoqiao Zhang, ed., *Directory of Chinese Agricultural and Related Organizations* (in Chinese), cited in Scott Waldron, Colin Brown and John Longworth, "State Sector Reform and Agriculture in China," *China Quarterly*, no. 186 (June 2006), 287n41.

Yet knowledge of the many dimensions of China's food security is not yet complete, as indicated by the important gaps in the scientific literature on environmental and plant/pest disease threats, among other areas.

9.2.2 Budget Allocations for Research and Development

Waldron et al. estimate that the number of government staff in China's agricultural industry (including fisheries, forestry, and livestock) in 2002 was 4.1 million, a decline from 7.3 million in 1990.[7] The agricultural sector sustained a much smaller reduction in personnel than other economic areas such as construction and finance. Although the number of employees may seem large in comparison to agricultural sectors in the United States and European nations, it includes central-level staff, those in departments at the provincial and autonomous region-level, and staff in bureaus of counties and cities. Also, the number includes not only personnel in the Ministry of Agriculture, departments and bureaus of agriculture, but also "service" and "enterprise" units. Service units (*shiye danwei*) include nearly half of the agricultural personnel; they are engaged in research, testing, inspection, and service delivery.[8] Enterprise units include agricultural conglomerates as well as local-level state owned firms and shareholder enterprises.

In 2003, reports indicated that overall state spending on agriculture, in support of agricultural production, science and technology, capital construction, and poverty relief totaled $23.4 billion, about 7% of government spending then (as compared to 10% in 1990). The state spent an additional $827 million on price subsidies for grain, cotton, and edible oil.[9] Overall spending on agriculture exceeded that on national defense (as defined in public government statistical reports), but followed expenditures in the areas of education and health and capital construction. Yet the real growth rate of spending on agricultural research has slowed in China, a pattern noticeable globally.[10] In 2003, China's investment in agricultural R&D was just 0.4% of agricultural output, compared to 3% in economically developed countries, and 1% average internationally.[11] Moreover, the ratio of public to private sector spending is still far higher than in most other countries.[12]

---

[7] Ibid., 280.

[8] Ibid., 286.

[9] Ibid., 281.

[10] See P. G. Parday and N. M. Beintema, "Slow Magic: Agricultural R&D a Century after Mendel," IFPRI Food Policy Report. Washington, DC, 2001.

[11] Lubiao Zhang, "China's Campaign on New Socialist Countryside and Agricultural Policy Reform," Chinese Academy of Agricultural Sciences. Beijing: September 20, 2006.

[12] For information on spending in science and technology, see Ministry of Science and Technology, *2006 Report on Technological Development of China's Villages* (in Chinese). Beijing: China Agricultural Publishers, 2007.

9.2.3 Knowledge of China's Farming Population

Agricultural practices in traditional China as compared to those in other pre-modern societies were relatively efficient. Communist rule changed the organization of agriculture, and as discussed in Chapter 2, this had deleterious effects, especially on incentives of farmers. The Maoist regime did, however, start a technological revolution in agriculture, with development of mechanization and improvements in irrigation, rural electrification, fertilization, and crop breeding. The regime began the dissemination of new farming techniques, and by the early 1960s, more than 100,000 staff worked out of extension service stations.[13]

After economic reforms began in the late 1970s, farm families had new incentives to increase production and lower costs. Increased agricultural research and the improvement of agricultural infrastructure led to increases in productivity as well. Yet the small size of most farms inhibited mechanization improvements (there are about 220 million farms in 2009). Also, demographic changes presented problems. Already less well-educated than urban Chinese, the better educated farmers were more likely to join the out-migration from rural areas, leaving a disproportionate number of very young and very old farmers in rural areas, and women were more numerous than men. In fact, at least 70% of the farmers joining the migrant labor pool have been men.

The exodus of farm labor into the cities is a difficult problem for the state, both in rural and urban areas. Called *Nungmingong* (farmer laborers or migrant laborers), the movement responds to push-pull dynamics. Farmers may earn no more than 400 Yuan for work on one *mu* of land, while urban factor labor earns at least three times as much annually. The number of Chinese employed in farming fell by more than 80 million between 1996 and 2006, as measured by a national agricultural census. Correspondingly, the number of migrant workers rose to 130 million, a figure 60 million greater than a decade earlier.[14] A village head in Inner Mongolia reported that 35% of his fellow villagers now worked outside. He remarked: "The loss of farm laborers will have a much bigger impact on rural production."[15]

Much more technical assistance is available to farmers today to disseminate new techniques and assist in improvement of farming practices.[16] The Chinese extension system has five levels, from the national through provincial, municipal, and county to the township. At this lowest level, about 400,000 staff work from 50,000 extension units, and they have most frequent encounters with farmers, spreading information on improved methods of crop cultivation, including fertilization and new seed varieties. About one fifth of villages have farmers' service organizations, and

[13] Kang Chao, *Agricultural Production in Communist China, 1949–1965*. Milwaukee, WI: The University of Wisconsin Press, 1970, 89.

[14] Xinhua, "Farming Losing Its Appeal," *China Daily*, February 22, 2008, 3.

[15] Zhiming Xin, Jing Fu, and Ping Zhu, "Security of Food Calls for Serious Thought," *China Daily*, January 16, 2008, 7.

[16] For an overview of recent rural education and training programs, see Chenggui Li, "Reform of China's Agricultural Technology Structure." Unpublished paper. Beijing: 2007.

there are, nation-wide, about one million farmer technicians. But most villages lack farmers' organizations and easy access to extension assistance.

An important feature in reorganization of state agricultural services at all levels has been development of the enterprise philosophy and contract system. Technical contracts are made between farmers and technical and extension units, with the fee structure varying by wealth of village and profitability of crops grown.[17] Studies of technology extension indicate that the fee structure has had less of an impact on adoption of new agricultural technologies than personal and family characteristics (e.g., age and education levels) and perceived benefit of the technology.[18]

While dissemination of information on new agricultural technologies has spread apace, farmers in most areas have not yet changed their practices to reduce environmental degradation and threats to human health (especially from improper pesticide and fertilizer use). For example, a recent survey of farm households in the North China Plain found that only about 6% followed recommendations of extension agents regarding balanced input use.[19]

9.3 Challenges To Administrative Coordination

China's large population and its vast territory present obstacles to the effective implementation of policies pertaining to food security. In the Chinese bureaucracy, a system called *Tiao/kuai Guanxi* has developed to administer government functions. The *tiao* are lines describing vertical bureaucracies, such as the agricultural function from the center to the county. The *kuai* are "pieces" in reference to the horizontal coordinating bodies at different levels. As Lieberthal notes, this nests any official in a context with multiple bosses in different places, leading to his characterization of the Chinese political system as one of "fragmented authoritarianism."[20] In the case of a single problem area, such as public construction, typically there will be a "leadership relationship" (*lingdao guanxi*) between the official and the most important department (whether a "line" or a "piece"), and a "professional [implying non-binding] relationship" (*yewu guanxi*) with the other.[21]

The issues of food security are at a higher range of complexity, because they involve more than one line agency and more than one horizontal piece. We discuss this complexity in terms of both vertical and horizontal aspects of policy coordination.

---

[17] Frederick W. Crook, "China's Extension System," *Journal of Extension*, Vol. 30, no. 3 (Fall 1992), 3.

[18] Jianmin Cao, Ruifa Hu, and Jikun Huang, "Agricultural Technology Extension and Farmers' Modification of New Technology," *China's Soft Science* (in Chinese), no. 6 (2005), 60–66.

[19] Lin Zhen, Jayant K. Routray, Michael Zoebisch, Guibao Chen, Gaodi Xie, and Shengkui Cheng, "Three Dimensions of Sustainability of Farming Practices in the North China Plain," *Agriculture, Ecosystems and Environment*, Vol. 105, no. 3 (2005), 507–22.

[20] Kenneth Lieberthal, *Governing China*. New York: W. W. Norton, 1995, 169.

[21] Ibid., 169–70.

9.3.1 *Horizontal Coordination*

Two agencies of the central government play the key roles with respect to the subjects of this volume: the Ministry of Agriculture (MOA) and the Ministry of Environmental Protection (MEP). Both, however, share their prime function with a host of other central government ministries. For example, Lu et al. mention that 14 government departments share agricultural functions with MOA: eight have programs on quality and safety of farm produce, eight have units on agricultural investment, six have programs regarding farm processing and distribution, and five have offices regulating inputs to agricultural production.[22] One example of the replication of functions is the Department of Rural and Social Development lodged in the Ministry of Science and Technology (MOST).

MEP is a relatively new agency in Chinese government, and it attained ministerial status only in 2008. Virtually every other government ministry or administration has at least one environmental section, for example the Department of Environment and Natural Resources of MOA. MEP has very broad environmental protection functions, but other strong central government agencies have greater authority in environmental areas with a direct impact on food production. For example, the Ministry of Land and Natural Resources has greater power respecting the monitoring of changes to China's arable land, and the Ministry of Water Resources had greater power concerning threats to water sufficiency. Although as part of its elevation to ministerial status in 2008, the MEP has requested increases in budget and personnel,[23] it remains a relatively small agency, with less than one tenth the personnel of the US EPA, but with a far more expansive mandate.

In emerging areas such as the development of agricultural biotechnology, and in the recent crises concerning food safety, the State Council did establish coordinating mechanisms, for example the Biosafety Committee and the State Council task force on food safety. These two areas constitute evolving regimes, as discussed in Chapters 7 and 8, and they promise integration in the approach to national problems of food security, at least in terms of the central government's direction and coordination of public agencies.

The contaminated powdered milk crisis showed the limits of horizontal coordination. Even before this crisis, observers had drawn attention to overlapping of responsibilities and problems in law enforcement, because at least six government departments were involved. A United Nations report on the crisis suggested that China form a unified regulatory agency as "a single source of information." While some leaders have spoken in favor of having a single department responsible for the

[22] Liangshu Lu, Zhicheng Liu, Dongyan Wang, and Lizhi Zhu, "Agricultural Administration System Reform for Acceleration of Modern Agriculture Development," cited in Waldron et al., 2006, 282.

[23] Personal interview with division director, Ministry of Environmental Protection, Beijing, May 21, 2008.

entire food safety chain, as is the case in the United States, others believe it is more feasible to share responsibilities among agencies.[24]

9.3.2 *Vertical Coordination*

The problem of vertical integration is much more intractable, and mirrors the difficulty that China has faced in coordinating economic development activities from the center to provinces and local governments. The relatively simple case is displayed by Ma and Ortolano in their analysis of the implementation of water pollution control requirements. In many countries, the state environmental ministry plays the major role in actual implementation of environmental protection rules, but in China, although MEP drafts the rules, the local environmental protection bureaus (EPBs) are critical to their implementation. They state:

> (A)n EPB has two formal reporting relationships: *kuai-kuai* relations connect an EPB with the head of its local government, and *tiao-tiao* relations link an EPB with an environmental agency one level up in the hierarchy. Local government leaders control an EPB's budget and staffing, and thus they typically have a stronger influence on routine EPB actions than the next highest environmental agency in the hierarchy.[25]

There are about 2,900 EPBs in China, and they are critical to implementation success.

This implementation deficit has not been ignored by central government officials. For example, in the area of environmental protection, the MEP, NDRC, and Ministry of Finance have published a new monitoring plan for the remainder of the Eleventh Five-Year Plan (2006–2010). It would require $2.2 billion in investment; half would come from the central government and the rest from subnational governments and the private sector. In the next 5 years, $5 billion would be allocated to capacity building of environmental agencies, including developing an advanced environmental quality monitoring and warning system and toughening environmental law enforcement.[26] This plan may ease implementation logjams, and appears to be more practical than the "Green GDP" reports to which provinces and municipalities have objected.

As we reported in Chapter 8, food production and safety have been poorly monitored by local governments as well. There is little in the nature of the incentive system to encourage them to pay attention to the state of farming practices. Associations of farmers are not found in every village, and they are a weak force,

---

[24] See Zhe Zhu, "Food Safety Law Debate Goes On," *China Daily*, August 21, 2008, 2; Tian Lan,"China Needs Better Food Safety Laws: UN Report," *China Daily*, Octobere 23, 2008, 3; Xiaohuo Cui, "Thorough Food Check System Urged," *China Daily*, October 28, 2008, 4.

[25] Xiaoyang Ma & Leonard Ortolano, *Environmental Regulation in China: Institutions, Enforcement, and Compliance*. Lanham, MD: Rowman & Littlefield, 2000, 154.

[26] Xiaohua Sun, "Environmental Watchdogs to Sharpen Teeth," *China Daily*, April 16, 2008, 14.

unable to apply pressure on governments to improve outcomes. When more than one functional agency is involved, as is the case in most of the food security issues we have discussed, the problems of vertical coordination increase greatly.

A recent study by Tilt examines the variables influencing regulatory behavior of local government officials. The author studied pollution enforcement in an industrial township of rural Sichuan. The focus was a district EPB with limited funding and manpower, charged with implementing pollution regulations on zinc smelters and coal-washing plants, among other factories. Ordinarily, economic concerns – such as the survival of the firms and their contribution of revenue to local government – would dampen regulatory enthusiasm. But in the case studied by Tilt, local farmers complained to officials, and they also convinced local reporters to film a firm's environmentally degrading operations as well as the owner's statements that he was beyond the law, which were aired on local TV. Citizen complaints and media exposure – which Tilt calls broadly "civil society factors" – led to decisive and strict regulatory action by the EPB.[27] (As we note below, such grassroots pressures are not present in most of China's townships and villages.)

9.4 Modernization of China's Agricultural Infrastructure

Away from the east and southeast coast of China and its large cities, one does not have to drive far off major thoroughfares to encounter villages and rural households still resembling those in poor, developing areas of the Third World. As discussed in Chapter 3, the state has made large investments in the improvement of infrastructure (which also has reduced the supply of agricultural land). This investment has improved somewhat the access of rural residents to markets.[28]

Infrastructure needs in rural regions are not limited to the transportation system. They include irrigation and drainage improvements and watershed management; leveling land and improving forests and grasslands; developing sanitary water systems; and construction of health clinics and schools. A study by Zhang et al. of a sample of 216 representative townships in 2003 found that roads and bridges were at the top of the list of public goods projects between 1998 and 2003. Some 1,266 projects had been completed. Although national data indicated that $1.6 billion had been spent by central and provincial governments on roads, data collected by Zhang et al., from which they made extrapolations, indicated that only 43% of the total

[27] Bryan Tilt, "The Political Ecology of Pollution Enforcement in China: A Case from Sichuan's Rural Industrial Sector," *China Quarterly*, 192 (December 2007), 915–32.

[28] For a sophisticated treatment of the agricultural supply and distribution chain in China, see Jhihua Wu, *Research on China Food Logistics* (in Chinese). Beijing: China Agricultural Publishers, 2007.

budgeted investment funds spent in villages was likely to have come from upper-level governments.[29] Villagers themselves contributed labor to development projects, and village councils funded many of them.

Present and future investments in China's road system are expanding rapidly. The state plans 22,000 miles of new roads by 2010 and 44,000 miles by 2020, which eventually will penetrate as far as the Red Basin of Sichuan. The objective of state planning is to reduce the isolation of remote regions, and improve their access to national and international markets, which would increase their food security.[30] Yet much work remains to be done.

China's grain storage system has been modernized in recent years to counter the two granary crises reported on in Chapter 2. Given the high volume of China's grain production, much was put into storage where supplies could sit as long as 5 years. Previously, reported losses in storage in some facilities were as high as 5–10% because of plant pests and mildew.[31] This situation encouraged greater pesticide use, creating another problem. In a cooperative project with United National Industrial Development Organization, the State Grain Administration has now curbed use of bromomethane as a pesticide in storage facilities, because it not only left residue on food stuffs, but also contributed to ozone depletion, and it now uses phosphine fumigation instead.

The State Grain Administration also has phased out the bag handling system and large number of small port, rail, and storage facilities, and instead has contracted with multinational firms to develop bulk storage and handling facilities.[32] Such structures can provide storage in each facility of greater than 1 million metric tons, and incorporate ventilation, dust control, transfer, weighting, and monitoring systems.

There are four levels to grain storage in China: central, provincial, city, and county/township. About 60% of total storage capacity is under central government control, and these granaries are the most modern. The central grain reserve corporation purchases capacity from local granaries, but these are under the control of local governments (the provincial governor is fully responsible), and they earn storage fees from the central government. Most of the granaries at the county/township/village level have been privatized.[33] It is storage facilities at the subnational level that are most in need of renovation or replacement. In several regions of China, for example maize-growing areas of the Northeast, production has been constrained by shortage of transport and modern storage facilities.[34]

---

[29] Linxiu Zhang, Renfu Luo, Chengfang Liu, and Scott Rozelle, "Investing in Rural China," *The Chinese Economy*, Vol. 39, no. 4 (July-August 2006), 65.

[30] Edward Anderson, "Food Security in China," in Gerard J. Gill, John Farrington, Edward Anderson, Cecilia Luttrell, Tim Conway, N. C. Saxena and Rachel Slater, *Food Security and the Millenium Development Goal on Hunger in Asia*. London: Overseas Development Institute, 2003, 12.

[31] Jiao Wu, "More Efforts to Protect Grain Crop," *China Daily*, December 15-16, 2007, 2.

[32] See: www.rockwelllautomation.com

[33] Personal interview with grain storage specialist, State Grain Administration, Beijing, May 16, 2008.

[34] Hong Yang, "Trends in China's Regional Grain Production and Their Implications," *Agricultural Economics*, Vol. 19, no. 3 (December 1998), 309–25.

9.5 Poverty Alleviation and Food Security

Famine as a result of the Great Leap Forward in the late 1950s cost 20–30 million lives in China, and the memory of that event has driven efforts to reduce poverty and malnourishment. We introduced the problem as it related to policy implementation in the previous chapter. Here, we expand on that coverage by considering the significant reduction in poverty since the onset of economic reforms in the late 1970s, the nature of government poverty alleviation programs, and remaining challenges.

9.5.1 Poverty Reduction from the Late 1970s to the Present

China has had great success in reducing rural poverty since the late 1970s, and its reduction of poverty and malnourishment in absolute numbers as well as proportion of the population lifted out of poverty is one of the most successful stories in world history. According to estimates of the Food and Agricultural Organization (FAO), the number of undernourished people fell from 303.8 million (or 30% of the population) in 1979–1981 to 193 million (16% of the population) in 1990–1992, and to 119.1 million (9% of the population) in 1998–2000.[35]

Only lately has China developed careful measurements of poverty levels, and the definition of the poverty line has changed over time. By China's definition, the number of rural poor decreased dramatically, from 260 million in 1978 to 128 million in 1984, a reduction of more than half. Although the rate of decline slowed in the late 1980s, the fall quickened in the 1990s, and dropped to a level of 41 million in 1998. Comparing those defined as poor to the rural population as a whole, the percentage declined from 32.9% in 1978 to 15.1% in 1984 and fell to 4.6% in 1998.[36]

Since the 1990s, the World Bank has used the measure of $1.00 a day (in proportional purchasing power) as the poverty line. By this measure, there are much larger numbers of Chinese who would be considered absolutely poor, but the trend line follows that established by state figures. In 1990, some 280 million rural Chinese would be considered poor by World Bank standards (31.3% of the rural population), as compared to 124 million in 1997 (13.5%).[37]

In 2007, the definition of poverty aroused academic concern. A group of 30 economists led by Wang Xiaolu urged the government to raise its poverty line and include basic expenditures on education and medical services in the formal calculation.

[35] Ibid., 2.

[36] Linxiu Zhang, Jikun Huang, and Scott Rozelle, "China's War on Poverty: Assessing Targeting and the Growth Impacts of Poverty Programs," *Journal of Chinese Economic and Business Studies*, Vol. 1, no. 3 (September 2003), 302.

[37] World Bank, *World Development Report, 2000/2001: Attacking Poverty*. Washington, DC: World Bank, 2001.

Wang, a senior research fellow of the China Reform Foundation, is the lead author of the report *Alleviating Poverty through Development*. He argued that instead of the poverty line figure of 683 Yuan (about $95) needed to purchase a minimum amount of food and clothing for survival, the figure of 1,100 Yuan ($153) should be used. He stated: "Using the standards we propose, 80 million Chinese are still in poverty, rather than the official figure of 23 million (using the government standard)."[38] This estimate was supported by the China Development Research Foundation, whose survey of 4,041 poverty-stricken households of 72 villages in 2006 revealed that poor income in the agricultural sector and growing expenditures on health care and children's education were the major causes of poverty.[39]

Then, in late 2007, confusing the issue further, the World Bank raised its estimate of China's poor from 100 million to 300 million, after a new survey of prices altered the estimate of what $1.00 would purchase.[40] In early 2008, the State Council's Poverty Alleviation Office indicated it likely would raise the bar for China's impoverished from the per capita annual income of 1,067 ($152) to 1,300 Yuan ($186). This roughly 20% increase would increase the number qualifying as poor to about 80 million. The new figure is equivalent to $1.00, U.S. a day, which matched the international standard then.[41] But in mid-2008, the World Bank raised its poverty threshold from $1.00 a day to $1.25.[42]

The World Bank's 2009 report on poverty reduction describes China's accomplishments as enviable:

> Between 1981 and 2004, the fraction of the population consuming below this poverty line (888 yuan per per person per year at 2003 rural prices) fell from 65% to 10% and the absolute number of poor fell from 652 million to 135 million, a decline of over half a billion people. A fall in the number of poor of this magnitude over such a short period is without historical precedent.[43]

In short, there is consensus that the number of poverty-stricken Chinese has declined sharply, but no consensus on the number falling below the poverty line in 2009. We believe that the official figure of 23 million poor Chinese is quite low, almost ridiculously so, and side with the World Bank's estimate of 135 million. Indeed, the figure may be higher, not only because of increased food costs, but

---

[38] Jing Fu, "Call for New Definition of Poverty," *China Daily*, September 26, 2007, 14.

[39] Ibid. See also Jing Fu, "Work Needed to Better Livelihoods," *China Daily*, October 15, 2007, 3.

[40] Howard W. French, "Lives of Poverty, Untouched by China's Boom," *The New York Times*, January 13, 2008, A11.

[41] Xinhua, "Poverty Line to be Raised to International Standard," *China Daily*, April 14, 2008, 2.

[42] Jiao Wu, "Poverty Line May be Raised," *China Daily*, September 3, 2008. By year's end the poverty definition was adjusted. Those earning less than 786 yuan were in "absolute poverty," and those earning between 786 and 1,067 yuan were "low income." The total poor (all those in these two categories) then numbered 43.2 million. See Zhuoqiong Wang, "New Poverty Line Raises Number of Poor," *China Daily*, December 23, 2008, 2 and Xinhua, "Bumper Harvests Lift Farm Incomes," *China Daily*, December 29, 2008, 1.

[43] World Bank, *From Poor Areas to Poor People: China's Evolving Poverty Reduction Agenda*, 2009.

because poverty statistics do not uniformly include costs for medical treatment, children's education, and transportation expenses. If the urban poor are included, the number is much higher.

9.5.2 Government Poverty Alleviation Programs

Before the development of formal anti-poverty programs in 1986, the regime made a number of special provisions for poor rural areas. For example, at the beginning of economic reforms in the late 1970s, communes in poor areas pioneered decollectivization efforts, because they had the most to gain from agricultural privatization. Areas of substantial poverty received transfer payments from the central government and had opportunities to purchase grain at lower prices.[44]

The poverty alleviation program initiated in the mid-1980s was composed of three elements: a new institutional structure, increased government funding, and targeting of poor areas. The national-level institution was called the Leading Group for Poverty Reduction, in which a number of agencies participated. Its role was to supervise spending on poverty alleviation programs, coordinate efforts across several ministries and departments, and to be the central point of advocacy for the poor. Poverty Area Development Offices (PADOs) were established at both provincial and county levels and charged with administering funds from central and provincial governments.[45]

Initially, most of the funds flowing into poverty reduction programs were grants. Local PADOs approved applications for projects to improve rural conditions, and most of the funding did not need to be repaid. The grants were very popular in rural poverty areas; however, the funding amount did not change and was quickly eroded by inflation. In the late 1980s, low-interest loans became the chief funding vehicle. The PADOs made the loans at below-market rates of interest, directing them to local banks. The loans went to households to use in agricultural production and to meet other basic needs. The third funding vehicle was a Food for Work program, which addressed the perception that poverty reduction funds had not been used efficiently. These projects emphasized infrastructural development such as building roads and delivering drinking water. By the mid-1990s they had become the major source of anti-poverty funding efforts.[46] There are different estimates on the total funding for poverty alleviation programs from the late 1980s to the present.

---

[44] Albert Park, Scott Rozelle, Christine Wong, and Changqing Ren, "Distributional Consequences of Fiscal Reform on China's Poor Areas," *China Quarterly*, 147 (1996), 1004.

[45] Zhang et al., 2003, 306. Also see Zhong Tong, Scott Rozelle, Bruce Stone, Dehua Jiang, Jiyuan Chen, and Zhikang Xu, "China's Experience with Market Reform for Commercialization of Agriculture in Poor Areas," in Joaquim von Braun and Eileen Kennedy, eds., *Agricultural Commercialization, Economic Development, and Nutrition*. Baltimore, MD: Johns Hopkins Press, 1994, 119–40.

[46] Zhang et al., 2003, 306.

Zhang et al. report the spending of $630 million in 1990, $1.35 billion in 1996, and $1.87 billion in 1997.[47] Anderson, on the other hand (and reporting on a later period, in the early twenty-first century) indicates that the central government committed approximately $3.2 billion a year to poverty alleviation work.[48]

The targeting of areas for inclusion in the poverty alleviation programs was based on two statistics: gross income at the county level and net income per capita; and specialists acknowledge problems in the use of both measures.[49] Based on national data, the leading groups demarcated "nationally-designated" poor counties and "provincially-designated" poor counties, but this left a number of equally poor counties (by other measures) outside the poverty alleviation programs.[50]

Studies of the targeted counties in Sichuan Province demonstrated that the government poverty reduction programs did increase growth. Counties not included in the government programs grew at lower rates or not at all.[51] However, the directed efforts in poverty reduction of the state probably had less of an impact than general increases in economic development, spurred by government and other forms of investment. Zhang et al. note:

> (G)overnment spending on production-enhancing investments, such as agricultural R&D, irrigation, rural education and infrastructure (including roads, electricity, and communications) have contributed to both growth and poverty alleviation.... (T)he results show that although government anti-poverty investments do help reduce poverty, these investments actually have the smallest impact on poverty reduction when compared to any of the other investment programs.[52]

In the early twenty-first century, the state has developed a series of preferential policies to alleviate poverty in the countryside, including agricultural taxation reform and direct subsidies to insure gains from production of crops.[53] The reform campaign was labeled "New Socialist Countryside."[54] As discussed above, the main form of agricultural tax on farmers has been eliminated, and farmers have been

---

[47] Ibid., 307.

[48] Anderson, 2003, 13.

[49] Zhang et al., 2003, 307. Also see Albert Park and Sangui Wang, "China's Poverty Statistics," *China Economic Review*, Vol. 12 (2001), 394–98.

[50] Zhang et al., 2003, 308.

[51] Ibid., 313.

[52] Ibid., 314.

[53] In 2009, agricultural subsidies were increased on major crops, such as wheat, by 15.3%. This was one of several means to cushion poor against the global recession expected to affect rural areas more adversely than urban ones. See Xu Wang, "Agricultural Subsidies Increased," *China Daily*, October 21, 2008,1; Tingting Si, "Global Slowdown to Hit China's Hinterland Hard," *China Daily*, November 27, 2008, 16; and Zhuoqiong Wang, "Help-Poor Policies on Anvil," *China Daily*, December 8, 2008, 2.

[54] For reform elements, see Lubiao Zhang, "China's Campaign on New Socialist Countryside and Agricultural Policy Reform." Beijing: Chinese Academy of Agricultural Sciences, 2006. Also see Yinan Hu, "Minimum Allowance," *China Daily*, January 30, 2008, 2. Also see Jiao Wu, "Rural–Urban Divide Targeted," *China Daily*, October 13, 2008, 3.

made exempt from other taxes such as those in the slaughtering and animal husbandry industry. This has brought about a net reduction in rural taxation in the amount of about $17.4 billion.[55]

9.5.3 Challenges for Future Poverty Alleviation Policies

Although the state has significantly reduced poverty in rural areas of China, in absolute terms the number of poor is still quite large, at least 135 million (or 10% of the population). A clustering of five characteristics describes the rural poor. First, they tend to live in resource deficient rural areas: in upland regions of interior provinces in southwestern, northern, and northwestern China.[56] Second, the regions of rural poverty are also those of substantial gender inequality. Although women are responsible for from 50% to 60% of domestic food production, they tend to have insecure access to land, credit, and extension services; they work longer hours than men; and they are far less likely to migrate to cities where employment opportunities are greater than in rural areas.[57] Third, minorities are also disproportionately represented among the rural poor. Fourth, the rural poor are less likely to have access to clinics and other medical care, and thus they are more subject to diseases and more likely to die at early ages. Fifth, they have substantially reduced opportunities for education, with only half of the boys and fewer girls able to attend primary school and attain literacy.

As China's economy has rapidly expanded, the gap between urban and rural development has deepened, and overall income inequality has increased. Vice-minister of Agriculture Yin Chengjie pointed out in 2007 that per capita net income of farmers had increased by more than 6% in the previous 3 years, but farmers' incomes still lagged behind those of urban residents: "The ratio of urban and rural residents' incomes was 3.28:1 last year, up from 3.21:1 in 2004, and the difference in per capita income has reached 8,173 Yuan (about $1,135)."[58] From 1981 to 1995, the Gini coefficient rose from 28.8 to 38.8, a very large rise by international standards.[59]

---

[55] Xinhua, "Countryside, Rural People a Top Priority," *China Daily*, October 8, 2007, 1. Reduction in agricultural taxes increased greatly the fiscal burden on local governments, for which tax revenues constituted the greater part of income. Some economists estimated that governments at the county level and below had accumulated about 1 trillion yuan in debt. In early 2008, the Ministry of Finance pledged that it would increase transfer payments to these local governments. See Zhiming Xin, "Transfer Payments Set to Rise," *China Daily*, April 17, 2008, 13.

[56] Ibid., 314.

[57] Anderson, 2003, 7.

[58] Dingding Xin, "Farmers Earn More, but Still Lag Behind," *China Daily*, September 14, 2007, 3. The vice-minister referred to ministry figures showing that at least 210 million rural laborers had migrated to cities or townships in 2006; in the first 6 months of 2007, the number of migrant workers had increased by 8.1 percent: "This migration has caused problems, such as leaving just women and old farmers to cultivate farmland."

[59] Ibid., 6.

Poor regions in China have advanced little as compared to high growth areas, making quite glaring the situation not only of those beneath the poverty line but also of those just above it.

The state has made improvements in targeting the rural poor most in need of assistance. For example, in 2007, the State Council inaugurated a minimum living allowance system (called the *Dibao*) for rural areas. Li Liguo, the Vice Minister of Labor and Social Security announced that the system would cover more than 20 million people in rural areas of the country, and the initial allocation would be $397 million.[60] The system is designed to consider more than income figures and to include other criteria such as housing conditions, access to water and electricity, health services and education, as well as variation by area in local prices of essential commodities.

Yet a critical analysis of poverty policy suggests much remains to be done. Director of the State Council Development Research Center's Rural Economy Department, Han Jun, goes well beyond the commonly mentioned "three problems of agriculture"(*san nung*). He contends that educational expenses take a disproportionately high percentage (some 30%) of disposable income, while little training is available for poor farmers. Cooperative medical insurance reaches less than 10% of the rural population. Central government funds often are misallocated, and rural development funds reach villages most in need in insignificant amounts. Poor farmers experience difficulty in obtaining credit too. Overall, Han argues that a "new socialist countryside" is far from realization.[61]

National officials directed attention to poverty alleviation at the 17th National Congress of the Communist Party of China in October 2007. Party General Secretary Hu Jintao, in his keynote address, said:

> We will increase transfer payments, intensify the regulation of incomes through taxation, break business monopolies, create equal opportunities and overhaul income distribution practices with a view to gradually reversing the growing income disparity.[62]

Hu pledged to eliminate absolute poverty in a new pattern of income distribution in which middle-income people would be the majority. He also promised to insure educational equity for the poor and the establishment of a sound social security system. This was the first occasion when the subject of the "people's livelihood" was included officially in a party report.[63] One hopeful development pursuant to these pledges was action by the Ministry of Education to grant subsidies of 1,000 Yuan a year to 300,000 poor students from 22 provinces, autonomous regions and municipalities in central and western areas.[64]

---

[60] Fangchao Li, "Basic Stipend for Rural Areas," *China Daily*, August 14, 2007, 2.

[61] Editors, "Leading State Adviser Offers Frank Assessment of Rural Challenges," *China Development Brief*, Vol. X, no. 7 (July/August 2006), 3–5.

[62] Zhe Zhu, "Narrowing Wealth Gap High on Party's Agenda," *China Daily*, October 16, 2007, 5.

[63] Ibid.

[64] Ying Wang, "Poorest Rural Students to get Government Financial Aid," *China Daily*, December 22–23, 2007, 2.

Planners within the Chinese Communist Party projected that 55% of the population would be middle class by 2020, distributed on a 70:30 basis between urban and rural residents. The definition of middle class in this case was those having annual household incomes of between 60,000 Yuan ($8,200), and 200,000 Yuan ($27,500).[65] At the third Plenary Session of the 17th CCP Central Committee in October 2008, decision-makers announced the goal of doubling per capita disposable income of rural residents by 2020, and eliminating absolute poverty by then.[66]

Another part of the overall poverty alleviation reform was a plan to expand the pension network to cover 60% of farmers by 2010 and 80% by 2015. The pension system will be based on experience of pilot projects in more than 300 counties. The director of the rural insurance department in the Ministry of Human Resources and Social Security, Zhao Dianguo, said: "A pension system would not only solve the social insecurity problem, but also act as a stimulus for higher consumption."[67] Indeed, increased consumption is necessary as the Chinese economy matures.

The focus of poverty alleviation programs has been on rural areas, but the largest growth in poverty is accounted for by migrants from rural impoverished areas who have moved to China's cities but have not been able to find adequate employment, housing, and social services there. The number of migrants is estimated to range from 100 to 150 million. While the state has attempted to establish a minimum living allowance system for urban residents, this is not comprehensive of a large number in the migrant population, many of whom lack household registration data needed to become eligible for such service programs. As the economic downturn unfolded in 2008–2009, the number of jobless migrants increased to 20 million, and they faced a difficult situation because five to six million new migrants join the job market annually.[68]

Overall, then, while the state has significantly reduced the rural poverty population, it has not eliminated this problem. (It should be kept in mind that even economically developed nation-states, such as the USA, have not eliminated the poverty problem.) Meanwhile, the focus of poverty alleviation has broadened to include the poor within the floating migrant population in China's cities.

9.6 Limited Opportunities for Public Participation

China's leaders have acquiesced to a limited amount of political liberalization during the reform period. Although citizens have no opportunity to elect members of the Politburo, ministers of the state or provincial governors, they may, in rural areas, vote for heads of village committees. Of equal importance, the regime in the

[65] Jiao Wu, "50% of People to be Middle Class," *China Daily*, December 27, 2007, 3.

[66] Xinhua, "Farmers' Income to be Doubled," *China Daily*, October 13, 2008, 1.

[67] Jing Fu, "Insurance for 80% Farmers in 7 Years," *China Daily*, November 17, 2008, 3.

[68] Yingzi Tan, "20 Million Migrants Lost Jobs: Survey," *China Daily*, February 3, 2009, 1.

early twenty-first century has become more tolerant of protest, particularly when the target of protest is not the regime itself but ineffective government programs, corrupt and abusive local officials, and foreign interests. We treated NGOs in the previous chapter. Here we discuss briefly changes in the media, civil society, and expressions of mass protests.

9.6.1 Media Reportage of Environmental and Food Security Issues

Most of the organs of mass communication in China are controlled and operated by the government. Those not sponsored by government agencies are supervised closely by them, and their products are subject to constant censorship. Nevertheless, more information is available concerning environmental problems, threats to food production, changes in food consumption, and the safety of food products than at any previous time in the history of the People's Republic.

Just in the area of the environment alone, by the start of the twenty-first century, China's 2,000 newspapers carried 47,000 articles on environment-related issues,[69] and the central television station CCTV had produced several programs on the environment. Television and radio are primary sources of news on issues such as environmental crises and food safety concerns, followed by newspapers, while the government itself is a distant third as a source of information.[70]

The latest source of information – the Internet – is perhaps the most influential in China today. About 253 million Chinese are connected to the Internet, the largest number in the world (overtaking the United States in 2008).[71] The Internet accommodates a large number of blogs and networks on the environment as well as food security issues.[72] Yet, the Internet is no more a secure medium than newspapers, television, and radio; blogs offensive to the regime have been shut down, and servers such as Google are carefully monitored. In short, there are limits to the ability of the media to accurately inform and to constrain state activity.

The media are largely responsible for growing awareness of social concerns.[73] Considerably more information has been available about environmental than food

---

[69] Bo Wen, *Greening the Chinese Media*. Washington, DC: China Environment Series, no. 2., Woodrow Wilson Center Press, 1998, 39.

[70] Elizabeth Economy, *The River Runs Black: The Environmental Challenge to China's Future*. Ithaca, NY: Cornell University Press, 2004, 163n50.

[71] Xinhua, "Nation has Largest Number of Netizens," *China Daily*, April 25, 2008, 1, and Baijia Liu, "Chinese World's Top Netizen Group," *China Daily*, July 25, 2008, 1. Yet, given China's large population, the proportion of Internet users is below the global average.

[72] See Jonathan Sullivan and Lei Xie, "Environmental Activism, Social Networks and the Internet," *China Quarterly*, no. 198 (June 2009), 422–32.

[73] For an analysis of how state-controlled media influence popular understanding of issues, see Alex Chan, "Guiding Public Opinion through Social Agenda-Setting: China's Media Policy since the 1990s," *Journal of Contemporary China*, Vol. 16, no. 53 (November 2007), 547–59.

safety issues, given the longer span of attention to the former. A recent survey by the China Environmental Awareness Program and the Chinese Academy of Social Sciences found that environmental protection was the fourth most important social issue, following healthcare, employment, and the income gap. While less than half of the 3,000 respondents (in 20 provinces) were familiar with concepts of biodiversity or greenhouse gas emissions, a majority had heard about concrete issues. Some 66% knew about trash sorting, 52% had heard about organic food, and 50% knew about the campaign to reduce use of plastic bags. Most respondents said they used water, electricity, and gas conservatively, but a majority declined to refrain from smoking in public. Finally, a majority said governments needed to pay more attention to the rural environment and issues such as safe drinking water.[74]

9.6.2 Impediments to the Development of Civil Society

The civil society concept is used extensively in the analysis of democratization movements globally. The media are one leg to the concept, and social groups are the second. There is a strong association between the autonomy of civic groups – such as labor unions, environmental organizations, and other voluntary groups – and the ability of people to select their leaders and make policy choices. A number of scholars have studied the evolution of social groups in China; while some are impressed with the growth in number of organized groups and the state's increasing tolerance for their quasi-autonomous status, none finds the condition of social groups in contemporary China to be "robust."[75]

What we noted in the previous chapter was that NGOs are relatively few in number. Estimates of the number vary widely, particularly if GONGOs and "grassroots" organizations operating outside the pale of the state, are included.[76] However, the number of NGOs that is registered following Chinese state requirements is small, just several hundreds. The largest single type of NGO is environmental, and studies of such groups indicate that in addition to the lack of government support, groups are deficient in technical capacity, funding, and internal organizational strength.[77]

---

[74] Wan Xiao, "People More Aware of Environmental Issues," *China Daily*, May 4, 2008, 3.

[75] See, for example, David Da-Hua Yang, "Civil Society as an Analytic Lens for Contemporary China," *China: An International Journal*, Vol. 2, no. 1 (March 2000); Ian Johnson, "The Death and Life of China's Civil Society," *Perspectives*, Vol. 1, no. 3 (2003); Rebecca Morse, "China's Fledgling Civil Society: A Force for Democratization? *World Policy Journal*, Vol. 18, no. 1 (2001); and Jude Howell, "New Directions in Civil Society: Organizing Around Marginalized Interests," in Jude Howell, *Governance in China*. Lanham, MD: Rowman & Littlefield, 2003.

[76] For example, E. Knup claimed that the number of grassroots organizations was larger than 200,000, without providing evidence for this number. See "Environmental NGOs in China: An Overview," *China Environment Series*, no. 1.

[77] See Jennifer Turner and Zhi Lu, "China's Environmental NGOs," in Yisheng Zheng, ed., *China Environment and Development Review* (in Chinese). Beijing: Social Sciences Academic Press, 2007, 349–69.

We pointed out intentions of the regime to loosen requirements (in 2007) and allow a larger number of NGOs to function, because they do play vital roles for the government, yet it seems unlikely that they will be able to influence environmental (or food security) policy significantly in the near future. Martens comments: "(A) substantial part of the Chinese population has as yet no access to contemporary consumption practices, let alone the opportunity to articulate an ecologically modernised demand for green products and services."[78]

9.6.3 Citizen Protests

In an environment of somewhat increased political liberalization, yet without clear channels to apply pressure on government agencies and political leaders, Chinese frequently have responded with ad hoc demonstrations, protests, and in some cases riots. For example, farmers have confronted factory owners, who often are supported by local government officials, when their fields and water systems are polluted with toxic contaminants,[79] and on occasion have been successful.

Conflicts over the use of land are perhaps the leading source of social unrest in China today. The Chinese Academy of Social Sciences reported that of about 60,000 messages sent by citizens to central media organizations in the first half of 2004, 36% concerned rural issues; of these, more than two-thirds concerned complaints about rural land use. The Academy's record of 130 mass confrontations between farmers and police in this year included two-thirds focusing on land use.[80]

Three recent demonstrations indicate changes in the nature of civic protests. In Chapter 7, we mentioned protests in Xiamen. Residents of a middle class residential area objected to plans, initially endorsed by city officials, to construct a petrochemical facility in their backyard. Their orderly demonstration led to suspension of the plan, and the likelihood it will be located elsewhere.[81] Second, protests by middle class residents of Shanghai to city plans to extend the Maglev rail line into their residential neighborhood seemed to be a sign of grassroots power.[82] The Chinese press referred to these two protests as "people's victories."

The most recent such protest occurred in Chengdu, the capital of Sichuan Province, in early May 2008. About 400–500 protesters walked calmly through the city's downtown area for several hours of a Sunday afternoon. They protested construction of a $5.5 billion ethylene plant some 35 km from the city core, saying

[78] Susan Martens, "Public Participation with Chinese Characteristics: Citizen Consumers in China's Environmental Management," *Environmental Politics*, Vol. 15, no. 2 (April 2006), 225.

[79] See Jing Jun, "Environmental Protests in Rural China," in Elizabeth Perry and Mark Selden, eds., *Chinese Society: Change, Conflict and Resistance*. New York: Routledge, 2000, 144.

[80] Cited in Yongshun Cai, "Civil Resistance and Rule of Law in China," in Elizabeth Perry and Merle Goldman, *Grassroots Political Reform in Contemporary China*. Cambridge, MA: Harvard University Press, 2007, 176.

[81] Sun Liangruo, *People's Daily*, December 17, 2007.

[82] Howard W. French, "Plan to Extend Shanghai Rail Line Stirs Middle Class to Protest," *The New York Times*, January 27, 2008, A12.

it would damage the "clean water and green mountains of Sichuan." The plant is a joint venture of the provincial government and PetroChina, a subsidiary of China National Petroleum Corporation, China's main oil producer. As in previous cases, protesters said such a project (already approved by the NDRC) required a public hearing and independent environmental assessment before proceeding. As in the two previous cases, most protesters were urban and middle class. The protest itself was organized through websites, blogs, and cell phone text messages.[83]

Indeed, the rising number of protest incidents leads some observers to suggest that China is on the road to democratization. Cautious analysts suggest that at the least, power holders now are more likely to consider policy innovations and institutional reforms.[84] Given the core value to the state of food security, addressing complaints and protests is essential to the maintenance of the regime's legitimacy. Yet limitations on opportunities for public participation in policy-making and policy review deprive leaders of information on public sentiment and are a challenge to political stability.

9.7 International Challenges

Not all of the challenges to China's food security are domestic in nature. Since its opening to the world in the late 1970s, China increasingly has become involved through trade and diplomatic ties with the other nations. This has increased opportunities for China's development while subjecting it to the requirements of bi- and multi-lateral agreements, international conventions, and treaties.

In this section we first consider the trade of China globally in agricultural products. We then treat issues related to China's entry to the World Trade Organization (WTO) in late 2001, such as changes to China's national standards. The third section discusses compliance requirements for other international conventions concerning food products.

9.7.1 China's Trade in Agricultural Products

Most of the food products upon which Chinese depend are produced and consumed domestically, a pattern that applies to many other countries. For instance, only 6% of rice and 17% of wheat production globally is exported.[85] Yet compared to the Maoist era, China's trade in agricultural products has increased exponentially.

[83] Edward Wong, "China Petrochemical Project Opposed," *New York Times*, Mayi 6, 2008, A19. As in the previous cases, Chinese media also featured the protest. See Zhiling Huang, "Chengdu People Walk to Express Concerns," *China Daily*, May 6, 2008, 4.

[84] See, for example, Kevin O'Brien and Lianjiang Li, *Rightful Resistance in Rural China*. New York: Cambridge University Press, 2006.

[85] Peter Oosterveer, *Global Governance of Food Production and Consumption: Issues and Challenges*. Cheltenham, UK: Edward Elgar Publishing, 2007, 4.

Table 9.1 Trade in major agricultural products (in millions, USD) (Adapted from Chen Liangbiao, "The Prospects of China's Food Security and Selection of Policy," in Zuo Mengxiao, *Crop Diversification in China*, FAO, 1998)

Type of product	Import	Export	Net
Cereal grains	1,989.47	2,147.98	158.51
Wheat	278.57	1.41	−277.16
Rough rice, rice	120.04	927.17	807.13
Maize	31.77	531.68	499.91
Sorghum	0.02	2.46	2.44
Other cereals	265.54	34.76	−230.78
Processed cereal products	70.05	368.10	298.05
Other produce	3,219.54	5,537.19	2,317.65
Vegetables	71.29	1,473.85	1,402.56
Fruits	241.78	433.31	191.53
Processed vegetables and fruit products	23.58	1,022.42	998.84
Coffee, tea	19.73	520.10	500.37
Sugar and sugared food	1,787.83	1,310.04	−1,477.79
Drinks, wine, and vinegar	75.33	450.10	374.77
Miscellaneous products	83.32	327.37	244.05

In 2000, China's total exports of food and live animal products amounted to $12.3 billion. It was the world's ninth largest exporter in that year of such products, responsible for 4.3% of world exports. It imported $4.8 billion of agricultural goods, making it the world's sixteenth largest importer, with about 1.6% of world imports.[86] In some areas such as fish production and consumption, it was the leader in the Asia-Pacific region. For example, in 1999, China was responsible for 33% of the world fish catch (41.5 billion tons), exporting product worth $715.5 million and importing $805 million.[87]

Table 9.1 displays the trade in major agricultural products of China in 1998:

Because this table includes data from 1998 only, it omits consideration of the major change in production of cereal grains in late 2003 and 2004. Shortfalls in cereal production in these years amounted to approximately 10 million tons (rice), 12 million tons (wheat), and 12 million tons (maize).[88] Initially, China drew from grain reserves to compensate for declines in production, but then had to import grains as these stores were depleted. This brought on a bulge of imports and a deficit in food products trade.

[86] Lee Ann Jackson, "Food Safety Standards Harmonization in the Asia-Pacific Region," in Reba Carruth, *Global Governance of Food and Agriculture Industries*. Cheltenham, UK: Edward Elgar Publishing, 2006, 328.

[87] Ibid., 329.

[88] Lester R. Brown, *Outgrowing the Earth*. New York: W. W. Norton, 2004, 146–47.

9.7.2 Issues Related to China's Entrance to the WTO

The share of agricultural products in China's international trade has declined from some 30% in 1980 to 16% in 2000.[89] Moreover, the proportion of food exports to total agricultural trade declined to 5% from 17% in 1980. As it prepared to enter the WTO, policy makers were aware that China's agricultural sector was less efficient than its industrial sector, and that initial adjustments could be expected.

As seen in Table 9.2., from its entry into WTO in late 2001–2007, China's farm trade doubled, reaching $60 billion in 2006 and it was expected to reach $70 billion in 2007.[90] However, a flood of foreign agricultural products that were more competitive than China's entered country markets, and following WTO accession, agricultural trade deficits were common. In 2004 the farm trade deficit was $4.64 billion, for reasons mentioned above. The deficit declined to $1.44 billion in 2005, but rose to $3.84 billion in 2006. In 2007, because of large increases in soybean and edible oil imports, the farm trade deficit reached $4.08 billion.[91] In 2008, there was a surplus in food trade of $1.7 billion. The expectation of food industry specialists was that imports would be normal in the next 5 years, and total about $147 billion.[92] The increase in imports has been driven by resource-intensive products (particularly, those requiring intensive use of land and water) such as soybeans, corn, cotton, edible oil, meat, and dairy products. Growth in exports has been driven by labor-intensive products such as vegetables, fruits, flowers, and aquaculture products,[93] where China has a comparative advantage.

China's international trade in grain is monopolized by the central government. The National Reform and Development Commission controls both the quantity of imports and exports. The Ministry of Commerce administers the cereal trade, but

Table 9.2 Growth of agricultural trade (in billions, USD) (Adapted by authors from Yao Xiangjin, Deputy Director General, Department of International Cooperation, Ministry of Agriculture, "Briefing on Trade and International Cooperation in Agriculture of China," October 10, 2007)

	2000	2001	2002	2003	2004	2005	2006
Value of imports	11.25	11.84	12.45	18.93	28.03	28.71	31.41
Exports	15.69	16.06	18.15	21.43	23.39	27.57	27.57
Surplus or deficit	4.44	4.22	5.70	2.50	–4.64	–1.14	–3.84

[89] Jikun Huang and Scott Rozelle, *China's Accession to WTO and Shifts in the Agriculture Policy*, Department of Agricultural and Resource Economics Working Paper No. 2. Davis, CA: University of California Davis, 2002.

[90] Zhiming Xin, "Farmers Help Cut China's Agro-trade Deficit," *China Daily*, January 9, 2007, 4.

[91] Huanxin Zhao, "Food Imports to Stay Steady," *China Daily*, March 11, 2008, 1.

[92] Xinhua, "$147b of Food to be Imported in 5 Years," *China Daily*, December 17, 2008, 3.

[93] Hongxing Ni, "Agriculture Trade Promotion in China." Beijing: Agricultural Trade Promotion Center, Ministry of Agriculture, 2007.

actual transactions are carried out by state trading enterprises. Two trading companies manage most of the trade. The most important is the China National Cereals, Oils and Foodstuffs Import and Export Corporation (COFCO). COFCO manages China's rice exports (and has a monopoly on rice imports), accounting for more than a tenth of world trade. It also is a large exporter of maize, however the primary maize exporter is the second largest international grain trading enterprise, Jilin Grain Group Importing & Exporting Co. (which is owned by the Jilin provincial government).[94]

The increase in global food prices of 2007–2008 put pressure on China's trade policy. Intent on protecting domestic price stability against the doubling of some cereal prices, in late 2007 (as discussed in Chapter 2), the central government abolished tax rebates for wheat, rice, corn, and soybeans. In early 2008, a 5.25% provisional export tariff was levied on 57 food stuffs, including wheat, corn, rice, and soybeans. The Ministry of Commerce alerted grain exporters that they would have export quotas for sales in 2008. These actions, predictably, increased the illegal exports of rice and wheat, because the domestic price, in the case of rice, was one fifth the global market price.[95] Government officials issued these reports while denying that China's continued importation of cereals had anything to do with global food price surges. A senior researcher of the State Grain Administration stated: "China is not a precipitator of the mounting increase in global wheat prices, but an important stabilizing factor for it."[96] The general manager of Chungliang (COFCO), the leading state trading company, said the company would not stop rice exports despite export restrictions by some major rice-producing countries such as Vietnam.[97]

China has had a sizable trade surplus for most of the economic reform era, expected to reach $254 billion for 2007, a 43% increase over the previous year.[98] In the twenty-first century, it has had the world's largest foreign exchange reserves, standing at $1.9 trillion in 2009. Before the tainted food products crises, agricultural trade was expected to balance out by the end of the 5-year transition to WTO membership.[99] Food safety crises may result in declining food product exports into the near-term.

Entry into the WTO obliged China to lower its agricultural tariffs, and the state claims that its average tariffs on farm products are only 15%, compared to a global

[94] Based on personal interview with grain policy specialist, State Council, May 19, 2008. See also "Major State Trading Companies – Food Exports" at http://www.fao.org/docrep/006/7y5109e/y5109e0g.htm (retrieved 7/23/2008).

[95] Ying Wang, "Customs Officials Nab Illegal Exports," *China Daily*, May 5, 2008, 3, and Jiao Wu, "Stiff Grain Export Rules to Remain," *China Daily*, June 20, 2008, 3.

[96] Yinan Hu, "Researcher Defends Wheat Policies," *China Daily*, February 19, 2008, 3. Also see Zi Chong, "China Not to Blame for Food Price Rise," *China Daily*, June 14–15, 2008, 4.

[97] Jing Fu, "Pledge Not to Stop Rice Exports Lauded," *China Daily*, April 25, 2008, 1.

[98] Ying Diao, "China's Trade Surplus Large but Slowing in '07," *China Daily*, May 8, 2007, 17.

[99] See Rural Development Institute, *Green Book on China's Rural Economy*. Beijing: Chinese Academy of Social Sciences, 2007.

average of 62%.[100] Notwithstanding these claims, some of China's trade partners believe it needs to reduce tariffs further.

A greater difficulty in adjusting to the WTO regime has been meeting requirements of the sanitary and phytosanitary measures for agricultural exports. The Agreement on Agriculture in the WTO, part of the initial rules at WTO's establishment in 1995, incorporated a Sanitary and Phytosanitary (SPS) agreement. The SPS protocol allows nations to continue to employ sanitary and phytosanitary standards, but they must be applied solely to protect human or animal health or plant life, and they must be based on scientific risk assessment. Standards preferably are to be consonant with guidelines issued by scientific standard-setting bodies: the Codex (of FAO and WHO's Alimentarius Commission), the International Office of Epizootics regarding animal health, and the International Plant Protection Commission regarding plant health.[101] China's products have encountered difficulties conforming to SPS measures of other countries, and in the judgment of trade partners, China has used its own SPS measures to discriminate against agricultural products of other nations.

China's Ministry of Commerce indicated that about 90% of China's exports of food products experienced foreign technical trade barriers in 2002 alone, with losses of $9 billion in potential sales.[102] Trade partners whose barriers curbed imports from China included the European Union, Japan, and the United States. The barriers affected labor-intensive agricultural products such as vegetables, fruits, livestock and poultry products, and seafood.

Analysts contend that the SPS difficulties in China's agricultural production are long-standing, but were revealed only in the consequence of accession to the WTO. Three factors account for most of the difficulties: (1) insufficient guidelines for agricultural and food production in the food quality regulatory and supervisory system; (2) implementation deficits in regulation and supervision of agricultural production and processing as well as noncompliance with regulations, which resulted, among other consequences, in misuse or abuse of chemical fertilizers, pesticides, and antibiotics; and (3) the very large number of quite small-scale farms, and the attendant problems of millions of farmers with limited education, little information about SPS standards, and less incentive to comply.[103]

Among China's trade partners, the United States probably has been the most critical of China's use of SPS as a barrier to agricultural and animal imports. A recent USDA report alleges that "China's phytosanitary and veterinary import standards are sometimes based on dubious scientific principles and are not always consistently applied."[104] China has restricted access for American exports of softwood lumber,

[100] Wei Jiang, "More Cuts in Farm Tariff 'Not Feasible'," *China Daily*, June 20, 2007, 2.

[101] Oosterveer, 2007, 73.

[102] Fengxia Dong and Helen Jensen, "China's Challenge: Conforming to Sanitary and Phytosanitary Measures for Agricultural Exports," *Iowa Agricultural Review*, Vol. 10, no. 2 (Spring 2004), 2.

[103] Ibid., 4.

[104] USDA, *Foreign Trade Barriers*. Washington, DC: 2002, 54.

fresh potatoes and fruits, livestock and seafood products. On occasion, China has denied access to its market of US agricultural products, alleging that they have "quarantine diseases" that are abundant in China's fields and farms.[105]

China's trade policy, its foreign policy, and concerns for food security as global cereal prices escalated intersected in the decision of the state to cultivate food crops abroad. A company in Heilongjiang Province has grown rice in the Russian Far East since 2004. By 2007, its operations extended to 42,000 ha. Since 2005 a Chongqing seed company has been growing rice in Laos. At the China-Africa Summit in November 2006, China gained tentative approval to establish ten agricultural centers in Africa. The first result of this episode of agricultural diplomacy was the agreement of the Chongqing company to cultivate 300 ha in Tanzania, beginning in 2009. The company will plant its rice seed varieties, process the rice, and sell part of the grain in China. Also, it will sell seeds to local farmers, purchase and then sell their harvests.[106] Of interest also are experiments with "outgrowing," defined as a system where native farmers control their own land but contract production to a Chinese company, studied by Brautigam and Tang.[107] These farmers' activities abroad occasioned a report in the *Financial Times* that China was "hoarding" overseas farmland, which officials quickly denied.[108]

China has benefited greatly from its membership in the WTO, but there have been costs to this membership as well.

9.7.3 Compliance with Other International Conventions

Treatment of agricultural biotechnology in Chapter 7 and food safety in Chapter 8 made mention of several other international protocols which increasingly affect China's behavior. Indeed, since the late 1970s China has become actively involved in international diplomacy and has ratified a host of conventions bearing on environmental change and food security, in addition to those mentioned above. We list a number of environmental conventions (China has ratified nearly 50) with direct bearing on food security:

- Convention on the International Trade in Endangered Species (CITES), which pledged China to ban the import and export of endangered species listed in the CITES annexes (including a number of wild plants and animals for which demand had increased)
- Montreal Protocol on Substances that Deplete the Ozone Layer

---

[105] Personal observation of the authors.

[106] Xinhua, "Firm Will Grow Rice in Africa," *China Daily*, May 9, 2008, 13.

[107] Deborah A. Brautigam and Xiaoyang Tang, "China's Engagement in African Agriculture: 'Down to the Countryside'," *China Quarterly*, no. 199 (September 2009), 686–706. Also see Deborah Brautigam, "Land Rights and Agricultural Development in West Africa: A Case Study of Two Chinese Projects," *Journal of Developing Areas*, Vol. 25, no. 4 (1992), 21–32.

[108] Xiaokun Li, "China Denies Hoarding Overseas Farmland," *China Daily*, July 4, 2008, 3.

- Convention to Combat Desertification
- United Nations Convention on the Law of the Seas
- Agenda 21, with its concept of sustainable development
- Kyoto Protocol on climate warming
- Convention on Biodiversity
- Convention on Persistent Organic Pollutants
- Cartageña Protocol on Biosafety

A characteristic of these conventions, treaties, and protocols is that they are an expression of "soft law," with limited international monitoring and little in the nature of enforcement from an international organization. They have brought about some change in behavior of the state – for example, identification of action plans to counteract invasive species. However, to the present the central government has not increased its spending to enable it to comply with the objectives of the international agreements into which it has entered.[109]

9.8 Conclusions

Because the subject of this volume is the relationship between environmental change and food security, we have explored in considerable detail the many dimensions of China's environmental crisis. Many scholars have treated threats to China's environment exhaustively. The difference in our approach has been the search for explicit linkage to China's food system – food production, distribution, and consumption – and its overall food safety. That has been the subject of the previous chapters.

In this chapter, however, we have focused directly on current and prospective challenges to the implementation of effective food security policies. We considered the state of knowledge concerning problems in food security, the match between state allocations and research and development needs, and remaining knowledge and training gaps. We next treated the staffing and training of the agricultural bureaucracy, and contrasted this with the large difficulties in improvement of farming practices and dissemination of up-to-date information to those who grow China's foods.

This discussion led to analysis of coordination in China's far-flung and complex administrative network as it applies to food policies, where we commented on the difficulties of both horizontal and vertical integration. China, like all other developing countries (and several economically developed countries as well) has an "implementation deficit," which is revealed in recent crises of contaminated foods. The gap in compliance is a more intractable problem than improvement of China's agricultural infrastructure. While effective enforcement requires coordination

[109] See Gerald Chan, *China's Compliance in Global Affairs*. Singapore: World Scientific Publishing, 2006, 166–67.

of human efforts, it is also dependent on a sufficiency of finance and China is fortunate to have gained wealth ample enough to modernize its food production and distribution systems.

A possible advantage to authoritarian rule in China is the ability to change policy direction quickly and to mobilize resources to resolve outstanding problems of the state. But we also considered a serious disadvantage to authoritarian polities – their unwillingness to nurture formation of a robust civil society, which would expand opportunities for public participation by enlisting the citizenry in the monitoring of the safety of China's food supply, among other subjects meriting enhanced public scrutiny. Only an authoritarian state in the process of withdrawal, for example Taiwan in the 1980s, is likely to empower individuals and groups (and particularly NGOs). China's party leaders give no indication of an intent to withdraw from rule, but have become somewhat more tolerant of groups, such as environmental NGOs, which do not challenge the legitimacy of the party-state.

We reviewed the current status and prospective plans for resolution of poverty problems in China, particularly in rural areas. The state has made great strides in poverty alleviation, primarily through its very rapid rate of economic growth which has lifted more than 500 million Chinese out of poverty, a truly colossal accomplishment. Still, a substantial minority of Chinese has difficulty affording nutritious food, and this remains a serious problem for the state, now in both urban and rural areas.

Finally, we considered briefly the international context in which China's food system is nested. Entrance into the WTO has benefited China, but trade in agricultural products and especially food has brought new challenges. Adaptation of Chinese food producers, especially the millions of farmers on quite small farms, to international standards will take time and the development of new incentives. As recent food safety crises highlight, however, these changes are essential both for the safety of China's domestic food supply and for the continued development of its export-based economy.

In the next chapter we summarize the argument and speculate briefly on what China's experience in the development of food security policies augurs for the future.

Chapter 10
Summary and Conclusions

Abstract The first section of the chapter summarizes each of the nine foregoing chapters. Then, the authors draw ten observations concerning: (1) the priority of food security; (2) the changing definition of food in China; (3) comparative advantage and food security, (4) difficulties in policy coordination, (5) the role of science, (6) the role of crises, (7) the *Nungmingong* (farmer-laborer) problem, (8) domestic challenges to food security, (9) international challenges to food security, and (10) environmental challenges to food security.

Keywords Mandate of Heaven (*Tianming*) • "Made in China" label • Comparative advantage • "Socialism with marketizing characteristics," Civil society • "Super (hybrid) rice," Food and Agriculture Organization (FAO) • Codex Alimentarius

Summary

We conclude this volume in two sections. First, we summarize the argument of the earlier chapters. Then we draw ten observations on what China's experience of environmental change and food security augurs for the future.

Chapter 1 introduced the concepts of food system and food security, defining the latter as the condition reached when a nation's population has access to sufficient, safe, and nutritious food to meet its dietary needs and food preferences. With 22% of the world's population, China's food situation is very important to global food security. We introduced the contents of the volume and then, in broad strokes, surveyed food security in ancient and dynastic (211 BC–1912) China. The focus was on environmental stressors, including population growth, natural disasters, pathogens and insect pests as well as on the responses of the state (for instance, development of systems for irrigation, flood control, storage, and transportation). Too, the chapter briefly introduced the Republican era from 1912 to 1949, and compared the environmental stressors and government responses at that time to those of the imperial period.

J.H. McBeath and J. McBeath, *Environmental Change and Food Security in China*, Advances in Global Change Research 35, DOI 10.1007/978-1-4020-9180-3_10,

Few nations have undergone the dramatic economic and political changes of China in its 60-year history under communist party rule. Chapter 2 reported on the new regime's establishment of a command economy with central planning of agricultural production and ideological control of plan targets and methods. After land reform and collectivization, the Maoist leadership launched two campaigns – the Great Leap Forward (GLF, 1958–1960) and Cultural Revolution (1966–1976), which had disastrous consequences for both environmental and food security. After Mao's death in 1976 and a brief succession struggle, Deng Xiaoping consolidated power and launched broad scale reforms. These introduced production incentives for farmers (the household responsibility system), loosened somewhat state control of the grain market, improved the country's infrastructure, and opened China to foreign trade and investment. The chapter then treated China's current food system. It described plant and animal food in the average diet, and remarked on China's self-sufficiency in grain. Yet there are important regional and income variations in access to food, while growing wealth has brought about changes in food preferences.

Chapter 3 presented information about the environmental stressors on food security. It began by reviewing the debate over how much arable land China has, and then examined the loss of land used for production of food crops because of population growth, urbanization, and economic development pressures. The effects of socioeconomic change have been almost uniformly negative so far as China's environment is concerned, and we analyzed degradation of land (erosion, deforestation, desertification, and land pollution), air pollution, and degradation of water (overuse of water for agriculture, pollution of rivers, lakes, and groundwater, and ocean pollution). The chapter concluded by examining seven different responses the state has made to environmental stressors: policy restricting conversion of arable land, the "one-child" policy, investment in irrigation systems, large-scale dam construction, the South–North Water Diversion Project, large-scale afforestation and reforestation campaigns, and the program to convert marginal agricultural lands to forests and grasslands.

Climate warming is a special case of environmental change, and it has been an issue in China for less than 2 decades. Chapter 4 described China's energy policy – its heavy reliance on coal and low energy efficiency, which have made China the world's largest emitter of greenhouse gases. The chapter also discussed contributions to climate warming of China's traditional agricultural system. Because the East Asian monsoon influences climate variability, we examined natural climate factors, and national and regional effects. Then, the chapter considered effects of climate change observed by scientists: changes in temperature, precipitation, surface evaporation, and sunshine duration. Also, the chapter considered the correlation between climate change and extreme weather events – floods, drought, heat waves, and typhoons. Through cooperative relationships with European institutions, Chinese meteorologists have used climate models to predict effects on rice, wheat, maize, and cotton production, and the chapter described some results of this research, as well as the difficulties in mitigation of climate change effects.

Chapter 5 introduced comprehensively the plant diseases and pests affecting agricultural production. It defined central concepts, such as ecological services, identifying the beneficial ecosystem effects of plant pathogens and insect pests. Using the most recent data, it discussed the broad range of economic damage done to food crops. Because migratory locusts, rice blast disease and wheat rust are major epidemics in China, we examined both their historical as well as contemporary impact on food security. Although thousands of plant diseases and insects infest China's staples, we focused on a little more than two dozen that have reached epidemic scales, organizing the discussion by the major food crops the diseases and pests threaten: rice, wheat, maize (corn), potatoes and soybeans. Then we separately treated rice false smut, wheat scab, and other post-harvest diseases and pests, because they not only affect crop yield but also endanger human and animal health. We considered farm cultural practices having an impact on diseases and pests as well as changes in government policy, including improvement of domestic transportation. We also discussed briefly the broadest environmental change affecting incidence and severity of disease and pest infestation – climate change. Finally, the chapter introduced the major control measures adopted to deal with plant pathogen and insect infestation: regulations and standards, cultural controls, biological control measures, and physical and chemical controls.

Although foreign species such as maize, wheat, and potatoes are now major staples in China's food system, attention recently has focused on alien invasive species that threaten China's food production, and Chapter 6 recounted this story. It distinguished species intentionally introduced to China from the uninvited, and discussed economic and non-economic effects. The chapter gave examples of plant, insect, fish, and other invasive species with impacts on food production. Also, it examined causes of species invasion – broadly, socioeconomic change and specifically, improved domestic transportation routes as well as rapid growth of international tourism and trade. All provide channels for increased numbers of alien species to enter China. The chapter pointed out China's efforts to reform laws and regulations to counter the threat. It recognized efforts of Chinese scientists, influenced by global standard-setting bodies, such as the World Conservation Union (WCU), to adopt the precautionary principle to eliminate threats. The chapter reviewed measures adopted to prevent or control harmful invasives: physical, biological, chemical, and integrated pest management. Finally, the chapter suggested future directions in research, outreach, and international collaboration to counter this new challenge.

Chapter 7 evaluated the biotechnological responses China has authorized to increase food production. The chapter explored motivations explaining China's aggressive pursuit of breakthroughs in agricultural biotech research, the public sector emphasis of China's approach, and state financial support that is greater than that of any other developing country. The chapter featured the development, impacts, and problems (e.g., new pests) of China's greatest GMO success – Bt cotton. Then, it examined as a special case Bt rice. Because rice is China's major food crop, and because global attitudes toward GMOs worsened while Bt rice was undergoing testing, the case has been very controversial. The chapter described the testing procedures under new regulations, considered the opportunity for production increases that Bt rice might open, and analyzed the several reasons advanced to delay or prohibit the commercialization of this important GM crop. The chapter concluded by illustrating some divisions in the elite as to whether Bt rice should be commercialized. It emphasized that, given limited information possessed by the public (and normal constraints in an authoritarian system on citizens' use of information if they had it), leaders face only two obstacles: their own lack of consensus and potential foreign opposition. This chapter concluded with analysis of China's biosafety "regime" (a network of laws, regulations, policies, and institutions directing resources to resolve problems). China's scientists and political leaders have developed a structure to manage GMOs, changed laws and regulations, and reined in behaviors threatening state objectives.

From 2007 to 2009, China encountered the greatest threat yet to the "Made in China" label, when foreign trade partners and Chinese themselves protested that some Chinese food products were unsafe for consumption. Chapter 8 addressed the legal and institutional framework China has developed to achieve food security. It described constitutional, legal, regulatory, and policy elements in the state's authority system. Then it classified the most relevant ministries and agencies by their function in the administrative process – food production, food consumption, and system-wide control. Because the state's devolution of authority to provinces, autonomous regions, municipalities and county/townships has weakened central authority, the chapter briefly examined manifestations of autonomy at the local level, and it reviewed attempts to improve integration and coordination system-wide. Indeed, it is local autonomy that is the ultimate explanation of most threats to both China's environmental and food security. The chapter described the role a non-governmental organization, Greenpeace, has played in embarrassing the regime because of its poor handling of food issues. The second case of a food regime applied the concept to food safety. In this food safety regime, leaders instantly created a new adaptable structure, revised numerous laws and regulations, and tightened enforcement systems. However, the new regime was unable to guarantee safety of all China's food products, as indicated by the 2008–2009 crisis of contaminated milk powder.

Chapter 9 adopted a different perspective than previous chapters. It both acknowledged China's success in providing sufficient food for its growing population, while pointing out "implementation deficits" retarding future plans to improve food security, with the proviso that all economically developing countries (and several economically developed nation-states as well) face similar difficulties. The chapter presented six challenges and questions: (1) Whether China has invested sufficiently in its intellectual establishment, agricultural research and development, and improvement of farming practices, (2) Whether administrative coordination, both vertically and horizontally, is adequate; (3) Is the food production infrastructure effective ? (4) Has the state sufficiently improved access of both rural and urban Chinese to food, and specifically, have poverty alleviation programs identified those most in need of assistance and helped them? (5) Whether the authoritarian regime has created sufficient opportunities in public participation to enlist the energies of the citizenry in monitoring the quality of the environment and food products (while providing an escape valve for lack of other meaningful participation opportunities and popular liberties); and (6) Whether China has adjusted and adapted to global standards, required to match its increasing participation in international government organizations (such as the WTO) as well as environmental conventions and agreements.

10.1 Observations

We conclude this volume by drawing ten observations on our topic.

10.1.1 Priority of Food Security

As mentioned in several places in the text, China's leaders have given a high priority to food security.[1] Consistently, they have emphasized the importance of agriculture, while this sector of the economy declines as a share of China's GDP. Although we pointed out the lack of resources allocated to the improvement of farming practices at the local level, the central government regularly spends vast sums on irrigation, to the extent that over half of China's grains (and specifically, most of its important food crop, rice) come from irrigated fields. In fact, agricultural production commands more of China's scarce water than either industrial production or household consumption. The regime also has greatly increased spending on rural roads, railroads, ports and harbors, storage facilities, and other agricultural infrastructure.

Too, the central government has spent billions of yuan on projects reducing environmental stress on arable lands and countering environmental degradation. The sloping lands conversion program as well as afforestation and reforestation campaigns, discussed in Chapter 3, indicate the government's willingness to spend if it can be established that agricultural productivity will benefit. Critics of China's environmental protection efforts say it spends too little, less than 1.5% of GDP, when economically developed countries spend more than that percentage. We agree that China needs to allocate a greater proportion of its resources to addressing environmental problems, but are impressed with the change in attitude of leaders regarding environmental problems and increased government spending on them, in the last decade.

---

[1] For a treatment of the "core interests" of the state, see Theda Skocpol, *States and Social Revolutions*. Cambridge: Cambridge University Press, 1979; and John Dryzek, David Downes, Christian Hunold, and David Schlosberg with Hans-Kristian Hernes, *Green States and Social Movements*. New York: Oxford University Press, 2003.

China has not needed to establish a new priority of food security, for this has been a state objective from early times. The philosopher Mencius (fourth century, BC) said that nourishing the people was the first principle of government.[2] Although by no means a liberal democrat, Mencius believed that ultimate sovereignty lay with the people,[3] and the phraseology of his time was that rulership was "granted by Heaven." Later, the Mandate of Heaven (*Tianming*) expressed the belief that natural calamities (including extreme weather events such as famines, pestilence and floods, causing malnourishment and starvation of the people) were products of bad government, and justified the people's withdrawal of support for the regime.

China's long history has been punctuated by natural disasters, but given the vast expanse of land and highly diverse biogeographical regions (and climatic conditions), disasters striking all parts of the country simultaneously have been extremely rare. The Great Leap Forward was such a period, as for 3 consecutive years (1959–1961) floods occurred over all of China. Human errors in the commune system, rapid deforestation and soil erosion accompanying the GLF, and destruction of flood-control systems compounded the effects of heavy rainfall.[4] A legacy of the policy failures of the Maoist era was increased attention to food security. Throughout the reformist era beginning in the late 1970s, China's leaders have pledged their support to increased food production. Unity of the leadership on the importance of agricultural production provides some insurance that food security threats can be managed into the near term. Indeed, this is a "core" function of the state.[5]

10.1.2 Changing Definition of Food

The Chinese state's definition of food centers on grains, in the order of rice, wheat, corn, and species of the millet family. Potatoes are considered the "fifth grain." Vegetables are included along with eggs and meat (pork, chicken, and mutton), and fish, but traditionally they have comprised a smaller part of daily food. As discussed in Chapter 2, diets of urban Chinese, whose incomes are higher than residents of rural areas, have changed. Now, staples are somewhat less important, while vegetables, fruit, meat, and increasingly dairy products compose a larger portion of table fare.

The official definition of food has been slow to change, and this has influenced the debate over arable land (discussed in Chapter 3). As critics such as Smil contend,[6]

---

[2] Kung-chuan Hsiao, *A History of Chinese Political Thought*. Princeton, NJ: Princeton University Press, 1979, 155. *Mencius* remarked: "A sage governs the kingdom so as to cause pulse and grain to be as abundant as water and fire."

[3] Ibid., 158.

[4] Kang Chao, *Agricultural Production in Communist China*. Milwaukee, WI: University of Wisconsin Press, 1970, 129–30.

[5] Dryzek et al., 2003.

[6] Vaclav Smil, "Research Note: China's Arable Land," *China Quarterly*, 158 (June 1999), 414–29.

China's conception of arable land follows the definition of food: land appropriate for the cultivation of grains such as rice, wheat, and maize. Land currently used for fruit orchards, ponds for aquaculture, truck gardens, and other food production purposes does not usually enter the arable land column of official reports.

In his 1995 book *Who Will Feed China?* Lester Brown made these issues controversial by predicting that China would be unable to provide sufficient staples for its population by 2020.[7] It would then need to purchase large quantities of grain in the international market, thereby driving up food prices globally. Both Chinese and foreign experts have challenged this provocative prediction. The regime's figures (the amount of 130 million hectares in the government's 2007 statistical yearbook, or the commonly used statistic of 121.5 million hectares in other government reports) demonstrate that arable land today is sufficient to meet current grain needs. Nevertheless, this debate and pressure to convert farm land to other purposes have led the regime to draw a line in the sand: arable land cannot be allowed to drop below 120 million hectares.

10.1.3 Comparative Advantage and Food Security

China's grain production in 2008 was 510 million metric tons, a bumper harvest despite a larger than average number of extreme weather events – floods, drought, and heat waves (which some observers called early signs of climate warming). At this level of production, China met approximately 95% of its needs for grains. Nevertheless, officials continued to push for increases in productivity, and expressed what we call the "rhetoric of grain security": that complete self-sufficiency in grain production, rising to accommodate increases in population, is necessary to protect China in an uncertain world.

In fact, China has not revised its food security policy to take into consideration agricultural comparative advantages in the international economy. A number of experts have argued that China should increase the importation of food products that require the intensive use of land (for example, wheat, maize, and rice), because as noted throughout this volume China has a significant and increasingly serious arable land deficit.[8] While this redirection of policy would result in large increases in the importation of grain, this deficit would be partly offset by exportation of other foods for which China has a comparative advantage, such as fruits and vegetables; and China's export of manufactured goods would more than make up for a deficit in agricultural trade.

China is close to self-sufficiency in rice and sorghum, but in recent years has imported wheat, maize (mostly for animal feed), and soybeans (also for livestock

---

[7] Lester Brown, *Who Will Feed China?* New York: W. W. Norton, 1998.

[8] See, for example, Chenggui Li and Hongchun Wang, "China's Food Security and International Trade," *China and World Economy*, no. 4 (2002), 32.

feed and for oil and processed foods). We believe that for a country as populous as China, and with the world's largest number of small farms (about 220 million, in 2009), continued self-sufficiency in food production is an unrealistic objective. China's population will increase until, as experts estimate, it will peak at 1.6 billion in 2030. China's farms are too small to allow productivity increases through mechanization, and increasing fertilizer and pesticide inputs would be counterproductive. GMOs have not yet demonstrated large yield increases. Hybrids promise increased productivity, but only with improved farming practices (including disease/pest control).

As noted in Chapter 9, for most of its history under communist rule, China has exported more food products than it has imported. (On occasion, such as during the GLF, this worked to the detriment of the population.) Upon entry to the WTO in late 2001, China lowered agricultural tariffs, and foreign food products became more competitive in the domestic marketplace. For most of the first 9 years of the twenty-first century, China has imported more cereals than it has exported. Yet China's overall global trade surplus continues to outpace that of any major trade partner. Clearly, China can afford to import more cereals than it has in recent years, without risking a current account deficit.

As China's food production costs rise, its domestic cereals will be more expensive than products from trading partners such as the USA, Australia, and Canada, which enjoy greater economies of scale, have better control over disease and pest problems, and suffer from less environmental damage. Following the principle of comparative advantage, China should import more cereals instead of fewer, compensating for this increase by further specialization in areas such as basic manufacturing. This adaptation, however, is unlikely to occur quickly because of the logjams in domestic transportation and storage systems.

10.1.4 Contradictions of Policy

The Chinese Leninist party-state survived the dissolution of the Soviet Union and the "color revolutions" of Central and East Europe because, since the late 1970s, it has adapted to changes in its domestic and international environment. Political liberalization has been limited yet significant. The CCP now has a few members whose credentials are based on business success; farmers can elect village chiefs, and while a majority has party membership some challenges to local party rule have succeeded. Economic liberalization has been much more deep and thoroughgoing than at any previous time in China's long history. The result in the early twenty-first century is an anomaly: what leaders call "socialism with marketizing characteristics." This hybrid form has inherent contradictions, and they make more complex the pursuit of food security. We discuss four.

First, the regime increased individual incentives through establishment of the household responsibility system, but it did not privatize land, which is still owned by the state or collectives. While farmers had 30-year leases on the lands they farmed, they lacked secure title or property rights to land: They could not sell or transfer the land they used; nor could they buy additional land. Meanwhile, the state devolved land control functions to local governments, which had incentives to transfer land to the most profitable activities. This meant that local governments allowed conversion of collective or state land from agricultural production to industrial, commercial, and in some cases residential, uses. This is the main factor explaining the loss of arable land in the last 2 decades. It also explains the increased levels of rural unrest as farmers protested, sometimes violently, the expropriation of lands they used and their physical displacement from the land.[9] In late 2008 the party endorsed the formation of markets to allow farmers to sub-contract, lease and exchange their land use rights, and to pool their land in larger shareholding entities; however, it is too soon to say whether this will release tensions in this area.

Second, as most observers of China's environmental crisis have pointed out, economic development pressures and urbanization seriously threatened environmental values. The relentless pressures to grow the economy contradicted the spirit of environmental laws and policies of the central government. Central leaders were willing to look the other way, or to appeal to the environmental Kuznets curve: after China became rich it would clean up the polluted landscape. As we emphasized in Chapters 3 and 4, however, the environmental crisis now is affecting food production. This means that officials outside the Ministry of Environmental Protection (MEP), such as in land, water, even planning ministries, lament environmental damage. The contradiction of rapid economic development and rapid environmental degradation has become manifest, as it has influenced food security.

Third is the contradiction between centrifugal and centripetal forces in the administration of food production, control, and consumption policies. Devolution of authority from the center to provinces, autonomous regions, municipalities, counties and ultimately townships and villages has seriously diluted policy intent and outcomes. For example, local governments have become more autonomous, not just in converting arable land to other uses, but also in failing to control production of contaminated products by local factories operating in a regulatory void (as noted by the most recent crisis in 2008–2009, reported on in Chapter 8). Attempts by the center to reassert control, as we shall see, generally have been short-lived and ultimately unsuccessful.

Finally, the need for a vigilant citizenry contradicts the survival and controlling instincts of the party-state. China has allowed non-governmental organizations to form (discussed in Chapters 8 and 9), the media are somewhat freer,[10] and protests that do not threaten party-state survival (as described in Chapter 9) are permitted.

---

[9] Kevin J. O'Brien and Liangjiang Li, *Rightful Resistance in Rural China*. New York: Cambridge University Press, 2006.

[10] See Yuezhi Zhao and Wusan Sun, "Public Opinion Supervision: Possibilities and Limits of the Media in Constraining Local Officials," in Elizabeth J. Perry and Merle Goldman, *Grassroots Political Reform in Contemporary China*. Cambridge, MA: Harvard University Press, 2007.

Yet the regime still censors the media, closes down NGOs, and rounds up dissidents. This is inauspicious for the formation of grassroots watchdog organizations such as consumer advocacy groups, because in the areas of food production, consumption, and control they likely would criticize state policy and regulations, ineffective enforcement, and political corruption.

10.1.5 The Role of Science

China's elite in the early twenty-first century has a technocratic cast. Certainly as compared to the status of intellectuals in the Maoist era, scientists have moved closer to power. The agricultural sciences in particular are useful to the regime to attain its core goals of economic development and food security and to avoid legitimacy crises. Indeed, scientific accomplishments, such as the sequencing of the rice gene, benefited national pride. Throughout we have emphasized the role that scientific research has played in increasing agricultural productivity, as witnessed in the development of "super (hybrid) rice." The state relies on scientists to ward off new threats to food production from domestic and invasive species (see Chapters 5 and 6), and to ascertain risks and recommend controls for rogue products in the food production system (see Chapter 8). As nuclear scientists gained prestige for China by developing weapons, so agricultural scientists enhance China's status through biotechnological research and development of GMOs, such as Bt rice, wheat, and maize (see Chapter 7). In return, the regime has handsomely funded these research enterprises, making China the leader among developing nations in this area.

At the 17th Congress of the CCP, party secretary Hu Jintao acknowledged the new status of China's intellectual establishment by calling for balanced "scientific development." This may simply be lip service of the regime in response to criticism of large-scale development projects, such as the Three Gorges Dam or the South–North Water Diversion Project. However, there is evidence that the state is paying greater attention to the costs and benefits of changes to China's environment, especially as it bears on food security. We noted in Chapter 3 that the central government is now systematically investigating land resources; MEP has launched surveys of environmental pollution; even NGOs, such as Ma Jun's Institute of Public and Environmental Affairs, have developed methods to estimate and publicize the sources of water pollution (but with a focus on foreign multinationals). In Chapter 4 we discussed the increase in monitoring nation-wide of changes in temperature and precipitation associated with climate change.

However, science does not drive policy investigation and formation, and in several areas related to food production, consumption, and control, scientific information clearly is inadequate. In Chapter 5, for example, we indicated the continuing lack of knowledge of plant pests and diseases, which affects food production.

10.1.6 The Role of Crises

The role of science in the policy-making process and regime formation typically expands when nations encounter crises to their environments. Karen Liftin opines that: "The ability of scientific knowledge to foster environmental action can be greatly enhanced by the emergence of an apparent crisis."[11] Our question is whether China's recent crises have had a positive impact on both the structure and behavior of the state in addressing food security problems.

Significant deforestation in the Yangtze and Yellow River watersheds has increased flooding, and in 1998, the worst flooding in 50 years took 4,000 lives.[12] This crisis did not bring about a change in the structure of decision-making, but as we noted in Chapter 3, it was responsible for the development of several large nation-wide campaigns of afforestation and reforestation. A second crisis with a more direct impact on China's food system occurred in 2007, when China's trade partners reported that some of her exported food products were unsafe for human (and animal) consumption. The regime's response to this crisis was nearly instantaneous. As discussed in Chapter 8, it led to the execution of the head of the State Food and Drug Administration, reform of laws and regulations, and a 4-month campaign to increase inspections and improve enforcement. But then, a year later, milk powder contaminated with melamine took six lives and threatened 300,000, and public skepticism increased at the same drill by the state: executions and long prison terms for the main culprits, more reforms of regulations, another 4-month campaign of administrators to insure that food products were free of non-edible substances and excessive levels of additives.

During the Maoist era, campaigns frequently were used to mobilize the population to achieve objectives, especially ideological goals, of the regime. Limited political liberalization of the reform governments since 1978 has reduced the role of campaigns and particularly has eliminated most ideological campaigns. Instead, as China marketizes, its leaders increasingly have employed nationalist themes, both to include most of the population and to increase regime legitimacy.[13] We estimate that in the short-term, the campaigns of afforestation and reforestation on the one hand and the first campaign to improve product safety on the other were successful. The difficulty lies in sustaining campaigns over a long period. The first food safety campaign lasted just 4 months, and once completed, the administrative

---

[11] Karen Liftin, "Eco-regimes: Playing Tug of War with the Nation-State," in Ronnie D. Lipschutz and Ken Conca, eds., *The State and Social Power in Global Environmental Politics*. New York: Columbia University Press, 1993, 101.

[12] Cynthia W. Cann, Michael C. Cann, and Shangquan Gao, "China's Road to Sustainable Development," in Kristen A. Day, ed., *China's Environment and the Challenge of Sustainable Development*. Armonk, NY: M. E. Sharpe, 2005, 7.

[13] See Suisheng Zhao, "The Rise of State-Led Pragmatic Nationalism: An Instrumental Response to the Decline of Communism in China," in his *A Nation-State by Construction: Dynamics of Modern Chinese Nationalism*. Stanford, CA: Stanford University Press, 2004, 209–47.

and enforcement systems seemed to return to normal. In an era when food fears have become global,[14] and the authoritarian state continues to control many food prices, company management and the flow of information, further food safety crises in China are inevitable.

In the absence of structural changes to improve coordination and implementation in the Chinese bureaucracy, we suspect that the momentary changes of behavior occurring in response to crises will be insufficient. Other, larger scale crises – for example, mitigating and adapting to climate change – will soon occur, and we are not sanguine that China's success in administering short-term fixes will be sufficient in addressing near- and longer-term environmental crises that directly impinge on food security.

10.1.7 Domestic Economic Challenges: The Nungmingong (Farmer-workers)

Nearly 150 million farmers have left rural areas to improve their lot in life. They are called the *Nungmingong* (or farmer-workers, and sometimes migrant workers). Economic forces drove them from the land, for as we reported in Chapter 9, they make three to four times more in factory work on average than they did cultivating the soil. Most studies object to the unfair treatment of this large group. Researcher Han Jun points out that *Nungmingong* are exploited: "They earn roughly one quarter what city workers make for doing the same work." They also are not eligible for health, education, and other services available to urban residents, because they lack *hukou* (household registration) in cities.[15] Yet what is often ignored is that *Nungmin Gong* retain leasehold rights to rural farmland, an important guarantee and form of security.

Over two-thirds of the farmer-workers are male, and most are relatively young (18–40). Their migration to urban areas removes a highly productive factor from agricultural work. Left to cultivate the fields are women, children, and old people. The women and elderly are likely to be ill-educated. (About one-half of China's illiterate population resides in rural areas.[16]) They are likely to be less proficient in the cultural practices of farming, and few government programs are directed to their training. Clearly, the challenge to the central government is to rebalance the incentive system to increase the rewards of farming, which would pull the *Nungmin Gong* back to their homes. This seems improbable, however, as increasing prices for cereals produced domestically would have large inflationary impacts. The absence of incentives for the farmer-workers will have a dragging effect on future agricultural productivity.

---

[14] Alison Blay-Palmer, *Food Fears: From Industrial to Sustainable Food Systems*. Aldershot, Hampshire, UK: Ashgate, 2008.

[15] Editors, "Leading State Adviser Offers Frank Assessment of Rural Challenges," *China Development Brief*, Vol. X, no. 7 (July/August 2006), 4–5.

[16] See Chenggui Li, "Reform of China's Agricultural Technology Structure," Unpublished paper, 2007.

A related challenge is the rising inequality between urban and rural populations. The number of Chinese described as impoverished has declined significantly, in some accounts by 500 million, making this the largest poverty reduction campaign in world history. Yet the rural poor still number about 135 million, and thus comprise 10% of the national population. Rural migrants have flooded China's cities and increased the urban welfare burden.

10.1.8 Domestic Political Challenges to Food Security

To the present, China has not made large changes to its system of governance that would facilitate resolution of near- and longer term food security problems. Indeed, the hierarchical, top-down approach of the communist party echoes important practices and routines of the dynastic era. There are advantages to this historically contingent (path dependent) approach, because the costs to the regime of bureaucratic reorganization (and the likely short-term costs of democratization) would be very high.[17]

We have indicated throughout, however, the disadvantages of an authoritarian system in addressing environmental degradation and food security issues. At the least, the lack of civil society development will inhibit the regime's monitoring of local governments' performance in regulating the millions of small farms and factories which threaten both environmental and food security values. Currently, China suffers both from large implementation deficits and regulatory voids, which imperil efficient administration of environmental and food safety rules.

Some observers are optimistic that the combination of local elections, growth in NGOs, and rise in civil protests will lead to democratization. China scholar Lowell Dittmer believes that there are "signs that China will follow its East Asian neighbors with some form of representative democracy."[18] We do not share this optimism, because there have been no signs that the authoritarian elite is interested in withdrawing from or sharing power. Indeed, in his opening address to the 17th National Congress of the CCP, President Hu Jintao advocated only incremental political changes. He reasserted the leadership's intent to maintain the party's monopoly on power, remarking that it must remain "the core that directs the overall situation and coordinates the efforts of all quarters."[19]

---

[17] Ronnie D. Lipschutz with Judith Mayer, *Global Civil Society & Global Environmental Governance*. Albany, NY: State University of New York Press, 1996, 45. See also Paul Krugman, *Peddling Prosperity – Economic Sense and Nonsense in the Age of Diminished Expectations*. New York: W. W. Norton, 1994.

[18] Lowell Dittmer, "Conclusion – China's Reform Deepening," in Lowell Dittmer and Guoli Li, *China's Deep Reform: Domestic Politics in Transition*. London: Rowman & Littlefield, 2006, 505.

[19] Quoted in Joseph Kahn, "China's Leader Closes Door to Reform," *The New York Times*, October 16, 2007, A15.

10.1.9 International Challenges to Food Security

Some reform-minded leaders in China see the country's recent entrance into the WTO, and its increased participation in international treaties, conventions, and protocols, as extremely beneficial to the development of domestic environmental and food security. Certainly, as we noted in Chapters 4, 6 and 7, international conventions, such as the Convention on Biodiversity Conservation, the Kyoto Protocol, and the Cartageña Protocol, have brought external pressure to bear on the state to follow international guidelines in the prevention and mitigation of invasive species, curbing greenhouse gas emissions, and in insuring the safety of new GMOs before they are introduced to the marketplace. Specialized agencies of the United Nations – the Food and Agricultural Organization (FAO) and the World Health Organization (WHO) – and their subordinate unit, the Codex Alimentarius, provide international regulations on food safety, but harmonization of quality and safety standards is incomplete (and enforcement by global institutions is lacking).[20] The WTO itself has rules on phytosanitary and safety standards, which China has obligated itself to follow, notwithstanding the flexibility any country has in determining the scientific basis on which quarantine lists are developed.

In the early twenty-first century, there is an increasingly complex web of international institutions, global NGOs, and regulations on food safety, and this network has been brought into play in the cases of food safety in China we have discussed. To defend the "Made in China" label of its exports, the Chinese state did immediately address problems of its food products. In the estimates of critical observers, the response fell short of the expectation that Chinese products on the shelves of supermarkets in economically-developed nations would be no less safe than products from Europe or North America.

China's rise to global economic power in the last generation corresponds with a sharp increase in globalization. Yet rich and powerful nation-states, such as China, remain sovereign agents in policy-making and especially in the implementation and enforcement of policy. As Peter Oosterver notes: "(P)romising approaches to identify global food governance arrangements will have to recognize that most food production practices, as well as certain food consumption practices, remain structured within the space of place."[21] Thus, while international pressures have assisted in the modernization of China's food system, it is domestic forces that will play the most important roles in mandating improvements, if any, in production, consumption, and control of food in China tomorrow and in the near-term.

---

[20] Reba A. Carruth, "Socio-economic Foundations of Food Safety Regulations and the Governance of Global Agri-food Industries," in Reba A. Carruth, ed., *Global Governance of Food and Agriculture Industries*. Cheltenham, UK: Edward Elgar, 2006, 26–28.

[21] Peter Oosterver, *Global Governance of Food Production and Consumption: Issues and Challenges*. Cheltenham, UK: Edward Elgar, 2007, 228.

Nevertheless, we remain impressed with the changes in China during the 30-year period of reform, which we have observed closely. Problems of environmental change and food security today are commonly recognized among the elite. Increasingly, citizens include them in their evaluation of government performance. Food security is a core interest of the state, and China has succeeded in feeding its large population.[22]

10.1.10 Environmental Challenges to Food Security

The concerns about food security are for the near- and mid-term, and not the present. China's environmental challenges are far greater, and if unresolved will advance the date when food sufficiency becomes problematical. China is an old and remarkably resilient land, but some adverse environmental change already has become irreversible: for example, loss of arable land to cities and factories, which cannot be restored; erosion and desertification of millions of hectares; pollution of lands by chemical factories, mining, and runoff of fertilizers and pesticides; and the melting of glaciers.

The state has been successful in redressing reversible changes, such as in its massive re- and afforestation campaigns. It has allocated increased resources to environmental protection and restoration, as national income rises, yet these have not been sufficient to protect biodiversity or ecosystem services, or counteract in the short-term the results of decades of land, water, and air pollution. It has made important institutional changes, such as elevating the environmental protection function to ministerial status in 2008 and increasing somewhat its staff, but as many observers point out, deficiencies remain in implementation and enforcement of environmental law.[23]

The evolution of state-society relations will be critical to effectively meeting environmental challenges. Increasingly, party leaders have directed attention to environmental issues and on occasion have acted to advance environmental goals. The size of the high school and college-educated population is increasing, and environmental awareness is growing apace, but not rapidly enough to represent a movement. Although more environmental news now is carried in the press and on electronic media, still a large segment of the public is uninformed about pressing environmental problems (and the state still controls the media). And, as we have

---

[22] This is not a comparative study, but it is worth mentioning that the food policies of China have been more successful than those of India, which has a smaller population, a greater percentage of arable land, and a democratic government. See: Somini Sengupta, "In Fertile India, Growth Outstrips Agriculture," *New York Times*, June 22, 2008, A7i.+

[23] In addition to sources in Chapters 8 and 9, also see Canfa Wang, "Chinese Environmental Law Enforcement: Current Deficiencies and Suggested Reforms," *Vermont Journal of Environmental Law*, Vol. 8, no. 2 (Spring 2007), 159–94.

mentioned repeatedly, China lacks a robust civil society of autonomous social organizations to pressure the government. We think it unlikely that the national elite will forge the link between food security and environmental degradation on its own, because its interests are in short-term survival. Because pressures from all sources – international governmental organizations and NGOs, other nation-states, domestic NGOs, the mass public – are weak, we are of two minds as to whether China will be able to respond effectively to environmental challenges before they imperil food security in the mid-term.

Selected Bibliography

Chinese-Language Sources

Bianju Weiyanhui (ed) (2007) Qixiang Bianhua Guojia Pingu Baogao (China's National Assessment Report on Climate Change). Kexue Chubanshe (Science Publishing), Beijing

Chen BM, Zhou XP (September 2004) Zhongguo Jinqi Gengdi Ziyuan yu Liangshi Zonghe Shengchan Nengli de Bianhua Taishi (Changes of agricultural resources and grain comprehensive production capacity of China in recent years). Ziyuan Kexue (Resour Sci) 26(5):38–45

Dong ZP, Jiang JY (2007) Xiaomai Bingchongcaohai Fangzhi Caise Tupu (Wheat pictorial guide to diseases, insects and weed and control). Zhongguo Nongye Chubanshe (China Agricultural Publishing Co.), Beijing

Guojia Huangjing Baohu Zongju Zhongguo Luxing (Shengwu Duoyangxing Gongyue) Gongzuo – Xietiaocu Bangongshi. (Biodiversity Conservation Office, SEPA, China. In: Shengwu Duoyangxing Baohu yu Fupin Yantao Hui (Proceedings of the Workshop on Biodiversity Conservation and Poverty Alleviation). Zhongguo Huangjing Kexue Chubanshe (China Environmental Sciences Press), Beijing

Guojia Linyequ (State Forestry Administration). Zhongguo Linye 2006 Nianjian (China Forestry yearbook 2006). Zhongguo Linye Chubanshe (China Forestry Publishers), 2006, Beijing

Guojia Linyequ (2007) Zhongguo Linye Fazhan Baogao (2007 Report on development of China's forestry). Zhongguo Linye Chubanshe (China Forestry Press), Beijing

Guojia Tongjiqu (2007, 2008) (National Bureau of Statistics of China). Zhongguo Tongji Nianjian (China Statistical Yearbook, 2007). Zhongguo Tongji Chubanshe (China Statistics Press), Beijing

Guowuyuan Nanshui Beidiao Gongcheng Jianshe Weiyuanhui Bangongshi (2007) (Office of Engineering and Construction, South–North Water Diversion Project, State Council). Zhongguo Nanshui Beidiao Gongcheng (China's South-to-North Water Diversion Project). Gongren Chubanshe (Labor Publishing Co.), Beijing

He BS, Huan S (eds) (2005) Shengtai Huanjing Anquan yu Guanli (Security and management of the ecological environment). Huanjing Kexue yu Gongcheng Chuban Zhongxin (Environmental Science and Engineering Publishing Center), Beijing

Hsu H, Wang J, Zhang S, Wang CY (eds) (2004) Wailai Wuzhong Ruqin – Shengwu Anquan – Zhuanji Ziyuan (Alien Species – Biosafety – Genetic Resources). Kexue Chubanshe (Science Publishing Co), Beijing

Huang GY, Cai YL (January 2002) "Fujian Sheng Gengdi Ziyuan Taishi yu Liangshi Anquan Duice (Current situation of cultivated land resources and food security in Fujian Province). Ziyuan Kexue (Resour Sci) 24(1)

Kang GL, Wu J (2006) Ziyuan Kaifa Huanjing Zhongjinshu Wuran yu Kongzhi (Heavy metal pollution and control in mining environment). Zhongnan Daxue Chubanshe (Central South University Press), Changsha

Jiang W, Chen F (March 2000) Woguo Beifang Diqu Fangzhi Shamohua de Falu Zikao (Thoughts on laws of controlling desertification for the north region of China). Ganhanqu Ziyuan yu Huanjing (J Arid Land Resour Environ) 14(1):34–38

Ju T (ed) (2006) Zhongguo Huanjing Baohu yu Kechixu Fazhan (Environmental pollution and sustainable development in China), vol. 10. Kexue Chubanshe (Science Publishing), Beijing

Kexue Jishu Bu (Ministry of Science and Technology) (2007). 2006 Zhongguo Nongcun Keji Fazhan Baogao (2006 Report on technological development of China's villages). Zhongguo Nongye Chubanshe (China Agricultural Publishing Co.), Beijing

Li GP (1994) Zhongguo Gudai Nongye (Ancient agriculture in China). Taiwan Shangwu Yengshuguan (Taiwan Commerce Publishing House), Taipei

Li KR (ed) (1992) Zhongguo Qihou Bianhua ji qi Yingxiang (Climate change in China and its impacts). Haiyang Chubanshe (Oceans Publishing), Beijing

Li M (2007) Zhongguo Liangshi Anquan Zhili Fanglu (China's plan for the control of grain security). Zhongguo Nongye Chubanshe (China Agricultural Publishing Co.), Beijing

Li ST, Xu XY (eds) (2004) Nanshui Beidiao yu Zhongguo Fazhan (The south-north water diversion and China's development). Jingji Kexue Chubanshe (Economic Science Publishing), Beijing

Li XY, Yang J, Wang LX (September 2004) Gancaoquo Tudi Shamohua de Renwei Qudong Zuoyong Fenxi – Yi Talimu Heliluyu Weili (Qualitative analysis on driving role of human activity on the land desertification in arid areas: the case of the Tarim River basin). Ziyuan Kexue (Resour Sci), 26(5):30–37

Li ZC, Zeng SM (2000) Zhongguo Xiaomai Xiubing (China wheat rust diseases). Zhongguo Nongye Chubanshe (China Agricultural Publishing Co.), Beijing

Li ZG (2007) Zhongguo Nongye Keji Tizhi Gaige (Reform of China's agricultural technology system). Beijing. Unpublished paper

Li ZY, Xie Y (eds) (2002) Zhongguo Wailai Ruqin Zhong (Invasive alien species in China). Zhongguo Linye Chubanshe (China Forestry Publishing House), Beijing

Liao YS (2005) Zhongguo de Guangai Yongshui yu Liangshi Anquan (China's irrigation for food security). Zhongguo Shuili Shuidian Chubanshe (China's Water Conservancy and Hydropower Publishing Co.), Beijing

Lin Y, Lu L (June 2006) Zhongguo Liangshi Chanliang Bodong Fenxi (A study on the fluctuation of China's grain yield). Jilin Nongye Daxue Xuebao (J Jilin Agric Univ) 28(3):346–50

Liu CS (ed) (2006) Zhongguo Tudi Ziyuan Janlu yu Quyu Xietiao Fazhan Yanjiu (Study of the strategy of land resources and regional harmonious development in China). Qixiang Chubanshe (Climate Publishing Co.), Beijing

Liu YQ (ed) (2005) Ganhan Zaihai dui Woguo Shehui Jingji Yingxiang Yanjiu (Research on the socio-economic impact of drought disasters on our country). Zhongguo Shuili Shuidian Chubanshe (China Water Conservancy & Hydropower Publishing Co.), Beijing

Nongyebu Yuyequ (Ministry of Agriculture, Fisheries Bureau) (2006) Zhongguo Yuye Nianjian (China fisheries yearbook, 2006). Zhongguo Nongye Chubanshe (China Agricultural Press), Beijing

Qian XP, Wang XQ (December 2005) Guwu Gongji Hanshu he Qihou Bianhua de Yingxiang Fenxi (Influence analysis of climate change on grain supply function and yield response in China) Zhongguo Nongye Qixiang (Chin J Agrometeorol) 26:14–19

Qin YF (July 2004) Woguo Shidi Ziyuan Baohu Lifa de Biyaoxing (Legislative necessity for wetlands protection in our country). Xueshu Jiaoliu (Acad Exch) 124(7):29–32

Ren GY, Xu MZ, Chu ZY, Guo J, Li QX, Liu XN, Wang Y (December 2005) Jin 54 Nian Zhongguo Dimian Qiwen Bianhua (Changes of surface air temperature in China during 1951–2004), Qihou yu Huanjing Yanjiu (Climatic Environ Res) 10(4):717–727

Ren GY, Guo J, Xu MZ, Chu ZY, Zhang L, Zou XK, Li QX, LiuXN (December 2005) Jin 50 Nian Zhongguo Dimian Qihou Bianhua Jiben de Tezheng (Climate changes of China's mainland over the past half century). Qixiang Xuebao (Acta Meteorol Sin) 63(6):942–956

Su YZ, Yuan YL (eds) (1999) Shangdo Huangchong (Shangdong locust). Zhongguo Nongye Keji Chubanshe (China Agricultural Technology Press), Beijing

Tang GL, Ren GY (December 2005) Jinbainian Zhongguo Dibiao Qiwen Bianhua Qushi de zai Fenxi (Reanalysis of surface air temperature change of the last 100 years over China). Qihou yu Huanjing Yanjiu (Climate Environ Res) 10(4):791–798

Tang SC (2004) Liangshi Anquan yu Gengdi de Guanxi (Relationship between grain safety and cultivated land). Shengtai Huanjing (Ecol Environ) 13(1):149–150

Tao FL, Yokozawa M, Hayasha Y (December 2005) Jiyu Jijie Qihou Xinxi de Xiaomai he Yumi Shengchan Yuce (Forecasting wheat and maize production based on seasonal climate information in China). Zhongguo Nongye Qixiang (Chin J Agrometeorol) 26:24–30

Wang H (ed) (2006) Zhongguo Shui Ziyuan yu Kechixu Fazhan (Zhongguo Kechixu Fazhan Zonggang. Di 4 Juan) (China's water resources and sustainable development: Volume 4. The overview of China's sustainable development). Kexue Chubanshe (Science Publishing), Beijing

Wu Q, Chaolunbagen G. Wang, Gao K (December 2005) Neimenggu Tudi Zhengli yu Shuitu Ziyuan Kechixu Liyung Yanjiu (Study on sustainable use of water and soil resources and land arrangements in Inner Mongolia). Ganhanqu Ziyuan yu Huanjing (J Arid Land Resour Environ) 19(7):166–173

Wu SH, Li RS (March 2002) Zhongguo Gengdi yu Weilai 30 Nian Liangshi Xuqiu, Baojiang ji Duice (Food demand guarantees and countermeasures for China in the next 30 years). Dili Kexue Jinjhan (Prog Geogr) 21(2):121–129

Wu ZH (2007) Zhongguo Liangshi Wuliu Yanjiu (Research on China food logistics). Zhongguo Nongye Chubanshe (China Agricultural Publishing Co.), Beijing

Xiao HF, Wang J (eds) (2007) Woguo Liangshi Zonghe Shengchan Nengli ji Baohu Jizhi Yanjiu (Study on China's grain production capacity and the supporting mechanism). Zhongguo Nongye Chubanshe (China Agricultural Publishing Co.), Beijing

Xie Y (ed) (2007) Shengwu Ruqin yu Zhongguo Shengtai Anquan (Bioinvasion and ecological security in China). Heibei Kexue Jishu Chubanshe (Hebei Science and Technology Press), Shijiajuang, Hebei

Xin HP (2003) Dadou Bingchonghai Fongzhi Caise Tupu (Soybean diseases and insects, their protection and pictorial guide). Zhongguo Nongye Chubanshe (China Agricultural Publishing Co.), Beijing

Xu HG, Zhang S (eds) (2004) Zhongguo Wailai Ruqin Wuzhong Bianmu (Inventory: invasive alien species in China). Zhongguo Huanjing Kexue Chubanshe (China Environmental Sciences Press), Beijing

Xu JY, Ren BQ, Huang HW, Zhang YL (January 2003) Liangshi Baohu Zhengce he Liangshi Anquan Wenti de Shencemg Zikao (Thoughts on the policy of protection and safety of food. Cai Jing Luncong (Coll Essays Finance Econ) 1:20–25

Xu JX, Zhang ZZ, Liu F (October 2005) Qianyi Liangshi Anquan de Shui Ziyuan Baozhang (Analysis of water resources' assurance of grain production security). Shuili Keji yu Jingji (Water Conserv Sci Technol Econ) 11(10):611–614

Xu M, Ren G (June 2004) Jin 40 Nian Zhongguo Qixiang Shengzhangqi de Bianhua (Change in growing season over China: 1961–2000). Yingyong Qixiang Xuebao (J Appl Meteorol Sci) 15(13):306–312

Xue DY (ed) Zhongguo Shengwu Yichuan Ziyuan Xianzhuang yu Baohu (Status quo and protection of bio-genetic resources in China). Zhongguo Huanjing Kexue Chubanshe (China Environmental Sciences Press), Beijing

Xue DY (ed) (2005) Zhuanjiyin Shengwu Fengxian yu Guangli – Zhuanjiyin Shengwu yu Huanjing Guoji Yantaohui Lunwenji (Risks and regulations of genetically modified organisms – Proceedings of the international workshop on GMOs and environment). Zhongguo Huanjing Kexue Chubanshe (China Environmental Sciences Press), Beijing

Xue DY (ed) (2006) Zhuanjiyin Shengwu Huanjing Yingxiang yu Anquan Guanli – Nanjing Shengwu Anquan Guoji Yantaohui Lunwenji (Environmental impacts and safety regulations of genetically modified organisms – Proceedings of the international workshop on biosafety). Zhongguo Huanjing Kexue Chubanshe (China Environmental Sciences Press), Beijing

Yang JS, Wen LD, Li LF (eds) (2007) Changjiang Baohu yu Fazhan Baogao (Yangtze conservation and development report). Changjiang Chubanshe (Yangtze Publishing Co.), Wuhan

Yu ZG, Hu XP (2007) Woguo Liangshi Anquan yu Gengdi de Shuliang he Zhiliang Guanxi Yanjiu (Research on the relation of food security in China and cultivated land in quantity and quality). Beijing. Unpublished paper

Yuan SQ, Liu DC, Zhao Y (January 2007) Shandong Sheng Gengdi Ziyuan Anquan Wenti Yanjiu (Study on cultivated land resource security in Shandong Province). Dili yu Dili Xinxi Kexue (Geogr Geo-Inform Sci) 23(1):55–58

Zhang H, Sun YP (1993) Wenshi Xiaoying dui Woguo Shuangjidao Qihou Shengchanli de Yingxiang (The impact of the greenhouse effect on climatic productivity of double-crop rice in China). In: Zhang Y, Zhang PY, Zhang H, Lin Z (eds) Qihou Bianhua ji qi Yingxiang (Climate change and its impact). Zhongguo Kexueyuan Dili Yanjiuzuo, Chuanqiu Yanjiu Xilu Wenji. Di Yi Ji (Institute of Geography, Chinese Academy of Sciences, Global Change Study no. 1, Series Publication), Beijing, pp 131–138

Zhi L, Li N, Wang J, Kong F (March 2004) Xibu Tuigeng Huanlin Jingji Bushan Jizhi Yanjiu (A discussion on the economic compensation system for conversion of cropland to forestland in western China). Linye Kexue (Sci Silvae Sini) 40(2):2–8

Zhongguo Huanjing yu Fazhan Guoji Hezuo Weiyuanhui (CCICED) (2007) Zhongguo Huanjing yu Fazhan: Shiji Tiaozhan yu Zhanlue Jueze (The environment and development in China: challenges and strategic choices of the century). Zhongguo Huanjing Kexue Chubanshe (China Environmental Sciences Press), Beijing

Zhongguo Nongye Keshueyuan (China Academy of Agricultural Sciences), Zhiwu Baohu Yanjuyuan (Plant Protection Institute) (1995) Zhongguo Nongjouwu Bengchonghai (China diseases and insects on agricultural crops). Zhongguo Nongye Chubanshe (China Agricultural Publishing Co.), Beijing

Zhonghua Renmin Gongheguo Nongyebu (People's Republic of China, Ministry of Agriculture), Zhongguo Nongye Nianjian Bianji Weiyuanhui (China Agricultural Yearbook Editorial Committee)(2006) 2006 Zhongguo Nongye Nianjian (2006 China agricultural yearbook). Zhongguo Nongye Chubanshe (China Agricultural Publishing Co.), Beijing

Zhonghua Renmin Gongheguo Nongyebu (People's Republic of China, Ministry of Agriculture), Zhongguo Nongye Nianjian Bianji Weiyuanhui (China Agricultural Yearbook Editorial Committee) (2008) 2008 Zhongguo Nongye Nianjian (2008 China agricultural yearbook). Zhongguo Nongye Chubanshe (China Agricultural Publishing Co.), Beijing

Zhongguo Shehui Kexueyuan Huanjing yu Fazhan Yanjiu Zhongxin (Chinese Academy of Social Sciences, Center for Environment and Development) (2007). Zhongguo Huanjing yu Fazhan Pinlun (China environment and development review). Shehui Kexue Wenxian Chubanshe (Social Sciences Academic Press), China, Beijing

Zhou D, Shen B (eds) (2005) Shui yu Shehui Jingji Fazhan de Huxiang Yingxiang ji Zuoyung (The mutual influence and impact of water and socio-economic development). Zhongguo Shuili Shuidian Chubanshe (China Water Conservancy and Hydropower Publishing Co.), Beijing

Zhou YQ, Ren GY (December 2005) Huabei Diqu Dibiao Qiwen Guance Zhong Chengshihua Yingxiang de Jiance he Dingzheng (Identifying and correcting urban bias for regional surface air temperature series of north China over period of 1961–2000). Qihou yu Huanjing Yanjiu (Climate Environ Res) 10(4):743–753

Ziran zhiYou (Friends of Nature) (ed) (2008) Zhongguo Huanjing de Weiji yu Zhuanji (Crisis and turning point for China's environment [2008]). Shehui Kexue Wenxian Chubanshe (Social Sciences Academic Press), Beijing

English-Language Sources

Adu-Gyamfi JJ (ed) (2002) Food security in nutrient-stressed environments: exploiting plants' genetic capabilities. Kluwer, Dordrecht, The Netherlands

Agrios GN (2005) Plant pathology, 5th edn. Elsevier Academic, Burlington, MA

Alston J, Pardey P, Roseboom J (1998) Financing agricultural research: international investment patterns and policy perspectives. World Dev 26(6):1057–1072

Anderson EN (1988) The food of China. Yale University Press, New Haven, CT

Anderson K, Yao S (2003) China, GMOs and world trade in agricultural and textile products. Pacific Econ Rev 8(2):157–169

Ash RF, Edmonds RL (December 1998) China's land resources, environment and agricultural production. China Q 156:836–879

Baum R (ed) (1980) China's four modernizations: the new technological revolution. Westview Press, Boulder, CO

Baumuller H (2003) Domestic import regulations for genetically modified organisms and their compatibility with WTO rules: some key issues. International Institute for Sustainable Development, Winnipeg, ON

Baviera ASP, Liao S, Militante CV (1999) Food security in China and Southeast Asia. Philippine-China Development Resource Center, Manila, Philippines

Bennett J, Wang X, Zhang L (eds) (2008) Environmental protection in China: land-use management. Edward Elgar, Cheltenham, UK

Biodiversity Working Group of China Council for International Cooperation on Environment and Development (2001) Restoring China's degraded environment: the role of natural vegetation. China Environmental Sciences Press, Beijing

Biodiversity Working Group of China Council for International Cooperation on Environment and Development (2001) Conserving China's biodiversity II. China Environmental Sciences Press, Beijing

Brautigam D (1992) Land rights and agricultural development in West Africa: a case study of two Chinese projects. J Dev Areas 25(4):21–32

Brautigam D, Tang X (September 2009) China's engagement in African agriculture: 'down to the countryside'. China Q 199:686–706

Brown L (1995) Who will feed China? Wake-up call from a small planet. Worldwatch Institute, Washington, DC

Brown L (July 1, 1998) China's water shortage could shake world security. *World Watch* (July 1, 1998)

Brown L (2004) Outgrowing the Earth. W. W. Norton, New York

Buck JL (1973) Land and agricultural resources. In: Wu Y (ed) China: a handbook. Praeger, New York

Burns J (2003) 'Downsizing' the Chinese state: government retrenchment in the 1990s. China Q 175:623–656

Carpenter J, Felsot A, Goode T, Hammig M, Onstad D, Sankula S (2002) Comparative environmental impacts of biotechnology-derived and traditional soybean, corn, and cotton crops. Council for Agricultural Science and Technology, Ames, IA

Carruth R (ed) (2006) Global governance of food and agricultural industries. Edward Elgar, Cheltenham, UK

Cartier C (August 2001) 'Zone fever', the arable land debate, and real estate speculation: China's evolving land use regime and its geographical contradictions. J Contemp China 10(28):445–470

Chan A (November 2007) Guiding public opinion through social agenda setting: China's media policy since the 1990s. J Contemp China 16(53):547–559

Chan A, Madsen R, Unger J (1984) Chen village: the recent history of a peasant community in Mao's China. University of California Press, Berkeley, CA

Chan G (2006) China's compliance in global affairs: trade, arms control, environmental protection, human rights. World Scientific Publishing Co., Hackensack, NJ

Chang WYB, King B (1994) Centennial climate changes and their global associations in the Yangtze River. Climate Res 4:95–103

Chao K (1970) Agricultural production in communist China: 1949–1965. University of Wisconsin Press, Madison, WI

China–UK Collaboration Project (2004) Investigating the Impacts of Climate Change on Chinese Agriculture. Department for Environment, Food and Rural Affairs (Defra), London

China Daily, selected issues

China Development Brief (June 1, 2001) Thirsty cities and factories push farmers off the parched earth. 1(6):1–7
China Development Brief (August 10, 2005) Greenpeace targets retail giant in fight against GM rice. VIII(8):5–7
China Development Brief (February 15, 2006) Statistics: seeking truth from (tonnes of) facts. IX(2):1–6
China Development Brief Greenpeace (March 31, 2005) Zhejiang hotels, stand firm against paper giant. VII(3):1–2
China Development Brief (July/August 2006) Leading state advisor offers frank assessment of rural challenges. X(7):3–5
Conservation International-China (2007) Promoting the use of natural regeneration in China to protect biodiversity and mitigate global climate change. Beijing
Crook FW (Fall 1992) China's extension system. J Extension 30(3):1–3
Day KA (ed) (2005) China's environment and the challenge of sustainable development. M. E. Sharpe, Armonk, NY
Deng X, Hunag JK, Rozelle S, Uchida E (2006) Cultivated land conversion and potential agricultural productivity in China. Land Use Policy 23:372–384
Ding J, Wang R, Fan Z (1995) Distribution and infestation of water hyacinth and the control strategy in China. J Weed Sci 9(2):49–51
Ding J, Fu W (1996) Biological control: conservation of biodiversity by using biodiversity. Biodiversity 4(4):222–227
Ding J, Xie Y (2001) The mechanism of biological invasion and the management strategy. In: Schei, PJ, Sung W, Yan X (eds) Biodiversity working group/CCICED. Conserving China's Biodiversity (II). China Environmental Science Press, Beijing
Dittmer L (1974) Liu Shao-ch'i and the Chinese cultural revolution: the politics of mass criticism. University of California Press, Berkeley, CA
Dittmer L, Li G (2006) China's deep reform: domestic politics in transition. Rowman & Littlefield, London
Dong F, Jensen H (Spring 2004) China's challenge: conforming to sanitary and phytosanitary measures for agricultural exports. Iowa Agric Rev 10(2):138–144
Dryzek J, Downes D, Hunold C, Schlosberg D with Hernes HK (2003) Green states and social movements. Oxford University Press, New York
Du S, Lu B, Zhai F, Popkin B (2002) A new stage of the nutrition transition in China. Public Health Nutr 5(1A):169–174
Eberhard W (1977) A history of China, 4th edn. University of California Press, Berkeley, CA
Economy E (2004) The river runs black: the environmental challenge to China's future. Cornell University Press, Ithaca, NY
Editorial Committee (2008) China's strategy for plant conservation. Guangdong Science and Technology Press, Guangzhou
Elvin M (1973) The pattern of the Chinese past. Stanford University Press, Stanford, CA
Fairbank JK, Goldman M (2006) China: a new history, 2nd edn. Harvard University Press, Cambridge, MA
Fan S, Zhang L, Zhang X (2004) Reforms, investment, and poverty in rural China. Econ Dev Cult Change 52(2):395–421
Fang JQ, Xie Z (1994) Deforestation in preindustrial China: the Loess Plateau region as an example. Chemosphere 29:983–999
Fang JY, Su SL, Liu GH (1995) Carbon pools on terrestrial ecosystems of China and their global significance. In: Galloway J, Melillo J (eds) Asian change in the context of global climate change. International geosphere–biosphere programme. Natural and anthropogenic changes: impacts on global biogeochemical cycles. Cambridge University Press, Cambridge
Friends of Nature (2007) A warming China: thoughts and actions for the Chinese civil society. Beijing
Fu CB (1994) An aridity trend in China in association with global warming. In: Ztpp RG (ed) Climate biosphere interaction: biogenic emissions and environmental effects of climate change. Wiley, Chichester, UK, pp 1–17

Fuller F, Huang J, Ma H, Rozelle S (2006) Got milk? The rapid rise of China's dairy sector and its future prospects. Food Policy 31:201–215
Gale F (ed) (2002) China's food and agriculture: issues for the 21st century. USDA Economic Research Service, Washington, DC
Galloway JN, Melillo JM (eds) (1998) Asian change in the context of global climate change. International geosphere–biosphere programme. Cambridge University Press, Cambridge
Garside R (1982) Coming alive: China after Mao. Mentor, New York
Gong GF, Hameed S (1991) The variation on moisture conditions in China during the last 2000 years. Int J Climatol 11:271–283
Green N (2009) Positive spillover? Impact of the Songhua River benzene incident on China's environmental policy. Wilson Center, China Environmental Forum, Washington, DC
Greenhalgh S (June 2005) Missile science, population science: the origins of China's one-child policy. China Q 182:253–276
Greenhalgh S, Winckler EA (2005) Governing China's population: from Leninist to neoliberal biopolitics. Stanford University Press, Stanford, CA
Greenpeace China (November 2004) Investigative report on APP's forest destruction in Yunnan. Hong Kong
Greenpeace International (2007) Sharing the blame: global consumption and China's role in ancient forest destruction. Amsterdam
Gregory PJ, Ingram SI, Brlacich M (2005) Climate change and food security. Philos Trans R Soc 360:2139–2148
Gupta A, Falkner R (November 2006) The influence of the Cartageña protocol on biosafety: comparing Mexico, China and South Africa. Global Environ Polit 6(4):23–55
Hamburger J (2002) Pesticides in China: a growing threat to food safety, public health, and the environment, China environment series, issue 5. Woodrow Wilson International Center for Scholars, Washington, DC, pp 29–44
Han NY, Jiang G, Li WJ (2002) Management of the degraded ecosystems in Xilingol biosphere reserve. Tsinghua University Press, Beijing
Harding AF (ed) (1982) Climate change in later prehistory. Edinburgh University Press, Edinburgh
Harkness J (December 1998) Recent trends in forestry and conservation of biodiversity in China. China Q 156:911–934
Harris RB (2008) Wildlife conservation in China: preserving the habitat of China's Wild West. M. E. Sharpe, Armonk, NY
Hartman DL (1994) Global physical climatology. Academic, San Diego, CA
Heilig GK (1997) Anthropogenic factors in land-use change in China. Popul Dev Rev 23(1):57–68
Ho SPS, Lin GCS (September 2004) Non-agricultural land use in post-reform China. China Q 179:758–781
Howell J (2000) New directions in civil society: organizing around marginalized interests. In: Howell J (ed) Governance in China. Rowman & Littlefield, Lanham, MD, pp 138–152
Hsiao HC (1979) A history of Chinese political thought. Princeton University Press, Princeton, NJ
Hsieh CM, Lu M (eds) (2004) Changing China: a geographic appraisal. Westview Press, Boulder, CO
Hsu HH, Gale F (eds) (November 2001) China's agriculture in transition. US Department of Agriculture, Economic Research, Washington, DC
Hu SY (2005) Food plants of China. The Chinese University Press, Hong Kong
Huang JK, Pray C, Rozelle S, Wang Q (January 25, 2002) Plant biotechnology in China. Science 295(5555):674–677
Huang JK, Rozelle S (2002) China's accession to WTO and shifts in the agriculture policy. Department of Agricultural and Resource Economics Working Paper No. 2. University of California Davis, Davis, CA
Huang JK, Hu R, Pray C, Qiao F, Rozelle S (2003) Biotechnology as an alternative to chemical pesticides: a case study of Bt cotton in China. Agric Econ 29:55–67

Huang JK, Rozelle S (2003) Trade reform: the WTO and China's food economy in the twenty-first century. Pacific Econ Rev 8(2):143–156
Huang JK, Hu R, van Meijl H, van Tongeren F (2004) Biotechnology boosts to crop productivity in China: trade and welfare implications. J Dev Econ 75:27–54
Huang JK, Zhu L, Deng X, Rozelle S (2005) Cultivated land changes in China: the impacts of urbanization and industrialization. Soc Photo-Opt Instrum Eng 5884, 58840I:1–15
Huang JK, Hu R, Rozelle S, Pray C (April 29, 2005) Insect-resistant GM rice in farmers' fields: assessing productivity and health effects in China. Science 308:688–690
Huang JK, Hu R, Rozelle S, Pray C (October 14, 2005) Debate over a GM rice trial in China. Science 310:231–233
Huang JK, Qiu H, Bai J, Pray C (2006a) Awareness, acceptance of and willingness to buy genetically modified foods in urban China. Appetite 46:144–151
Huang JK, Zhang HD, Yang J, Rozelle S, Kalaitzandonakes N (2008) Will the biosafety protocol hinder or protect the developing world: learning from China's experience. Food Policy 33:1–12
Huang Q, Dawe D, Rozelle S, Huang J, Wang J (2005b) Irrigation, poverty and inequality in rural China. Aust J Agric Resour Econ 49:159–175
Huang Q, Rozelle S, Lohman B, Huang J, Wang J (2006b) Irrigation, agricultural performance and poverty reduction in China. Food Policy 31:30–52
Huang R (1981) 1587, A Year of No Significance (Chinese version: Wanli Shiwunian, published by Shihuo Chubanshe, Taipei, in 1985). Yale University Press, New Haven, CT
Hyde WF, Belcher B, Xu J (2003) China's forests: global lessons from market reforms. Resources for the Future, Washington, DC
Hyun IT, Schreurs MA (2007) The environmental dimensions of Asian security. United States Institute of Peace Press, Washington, DC
Information Office, State Council, PRC (December 2007) China's energy conditions and policies. Beijing
Intergovernmental Panel on Climate Change (IPCC) (2007) Contribution of working group I to the fourth assessment report of the intergovernmental panel on climate change. Cambridge University Press, Cambridge
Jeon HK, Yoon SS (November/December 2006) From international linkages to internal divisions in China: the political response to climate change negotiations. Asian Surv XLVI(6):846–866
Jia H (June 2004) China ramps up efforts to commercialize GM rice. Nat Biotechnol 22(6):217–223
Jia S, Peng Y (2002) GMO biosafety research in China. Environ Biosafety Res 1:5–8
Jin S, Rozelle S, Alston J, Huang JK (August 2005) Economics of scale and scope and the economic efficiency of China's agricultural research system. Int Econ Rev 36(3):1033–1057
Johnson I (2003) The death and life of China's civil society. Perspectives 1(3):551–554
Joseph WA, Wong CPW, Zweig D (1991) New perspectives on the cultural revolution. Council on East Asian Studies, Harvard University, Cambridge, MA
Jun J (2000) Environmental protests in rural China. In: Perry E, Selden M (eds) Chinese society: change, conflict and resistance. Routledge, New York
Jussaume RA Jr (May 2001) Factors associated with modern urban Chinese food consumption patterns. J Contemp China 10(27):219–232
Kambara T, Howe C (2007) China and the global energy crisis. Edward Elgar, Cheltenham, UK
Karplus V (October 8, 2003) Let a thousand GM crops bloom. International Herald Tribune
Keeley J (April 2006) Balancing technological innovation and environmental regulation: an analysis of Chinese agricultural biotechnology governance. Environ Polit 15(2):293–309
Khan AR, Riskin C (June 2005) China's household income and its distribution, 1995 and 2002. China Q 182:356–384
Koohafkan P, Rey A, Antoine J (2005) Soil carbon sequestration in dryland farming systems. In: Lal R, Uphoff N, Stewart BA, Hansen DO (eds) Climate change and global food security. Taylor & Francis, London
Kung JK, Liu S (July 1997) Farmers' preferences regarding ownership and land tenure in post-Mao China: unexpected evidence from eight counties. China J 38:306–328

Lai, Elisa Chih-Yin (2009) Climate change impacts on China environment: biophysical impacts. Wilson Center, China Environmental Forum, Washington, DC
Lal R, Uphoff N, Stewart BA, Hansen DO (eds) (2005) Climate change and global food security. Taylor & Francis, Boca Raton, FL
Lardy NR (1994) China in the world economy. Institute for International Economics, Washington, DC
Lee HY (1978) The politics of the chinese cultural revolution: a case study. University of California Press, Berkeley, CA
Li C, Wang H (2002) China's food security and international trade. China World Econ 4:28–33
Li J, Liu J (2009) Quest for clean water in China's newly amended water pollution control law. Wilson Center, China Environmental Forum, Washington, DC
Li Y, Zhang H, Han Y, Xu D (1991) Preliminary investigation on American White Moth in Qinghuangdao. J Forest Sci Technol 4:29–30
Li Y (1998) China weeds. Agriculture Press, Beijing
Lieberthal K (1995) Governing China: from revolution through reform. W. W. Norton, New York
Liftin K (1993) Eco-regimes: playing tug of war with the nation-state. In: Lipschutz RD, Conca K (eds) The state and social power in global environmental politics. Columbia University Press, New York, pp 94–118
Lin E, Xiong W, Ju H, Xu Y, Li Y, Bai L, Xie L (2005) Climate change impacts on crop yield and quality with CO2 fertilization in China. Philos Trans R Soc B 360:2149–2154
Lin W, Somwaru A, Tuan F, Huang J, Bai J (2006a) Consumers' willingness to pay for biotech foods in China: a contingent valuation approach. AbBioForum 9(3):166–179
Lin W, Somwaru A, Tuan F, Huang J, Bai J (2006b) Consumer attitudes toward biotech foods in China. J Int Food Agribus Mark 18(1/2):177–203
Liu C, Zuo D (1997) Environmental issues of the three Gorges Project, China. Regulat Rivers: Res Manage 1:267–273
Liu YH, Lu DD, Cai F, Ge Q (eds) (2005) Research on human dimensions of global environmental change in China. China Meteorological Press, Beijing
Lindzen RS (March–April 1990) A skeptic speaks out. EPA J 16:13–16
Lipschutz RD, Mayer J (1996) Global civil society & global environmental governance. State University of New York Press, Albany, NY
Lo CWH, Chuang SS (2003) China's green challenges in the twenty-first century. In: Cheng JYS (ed) China's challenges in the twenty-first century. City University of Hong Kong Press, Hong Kong, pp 719–760
Lohmar B, Wang J, Rozelle S, Huang JK, Dawe D (2002) China's agricultural water policy reforms: increasing investment, resolving conflicts, and revising incentives. USDA Economic Research Service, Washington, DC, pp 1–27
Lomborg B (2007) The skeptical environmentalist. Cambridge University Press, Cambridge
Lu Q (1997) The importance of classical biological control to biodiversity protection. Chin Biodivers 5:224–230
Ma J (2004) China's water crisis. Eastbridge, Norwalk, CT
Ma T (2008a) Interconnected forests: global and domestic impacts of China's forestry conservation. Wilson Center, China Environmental Forum, Washington, DC
Ma T (2008b) Wielding the double-edged sword: the Chinese experience with agricultural genetically modified organisms. Wilson Center, China Environmental Forum, Washington, DC
Ma X, Ortolano L (2000) Environmental regulation in China: institutions, enforcement, and compliance. Rowman & Littlefield, Lanham, MD
MacFarquhar R (1990) The politics of China: eras from Mao to Deng. Cambridge University Press, Cambridge
Macilwain C (March 13, 2003) Chinese agribiotech: against the grain. Nature 422:111–112
Mackinnon J, Sha M, Cheung C, Carey G, Xiang Z, Melville D (1996) A biodiversity review of China. World Wide Fund for Nature (WWF) International, Hong Kong
Marshall Institute (1989) Scientific perspectives on the greenhouse problem. Washington, DC

Martens S (April 2006) Public participation with Chinese characteristics: citizen consumers in China's environmental management. Environ Polit 15(2):211–230
McBeath J, McBeath JH (2009) Environmental stressors and food security in China. J Chin Polit Sci 14:49–80
McBeath GA, Leng TK (2006) Governance of biodiversity conservation in China and Taiwan. Edward Elgar, Cheltenham, UK
McBeath J, Rosenberg J (2006) Comparative environmental politics. Springer, Dordrecht, The Netherlands
McBeath J, Wang B (April 2008) China's environmental diplomacy. Am J Chin Stud 15(1):1–16
McNeely JA (2001) Human dimensions of invasive alien species: how global perspectives are relevant to China. In: Biodiversity working group/CCICED. Conserving China's biodiversity. China Environmental Science Press, Beijing, pp 169–182
Meisner M (1986) Mao's China and after: a history of the People's Republic. The Free Press, New York
Ministry of Science & Technology et al., PRC (June 2007) China's scientific and technological actions on climate change. Beijing
Morse R (2001) China's fledgling civil society: a force for democratization? World Policy J 18(1):17–32
National Development and Reform Commission, PRC (June 2007) China's national climate change program. Beijing
New York Times, selected issues
Newell P (May 2003) Globalization and the governance of biodiversity. Global Environ Polit 3(2):56–71
Newell P (2003) Domesticating global policy on GMOs: comparing India and China. IDS Working Paper 206, Institute for Development Studies, Brighton, UK
Normile D (2004) Invasive species: expanding trade with China creates ecological backlash. Science 306(5698):968–969
O'Brien K, Li L (2006) Rightful resistance in rural China. Cambridge University Press, New York
Oi J (1989) State and peasant in contemporary China: the political economy of village government. University of California Press, Berkeley, CA
Oi J (September 1999) Two decades of rural reform in China: an overview and assessment. China Q 159:616–628
Oi J (1999) Rural China takes off: institutional foundations of economic reform. University of California Press, Berkeley, CA
Oosterveer P (2007) Global governance of food production and consumption: issues and challenges. Edward Elgar, Cheltenham, UK
Ou SH (1972) Rice diseases. Eastern Press, London
Pang J, Zou J (2008) China: climate change policy-making process. In: Kameyama Y, Sari AP, Soejachmoen MH, Kanie N (eds) Climate change in Asia: perspectives on the future climate regime. United Nations University Press, New York, pp 66–82
Parday PG, Beintema NM (2001) Slow magic: agricultural R&D a century after Mendel. IFPRI Food Policy Report, Washington DC
Park A, Rozelle S, Wong C, Ren C (September 1996) Distributional consequences of fiscal reform on China's poor areas. China Q 147:751–778
Park A, Wang S (2001) China's poverty statistics. China Econ Rev 12:394–398
People's Republic of China (2003) Initial national communication on climate change. Beijing
Perkins D (1969) Agricultural development in China, 1368–1968. Aldine, Chicago, IL
Perry E, Goldman M (2007) Grassroots political reform in contemporary China. Harvard University Press, Cambridge, MA
Popkin BM (1993) The nutrition transition in China: a cross-sectional analysis. Eur J Clin Nutr 47:333–346
Pray CE, Ramaswami B, Huang J, Hu R, Bengali P, Zhang H (2006) Costs and enforcement of biosafety regulations in India and China. Int J Technol Globalis 2(1/2):137–157

PRC State Council (August 17, 2007) China's food quality and safety. Beijing
Rawski E (1972) Agricultural change and the peasant economy of South China. Harvard University Press, Cambridge, MA
Reischauer EO, Fairbank JK (1960) East Asia: the great tradition. Houghton Mifflin, Boston, MA
Rice EE (1972) Mao's Way. University of California Press, Berkeley, CA
Rosenzweig C, Parry ML, Fischer G, Frohberg K (1993) Climate change and world food supply. Environmental Change Unit, Research Report No. 3. University of Oxford, Oxford
Ross L (1988) Environmental policy in China. Indiana University Press, Bloomington, IN
Rozelle S, Huang JK, Hu R (2004) Genetically modified rice in China: effects on farmers – in China and California. Giannini Found Agric Econ 2–6
Rozelle S, Huang J, Otsuka K (January 2005) The engines of a viable agriculture: advances in biotechnology, market accessibility and land rentals in rural China. China J 53:81–111
Saich T (March 2000) Negotiating the state: the development of social organizations in China. China Q 161:124–141
Saich T (2001) Governance and politics in China. Palgrave, New York
Saich T (2004) Governance and politics of China, 2nd edn. Palgrave, New York
Sanders R (February 2006) Organic agriculture in China: do property rights matter? J Contemp China 15(46):113–132
Schlesinger ME, Jiang X (1991) Revised projection of future greenhouse warming. Nature 350:219–221
Schram SR (1970) The political thought of Mao Tsetung. Praeger, New York
Schwartz J (Spring 2004) Environmental NGOs in China: roles and limits. Pacific Aff 77(11):28–49
Secretariat of the Convention on Biological Diversity (2001) Handbook of the convention on biological diversity. Earthscan, London
Shapiro J (2001) Mao's war against nature: politics and the environment in revolutionary China. Cambridge University Press, Cambridge, UK
Shine C, Williams N, Burhenne-Guilmin F (2000) Legal and institutional frameworks on alien invasive species: a contribution to the global invasive species programme global strategy document. IUCN Environmental Law Programme, Bonn, Germany
Shurtleff MC (1980) Compendium of corn diseases, 2nd edn. APS Press, St. Paul, MN
Simmons FJ (1991) Food in China: a cultural and historical inquiry. CRC Press, New York
Sinclair JB, Backman PA (2003) Compendium of soybean diseases, 3rd edn. APS Press, St. Paul, MN
Singer SF (ed) (1989) Global climate change: natural and human influences. Paragon House, New York
Smil V (1993) China's environmental crisis: an inquiry into the limits of national development. M.E. Sharpe, Armonk, NY
Smil V (1995) Feeding China. Curr China 94(593):137–153
Smil V (June 1999) Research note: China's arable land. China Q 158:414–429
Song Y, Chen D, Dong W (2006) Influence of climate on winter wheat productivity in different climate regions of China, 1961–2000. Climate Res 32:219–227
Spray SL, McGlothlin KL (2002) Global climate change. Rowman & Littlefield, Lanham, MD
State Environmental Protection Administration, PRC (2001) National biosafety framework of China. China Environmental Sciences Press, Beijing
State Forestry Administration, PRC (2001) Forestry in China. Beijing
State Forestry Administration, PRC (2001). Wildlife Conservation in China: Retrospect & Prospect. Beijing
Stevenson WR, Loria R, Franc GD, Weingartner DP (2001) Compendium of potato diseases, 2nd edn. APS Press, St. Paul, MN
Sullivan J, Xie L (June 2009) Environmental activism, social networks and the Internet. China Q 198:422–432
Tai, Pei-yu "Catherine", Ellis L (2008) Taihu Green Wash or Green Clean? Wilson Center, China Environmental Forum, Washington, DC
Tao F, Yokozawa M, Liu J, Zhang Z (2008) Climate change, land use change, and china's food security in the twenty-first century: an integrated perspective. Clim Change 71:138–151

Teets JC (June 2009) Post-earthquake relief and reconstruction efforts: the emergence of civil society in China? China Q 198:330–347
Tilt B (December 2007) The political ecology of pollution enforcement in China: a case from Sichuan's rural industrial sector. China Q 192:915–932
Tong CL, Hall CAS, Wang HQ (2003) Land use change in rice, wheat and maize production in China (1961–1998). Agric Ecosyst Environ 95:523–536
Tregear RR (1980) China: a geographical survey. Wiley, New York
Turner J, Otsuka K (2006) Reaching across the water. Woodrow Wilson International Center for Scholars, Washington, DC
Turner J, Lu Z (2007) China's Environmental NGOs. In: Zheng Y (ed) China environment and development review. Social Sciences Academic Press, Beijing, pp 349–369
Uchida E, Xu J, Rozelle S (May 2005) Grain for green: cost-effectiveness and sustainability of China's conservation set-aside program. Land Econ 81(2):247–264
Vassilos R, Franke T, Sanchez J (2008) Water situation in the North China plain. USDA Foreign Agricultural Service, Global Agriculture Information Network, Beijing
Vorosmarty C, Li C, Sun J, Dai Z (1998) Drainage basins, river systems, and anthropogenic change: the Chinese example. In: Galloway JN, Melillo JM (eds) Asian change in the context of global climate change. Cambridge University Press, Cambridge, pp 210–244
Waldron S, Brown C, Longworth J (June 2006) State sector reform and agriculture in China. China Q 186:277–294
Wang CF (Spring 2007) Chinese environmental law enforcement: current deficiencies and suggested reforms. Vermont J Environ Law 8(2):159–194
Wang F (1997) Occurrence and comprehensive prevention and cure of vegetable leaf miner. J Agric Sci Technol 2:73–81
Wang C, Yu Z, Wang D (2006) China: risk assessment and risk management. IUCN Regional Biodiversity Programme, Asia. http://www.rbp-iucn.lk/biosafetyi/CouStatuisi_China.htm. Cited 30 May 2006
Wang J, Xu Z, Huang J, Rozelle S (2005) Incentives in water management reform: assessing the effect on water use, production, and poverty in the Yellow River Basin. Environ Dev Econ 10:769–799
Wang J, Mendelsohn R, Dinar A, Huang J, Rozelle S, Zhang L (2007) Can China continue feeding itself? The impact of climate change on agriculture. The World Bank, Policy Research Working Paper 4470, Washington, DC
Wang Q, Zhang W (Mar/Apr 2004) China's potato industry and potential impacts on the global market. Am J Potato Res 81:101–109
Webster RK, Gunnell PS (eds) (1992) Compendium of rice diseases. APS Press, St. Paul, MN
Wei H (1997) Occurrence and continuous control of American rice water weevil in China. Plant Quarantine 11:60–62
Weise MV (1987) Compendium of wheat disease, 2nd edn. APS Press, St. Paul, MN
Wen B (1998) Greening the Chinese media. China Environment Series, no. 2, Woodrow Wilson Center Press, Washington, DC
Wexler R, Ying X, Young N (Sept 2006) NGO advocacy in China. China Development Brief
White LT, Li C (1990) Diversification among mainland Chinese intellectuals. In: Chang KY (ed) Mainland China after the thirteenth party congress. Westview Press, Boulder, CO, pp 143–171
Wisner B, Blaikie P, Cannon T, Davis I (2004) At risk: natural hazards, people's vulnerability and disasters, 2nd edn. Routledge, London
Wittfogel K (1957) Oriental despotism. Yale University Press, New Haven, CT
Womack B (ed) (1991) Contemporary Chinese politics in historical perspective. Cambridge University Press, New York
World Bank (2001) World Development Report, 2000/2001: attacking Poverty. World Bank, Washington, DC
World Bank (2009) From poor areas to poor people: China's evolving poverty reduction agenda. World Bank, Washington, DC

World Wildlife Fund (WWF) (2007) International climate change regime: a study on key issues in China. Environmental Sciences Press, Beijing
World Wildlife Fund International (2004) The Asia-Pacific Climate and Energy Programme. Giand, Switzerland
WWF-China Freshwater Programme (2005) Water for people, water for life. WWF, Beijing
Wu G, Lee W (eds) (1984) Managing the environment in China. Tycooly International, Dublin
Wu KM (April 2002) Impacts of Bt cotton on status of insect pests and resistance risk of cotton bollworms in China. In: Conference on resistance management for Bt-crops in China: economic and biological considerations. North Carolina State University, Raleigh, NC
Xie Y (2007) Alien Invasive Species in China. http://www.chinabiodiversityi.com/shwdyx/ruq/ruq2-4n.htm. Cited 16 Oct 2007
Xie Y, Li Z, Gregg WP, Li D (2000) Invasive species in China – an overview. Biodivers Conserv 10(8):1317–1341
Xie Y, Wang S, Schlei P (eds) (2004) China's protected areas. Tsinghua University Press, Beijing
Xiong W, Lin E, Ju H, Xu Y (2007) Climate change and critical thresholds in China's food security. Climate Change 81:205–221
Xu Z, Bennett MT, Tao R, Xu J (2004) China's sloping land conversion programme four years on: current situation and pending issues. Int Forest Rev 6(3–4):241–249
Xu Z, Xu J, Deng X, Huang J, Uchida E, Roselle S (2006) Grain for green versus grain: conflict between food security and conservation set-aside in China. World Dev 34(1):130–148
Yang DD (March 2000) Civil society as an analytic lens for contemporary China. China: Int J 2(1):33–49
Yang DL (1997) Beyond Beijing: liberalization and the regions in China. Routledge, London
Yang H (December 1998) Trends in China's regional grain production and their implications. Agric Econ 19(3):309–325
Yang J (2001) The alien and indigenous fishes of Yunnan: a study on impact ways, degrees and relevant issues, Biodiversity working group/CCICED. Conserving China's Biodiversity (II). China Environmental Science Press, Beijing, pp 161–168
Yang Y (2007) Pesticides and environmental health trends in China. Wilson Center, China Environmental Forum, Washington, DC
Yang Y, Turner J (2007) Food safety in China. Wilson Center, China Environmental Forum, Washington, DC
Ye D, Fu C, Chao I, Yoshino M (eds) (1987) The climate of China and global climate. Springer, Berlin
Yeh AGO, Li X (1997) An integrated remote sensing and GIS approach in the monitoring and evaluation of rapid urban growth for sustainable development in the Pearl River Delta, China. Int Plan Stud 2(2):193–210
Yi Q, Hua L (1997) Strengthening quarantine of plants introduced from Taiwan. Plant Quarantine 12:185–187
Yin R, Xu J, Li Z, Liu C (2005) China's ecological rehabilitation: the unprecedented efforts and dramatic impacts of reforestation and slope protection in Western China, Woodrow Wilson Center. China Environ Ser (7):17–32
Young N (June 2005) NGOs will have to 'negotiate the state' for some time yet. China Dev Brief IX(2):1–3
Young N (September 2005) Under Scrutiny. China Dev Brief IX(7):6–8
Zhang L, Huang J, Rozelle S (September 2003) China's war on poverty: assessing targeting and the growth impacts of poverty programs. J Chin Econ Bus Stud 1(3):301–317
Zhang L, Luo R, Liu C, Rozelle S (July–August 2006) Investing in rural China. Chin Econ 39(4):57–84
Zhang Q (2000) Agricultural biotechnology opportunities to meet the challenges of food production. In: Persley GJ, Lantin MM (eds) Agricultural biotechnology and the poor: Proceedings of an international conference. Consultative Group on International Agricultural Research, Washington, DC, pp 45–50

Zhang Z (1998) The economics of energy policy in China: implications for global climate change. Edward Elgar, Cheltenham, UK

Zhao S (ed) (2004) Chinese foreign policy: pragmatism and strategic behavior. M. E. Sharpe, Armonk, NY

Zhang Z (2004) A nation-state by construction: dynamics of modern Chinese nationalism. Stanford University Press, Stanford, CA

Zhen L, Routray JK, Zoebisch M, Chen G, Xie G, Cheng S (2005) Three dimensions of sustainability of farming practices in the North China Plain. Agric Ecosyst Environ 105(3):507–522

Zhong T, Rozelle S, Stone B, Jiang D, Chen J, Xu Z (1994) China's experience with market reform for commercialization of agriculture in poor areas. In: Von Braun J, Kennedy E (eds) Agricultural commercialization, economic development, and nutrition. The Johns Hopkins University Press, Baltimore, MD, pp 119–140

Zhou XP, Gu XK, Chen BM, Liu XW, Cui YM (2007) Correlation analysis on recent grain production and change of cultivated land resources in China. Beijing Normal University, School of Management, Beijing. Unpublished paper

Index

MIX
Papier aus verantwortungsvollen Quellen
Paper from responsible sources
FSC® C105338

If you have any concerns about our products,
you can contact us on
ProductSafety@springernature.com

In case Publisher is established outside the EU,
the EU authorized representative is:
Springer Nature Customer Service Center GmbH
Europaplatz 3, 69115 Heidelberg, Germany

Printed by Libri Plureos GmbH
in Hamburg, Germany